ELECTRICAL CIRCUIT ANALYSIS AND NETWORK THEORY

AUTHOR

Prof. Shouri Chatterjee

Department of Electrical Engineering
Indian Institute of Technology (IIT) Delhi

REVIEWER

Prof. M K Verma

Department of Electrical Engineering
Indian Institute of Technology (IIT) BHU

KHANNA BOOK PUBLISHING CO. (P) Ltd.
4C/4344, Ansari Road, Darya Ganj, New Delhi, India

All India Council for Technical Education
Nelson Mandela Marg, Vasant Kunj, New Delhi, India

Dear Readers,

To prevent the piracy, this book is secured with HIGH SECURITY HOLOGRAM on the front title cover. In case you don't find the hologram on the front cover title, please write us to at contact@khannabooks.com or whatsapp us at +91-99109 09320 and avail special gift voucher for yourself.

Specimen of Hologram on front Cover title:

Moreover, there is a SPECIAL DISCOUNT COUPON for you with EVERY HOLOGRAM.

How to avail this SPECIAL DISCOUNT:

Step 1: Scratch the hologram

Step 2: Under the scratch area, your "coupon code" is available

Step 3: Logon to www.khannabooks.com

Step 4: Use your "coupon code" in the shopping cart and get your copy at a special discount

Step 5: Enjoy your reading!

ISBN: 978-93-5538-338-9

Book Code: UG403EN

Electrical Circuit Analysis and Network Theory

by Shouri Chatterjee

First Edition: 2026

Originally Published by:

All India Council for Technical Education, Ministry of Education, GOI

Exclusively Published & Distributed by:

Khanna Book Publishing Co. (P) Ltd.
4C/4344, Ansari Road, Daryaganj,
New Delhi (INDIA)

Visit us at: **www.khannabooks.com**
Write us at: **contact@khannabooks.com**

To view complete list of books, Please scan the QR Code:

Printed in Bharat.

प्रो. म. जगदीश कुमार
अध्यक्ष
Prof. M. Jagadesh Kumar
Chairman

सत्यमेव जयते

अखिल भारतीय तकनीकी शिक्षा परिषद्
(भारत सरकार का एक सांविधिक निकाय)
(शिक्षा मंत्रालय, भारत सरकार)
नेल्सन मंडेला मार्ग, वसंत कुंज, नई दिल्ली–110070
दूरभाष : 011–26131498
ई–मेल : chairman@aicte-india.org

ALL INDIA COUNCIL FOR TECHNICAL EDUCATION
(A STATUTORY BODY OF THE GOVT. OF INDIA)
(Ministry of Education, Govt. of India)
Nelson Mandela Marg, Vasant Kunj, New Delhi-110070
Phone : 011-26131498
E-mail : chairman@aicte-india.org

FOREWORD

Engineers are the backbone of the modern society. It is through them that engineering marvels have happened and improved quality of life across the world. They have driven humanity towards greater heights in a more evolved and unprecedented manner.

The All India Council for Technical Education (AICTE), led from the front and assisted students, faculty & institutions in every possible manner towards the strengthening of the technical education in the country. AICTE is always working towards promoting quality Technical Education to make India a modern developed nation with the integration of modern knowledge & traditional knowledge for the welfare of mankind.

An array of initiatives have been taken by AICTE in last decade which have been accelerate now by the National Education Policy (NEP)2020. The implementation of NEP under the visionary leadership of Hon'ble Prime Minister of India envisages the provision for education in regional languages to all, thereby ensuring that every graduate becomes competent enough and is in a position to contribute towards the national growth and development through innovation & entrepreneurship.

One of the spheres where AICTE had been relentlessly working since 2021-22 is providing high quality books prepared and translated by eminent educators in various Indian languages to its engineering students at Under Graduate & Diploma level. For the second year students, AICTE has identified 87 books at Under Graduate and Diploma Level courses, for translation in 12 Indian languages - Hindi, Tamil, Gujarati, Odia, Bengali, Kannada, Urdu, Punjabi, Telugu, Marathi, Assamese & Malayalam. In addition to the English medium, the 1044 books in different Indian Languages are going to support to engineering students to learn in their mother tongue. Currently, there are 39 institutions in 11 states offering courses in Indian languages in 7 disciplines like Biomedical Engineering, Civil Engineering, Computer Science & Engineering, Electrical Engineering, Electronics & Communication Engineering, Information Technology Engineering & Mechanical Engineering, Architecture, and Interior Designing. This will become possible due to active involvement and support of universities/institutions in different states.

On behalf of AICTE, I express sincere gratitude to all distinguished authors, reviewers and translators from different IITs, NITs and other institutions for their admirable contribution in a very short span of time.

AICTE is confident that these out comes based books with their rich content will help technical students master the subjects with factor comprehension and greater ease.

(Prof. M. Jagadesh Kumar)

Acknowledgment

The author is grateful to AICTE for planning and publishing this book. In particular, I sincerely acknowledge the contribution of Dr. Amit Srivastava, Director, Faculty Development Cell of AICTE for the opportunity. The book was very thoroughly, thoughtfully and capably reviewed by Prof. M. K. Verma, Professor, IIT BHU. His contributions in improving the book are extremely noteworthy.

The book is primarily an outcome of teaching several batches of students over the years in the Circuit Theory course at IIT Delhi. Their questions and their lines of thought have inspired many aspects of this book. The individual units, aligned with the AICTE syllabus, originally took shape as hand-outs during the course. The exercises are the result of accumulating several years of tutorial and examination questions. The suggested project activities at the end of some of the chapters are ideas that second year students pursue while (or after) studying the course.

Several topics in the book were discussed at length with Prof. R.K. Patney, Prof. G.S. Visweswaran, Prof. B.K. Panigrahi, Prof. Ankesh Jain, and Prof. Anandarup Das, all at the Department of Electrical Engineering, IIT Delhi. Their contributions in the development of the book have been very important. Many thanks are due to my students, Aashish and Raghav, who have checked and re-checked many of the examples and questions.

Thanks are also due to many of my teachers. In particular, I would like to mention Prof. M. Anthony Reddy of IIT Madras, who taught me the value of clarity in reasoning. The subject of circuit theory has come to today's refinement through step-wise development of principles from the basic principles of Mathematics and Physics. The other teacher of mine who was a guiding force behind the book was Prof. Omar Wing [1928-2020] of Columbia University, New York (also the founding Dean of the Chinese University of Hong Kong). This book is modeled after his class of Network Theory at Columbia University, albeit re-tailored for an undergraduate audience.

One of the most important inspiring forces behind this book has been my former Ph.D. student, Dr. Nagarjuna Nallam. His constant push throughout the course of the development of the book is one of the reasons it is seeing the light of the day.

Last but not the least, the book would not have been possible without the support and patience of my family, my wife Uttara and my sons Dhimaan and Shivoham. They have borne my addiction to the book very very patiently.

Shouri Chatterjee
New Delhi
November 2022

Preface

Electrical Circuit Analysis and Network Theory has been written specifically with second year undergraduate students of Electrical Engineering and Electronics and Communications Engineering in mind. This book adheres to the AICTE model syllabus of two courses, namely PCC-EE01 Electrical Circuit Analysis for Electrical Engineering, and EC06 Network Theory for Electronics and Communications Engineering.

It is recommended that the contents of the book are taught and studied over the duration of one semester (roughly 45 one hour lectures). The table below gives a tentative budgeting of material for the teacher.

Unit	Title	Lecture Hours	Study Weeks
1	Fundamentals of circuit analysis	1-4	1
2	Network theorems	3-4	1-2
3	First and second order networks	8	3
4	Sinusoidal steady-state analysis	8	3
5	Laplace transforms	8	3
6	Fourier series and transforms	4	1
7	Two-port networks	6	2
8	Basic filters	3	1

There are eight units. Units 1 and 2 can be covered together in 4-8 lectures, depending on prior exposure of the students to circuit analysis. Units 3, 4 and 5 should be covered in 8 lectures each. The material in unit 6 can be delivered in 4 lectures, unit 7 can be budgeted 6 lectures, and lastly unit 8 is recommended no more than 3 lectures.

There are a few unique features of the book.

1. The book is reasonably rigorous in terms of the mathematics used. A student coming into the EE program from high school will appreciate the mathematical foundation of circuit theory.

2. The book contains guides to the use of hands-on numerical computational programs such as Octave and MATLAB®. These tools have deep relevance in Engineering and the Sciences today. The book exposes the student to these tools at an early stage in their undergraduate journey, to make them proficient in their usage. There are full programs in Octave that demonstrate circuit analysis, and analysis with Laplace and Fourier transforms. With enough practice in Octave, the diligent student can graduate to much wider realms of science and engineering.

3. The book also gives a preliminary introduction to SPICE, a very popular circuit simulation program.

The courses in Electrical Circuit Analysis and Network Theory lay the groundwork for several third year Electrical Engineering courses. Power engineering, power electronics, communications engineering, analog circuits, control engineering, filter design: the foundations for all of these fields are developed in the book.

The unit on sinusoidal steady state analysis brings out some interesting problems frequently put up by power engineers.

There are several illustrations of power electronic circuits through the book, including a harmonic analysis technique for a full-wave rectifier. Several questions on transient analysis are modeled after DC-DC converters, common in power electronics.

Introductions to Fourier series and Fourier transforms are included in the book. Several practical filter design examples for communication circuits are included in the text.

Fundamental concepts of power sources, as opposed to voltage and current sources, motivate analog circuits. Two-port network theory along with hints with how to design predictable feedback networks form the basis of modern analog circuit design.

The book provides a gentle but thorough study of linear differential equations and Laplace transforms. The section on Laplace transforms is foundational for the control engineer and many others.

Additional material for the inquisitive student is provided with the help of QR-codes through the book. These are extra topics that a student can branch into, and do not form the core text.

Outcome based education

For the implementation of an outcome based education the first requirement is to develop an outcome based curriculum and incorporate an outcome based assessment in the education system. By going through outcome based assessments evaluators will be able to evaluate whether the students have achieved the outlined standard, specific and measurable outcomes. With the proper incorporation of outcome based education there will be a definite commitment to achieve a minimum standard for all learners without giving up at any level. At the end of the program running with the aid of outcome based education, students will be able to arrive at the following outcomes:

PO-1: Engineering knowledge: Apply knowledge of mathematics, science, engineering fundamentals, and an engineering specialization to the solution of complex engineering problems.

PO-2: Problem analysis: Identify, formulate, review research literature, and analyze complex engineering problems reaching substantiated conclusions using first principles of mathematics, natural sciences, and engineering sciences.

PO-3: Design/development of solutions: Design solutions for complex engineering problems and system components or processes that meet the specified needs with consideration for public health, safety, culture, society, and the environment.

PO-4: Investigate complex problems: Apply research methods in the design of experiments, analysis and interpretation of data, to provide valid conclusions in the investigation of complex problems.

PO-5: Modern tool usage: Create, select, and apply appropriate techniques, resources, and modern engineering and IT tools for prediction and modeling of complex engineering activities with an understanding of its limitations.

PO-6: The engineer and society: Apply reason and contextual knowledge to assess societal, health, safety, legal and cultural issues and understand the consequent responsibilities relevant to professional engineering practice.

PO-7: Environment and sustainability: Understand the impact of the professional engineering solutions in societal and environmental contexts, and demonstrate the knowledge of, and need for sustainable development.

PO-8: Ethics: Apply ethical principles and commit to professional ethics, responsibilities and norms of engineering practice.

PO-9: Individual and team work: Function effectively as an individual, as a member or leader in diverse teams, and in multidisciplinary settings.

PO-10: Communication: Communicate effectively on complex engineering activities with the engineering community and with society at large. Comprehend and write effective reports and design documentation, make effective presentations, and give and receive clear instructions.

PO-11: Project management and finance: Learn the principles of engineering and management and apply these to one's own work, as a member and leader in a team, to manage projects in multidisciplinary environments.

PO-12: Life-long learning: Recognize the need for, and have the preparation and ability to engage in independent and life-long learning in the broad context of technological change.

Course outcomes

After completion of the course, the students will be able to:

CO-1: Analyze arbitrarily complex linear circuits at DC and AC optimally. Systematically analyze linear circuits under any given time-varying excitation, with and without the help of Laplace transforms.

CO-2: Develop an intuition of circuits. Be able to make predictions in terms of circuits.

CO-3: Develop a firm grip of Laplace and Fourier transforms. Develop a firm grasp of the relationship between Laplace, Fourier transforms, AC analysis, and frequency response. Be comfortable with circuit analysis with complex-numbers.

CO-4: Understand power in AC circuits. Understand three-phase AC systems.

CO-5: Visualize circuits as systems with input-output characteristics.

CO-6: Numerically analyze circuits.

	Expected mapping with program outcomes *(1: Weak correlation; 2: Medium correlation; 3: Strong correlation)*											
	PO-1	**PO-2**	**PO-3**	**PO-4**	**PO-5**	**PO-6**	**PO-7**	**PO-8**	**PO-9**	**PO-10**	**PO-11**	**PO-12**
CO-1	2	3	1	-	1	-	-	-	-	-	-	-
CO-2	2	2	2	3	1	-	-	-	-	-	-	1
CO-3	3	1	-	2	2	-	-	-	-	-	-	1
CO-4	3	1	1	1	2	1	-	-	-	-	-	-
CO-5	2	3	3	2	-	-	-	-	-	-	-	-
CO-6	1	1	1	1	3	-	-	-	-	-	-	-

Guidelines for Teachers

To implement an outcome-based education (OBE), the level of knowledge and skills of the students should be enhanced. Teachers are responsible for the proper implementation of OBE. Some of the responsibilities (not limited to) for the teachers in an OBE system are as follows:

- Within reasonable constraints, teachers should manage time to the advantage of all students.
- Teachers should assess students on certain defined criteria without considering any other potential ineligibility to discriminate them.
- Teachers should foster the learning abilities of the students.
- Teachers should ensure that all the students are equipped with knowledge and competence when they finish their education.
- Teachers should encourage students to improve their peak performance capabilities.
- Teachers should facilitate and encourage group work and team work.

- Teachers should follow Blooms taxonomy in every assessment.
- Students' learning should be connected with practical and real life situations.

Bloom's Taxonomy

Level	Teacher should check students' ability to	Student should be able to	Possible mode of assessment
Create	create	design or create	Mini project
Evaluate	justify	argue or defend	Assignment
Analyse	distinguish	differentiate or distinguish	Project/Lab methodology
Apply	use information	operate or demonstrate	Technical presentation
Understand	explain ideas	explain or classify	Presentation or seminar
Remember	recall	define or recall	Quiz

Guidelines for Students

Students should take equal responsibility in implementing an OBE. Some of the responsibilities for the students in an OBE system are as follows:

- Students should be aware of each UO before the start of a unit.
- Students should be well aware of each CO before the start of the course.
- Students should be well aware of each PO.
- Students should think critically and reasonably.
- Students should be well aware of their competence at every level of OBE.

Abbreviations, units, symbols

Units used		
Abbreviation	**Full form**	**Unit for**
V	volt	potential, voltage
A	ampere	current
s	second	time
rad	radian	angle, 2π radians make a circle
°	degree	angle, 360° make a circle
Hz	hertz	frequency, cycles per second
rad/s	radians per second	angular frequency
Ω	ohm	resistance, volt per ampere
S	siemens	conductance, ampere per volt
C	coulomb	charge
H	henry	inductance
F	farad	capacitance
W	watt	active power
V· A	volt ampere	apparent power, complex power
var	var	reactive power
dB	decibel	$10\log_{10} A$, A is a ratio of powers $20\log_{10} A$, A is an amplitude ratio

Note, 'mho' or ℧ is **not** an IEEE-standard unit for conductance.

List of symbols		
Symbol	**Description**	**Notes**
j	$\sqrt{-1}$	In EE, j is always $\sqrt{-1}$.
t	time	
T	period	of a periodic signal.
v, V	voltage	
i, I	current	In EE, i is always a current.
R	resistance	
C	capacitance	
L	inductance	
G	conductance	
M	mutual inductance	
Z	impedance, complex resistance	$Z = R + jX$
Y	admittance, complex conductance	$Y = G + jB$
ω	angular frequency	Unit is rad/s.
s	complex frequency, any complex number	Used for Laplace transforms. $s = \sigma + j\omega$
f	frequency	Unit is Hz.
P	power, active power	
S	complex power	$S = P + jQ$
$\mathbf{A}_i$	incidence matrix	
$\mathbf{A}$	reduced incidence matrix, also state-space matrix	depends on context
ϕ, θ	phase	
$\mathcal{L}$, $\mathcal{L}^{-1}$	Laplace transform, inverse	
$\mathcal{F}$, $\mathcal{F}^{-1}$	Fourier transform, inverse	
a_n, b_n C_n	Fourier series coefficients	Trigonometric Exponential, complex

Matrices are in bold. Symbols like $\overline{V}$ are phasors, or rotating vectors.

List of common abbreviations		
Acronym	**Description**	**Notes**
LHP	left half s-plane	Complex plane, $\mathrm{Re}[s] < 0$
RHP	right half s-plane	Complex plane, $\mathrm{Re}[s] > 0$
KVL	Kirchhoff's voltage law	KVL is always valid.
KCL	Kirchhoff's current law	KCL is always valid.
LTI	linear and time invariant	

Contents

1 Fundamentals of circuit analysis **1**

1.1 KCL and KVL . . . 3

1.1.1 Origins of KVL and KCL . . . 3

1.2 Device properties . . . 3

1.3 The node-voltage method . . . 8

1.3.1 The datum node . . . 8

1.3.2 Circuit examples for node-voltage analysis . . . 9

1.3.3 Circuit examples with dependent sources . . . 13

1.4 The mesh or loop-current method . . . 16

1.4.1 Circuit examples for mesh or loop analysis . . . 16

1.4.2 Choose the right loops . . . 20

1.4.3 Circuit examples with dependent sources . . . 21

1.5 Tellegen's theorem . . . 23

1.5.1 The incidence matrix . . . 23

1.5.2 Independent KVLs . . . 25

1.5.3 Tellegen's theorem . . . 25

1.6 Duality . . . 26

1.7 Unit summary . . . 29

1.8 Exercises . . . 31

2 Network theorems **37**

2.1 Tableau analysis . . . 39

2.2 Superposition theorem . . . 41

2.3 Thévenin and Norton theorems . . . 45

2.4 Source transformations . . . 49

2.5 Compensation theorem . . . 50

2.6 Reciprocity theorem . . . 52

2.7 Maximum power transfer theorem . . . 56

2.8 Unit summary . . . 60

2.9 Exercises . . . 62

3 First and second order networks **69**
3.1 First order differential equations . . . 71
3.1.1 Forms of the input . . . 71
3.1.2 Form of the solution . . . 72
3.1.3 Zero-input solution, free response . . . 72
3.1.4 Zero-state solution, forced response . . . 73
3.1.5 Complete solution . . . 73
3.1.6 Summary of steps . . . 74
3.2 Series RC and RL circuits . . . 74
3.2.1 Series RC circuit . . . 74
3.2.2 Initial and final conditions . . . 77
3.2.3 Series RL circuit . . . 78
3.3 Other first order RC and RL circuits . . . 79
3.3.1 Order of a circuit . . . 79
3.3.2 Step response of a first order circuit . . . 80
3.4 Second order networks: series, parallel RLC . . . 84
3.4.1 Series RLC circuits . . . 84
3.4.2 Over-damped . . . 87
3.4.3 Critically damped . . . 88
3.4.4 Under-damped . . . 89
3.4.5 Response metrics . . . 90
3.4.6 Parallel RLC . . . 92
3.5 Other second order circuits . . . 94
3.6 Unit summary . . . 98
3.7 Exercises . . . 99

4 Sinusoidal steady-state analysis **109**
4.1 Preliminaries . . . 111
4.2 The phasor . . . 114
4.3 Impedances and admittances . . . 117
4.3.1 Inductor . . . 117
4.3.2 Capacitor . . . 118
4.4 Mutual inductance . . . 122
4.4.1 The dot convention . . . 122
4.4.2 The transformer . . . 124
4.5 AC circuit analysis . . . 125
4.5.1 Network theorems . . . 129
4.6 Power computations with phasors . . . 131
4.6.1 Phasor convention . . . 131
4.6.2 Power in an element . . . 132
4.6.3 Maximum power transfer theorem . . . 135

4.6.4 Reactive power . . . 136
4.7 Three phase circuits . . . 138
4.7.1 Star-connected load . . . 139
4.7.2 Delta-connected load . . . 139
4.7.3 Line and phase . . . 140
4.7.4 Balanced loads . . . 141
4.7.5 Unbalanced three-phase networks . . . 143
4.8 Unit summary . . . 143
4.9 Exercises . . . 145

5 Laplace transforms **155**
5.1 The Laplace transform . . . 157
5.1.1 Convergence of the Laplace transform . . . 158
5.1.2 Important properties of the Laplace transform . . . 158
5.1.3 A few important Laplace transforms . . . 160
5.1.4 Initial and final value theorems . . . 165
5.2 The inverse Laplace transform . . . 166
5.2.1 Inversion integral . . . 166
5.2.2 Partial fraction expansions for inverse Laplace transforms 168
5.2.3 Partial fraction expansion with repeated poles . . . 170
5.3 Solving circuits with Laplace transforms . . . 173
5.3.1 Solving a differential equation . . . 173
5.3.2 Directly solving a circuit . . . 174
5.4 The circuit as a system . . . 180
5.4.1 Convolution . . . 180
5.4.2 Transfer function representation . . . 181
5.4.3 Poles and zeros . . . 182
5.4.4 Stability of a system . . . 183
5.5 A system in sinusoidal steady state . . . 184
5.5.1 Significance of poles and zeros . . . 185
5.5.2 Frequency response . . . 186
5.5.3 Series and parallel resonance . . . 188
5.6 Unit summary . . . 193
5.7 Exercises . . . 194

6 Fourier series and transforms **201**
6.1 Preliminaries . . . 203
6.1.1 Orthogonality of the trigonometric system . . . 203
6.1.2 Periodic signals . . . 203
6.2 Trigonometric Fourier series . . . 205
6.3 Waveform symmetries . . . 211

6.3.1 Even functions 211
6.3.2 Odd functions 212
6.3.3 Functions with half-wave symmetry 214
6.4 Exponential Fourier series 216
6.5 The square wave 217
6.5.1 Change in timing 217
6.5.2 Change in values of states 218
6.5.3 The sinc function 219
6.5.4 Discrete spectra 219
6.6 Steady state response of a network 221
6.6.1 Case study: oscilloscope probes 221
6.6.2 Case study: full-wave rectifiers 224
6.7 Power in a periodic signal 226
6.7.1 Series or shunt RC or RL loads 226
6.7.2 Any general load 228
6.8 The Fourier transform 229
6.8.1 Relationship with the Laplace transform 231
6.8.2 Properties of Fourier transforms 232
6.9 Unit summary 234
6.10 Exercises 236

7 Two-port networks **239**
7.1 Preliminaries 241
7.1.1 General idea behind two-port analyses 241
7.2 Impedance (Z) parameters 243
7.3 Conductance (Y) parameters 245
7.4 Star-Delta and Delta-Star conversion 247
7.5 Hybrid (H) parameters 252
7.6 Inverse hybrid (G) parameters 255
7.7 Input and output impodance 256
7.8 Transfer function 258
7.9 Transmission parameters 260
7.10 Relationship between variables 265
7.11 Circuit configurations 265
7.11.1 Series-series 265
7.11.2 Series-shunt 266
7.11.3 Shunt-series 267
7.11.4 Shunt-shunt 268
7.11.5 Overall remarks 268
7.12 Multi-port networks 270
7.13 Unit summary 271

7.14 Exercises 272

8 Basic filters **279**
8.1 Low-pass filters 281
8.1.1 High order Butterworth filters 282
8.1.2 Low pass filter transformation 284
8.2 High-pass filters 285
8.2.1 Low-pass to high-pass transformation 286
8.3 Band-pass filters 288
8.3.1 Low-pass to band-pass transformation 288
8.4 Band-stop filters 291
8.4.1 Low-pass to band-stop 292
8.5 Unit summary 295
8.6 Exercises 296

A Octave usage in this text **299**

List of Figures

1.1 Kirchhoff's voltage law (KVL) and Kirchhoff's current law (KCL). The sum of voltage drops around a loop is zero. The sum of currents leaving any node is zero. 4
1.2 (a) A resistor, (b) an inductor, and (c) a capacitor, with definitions of currents through them and voltage drops across them. 4
1.3 Definitions of the voltages and currents for a mutual inductance, as related to its equations given in (1.1) and (1.2). . . . 6
1.4 Several loops exist in any given circuit. The number of loops required to perform mesh or loop-current analysis is $B-(N-1)$, where there are B elements and N nodes in the circuit. The circuit shown has 13 elements, 10 nodes. 4 loop or mesh-current equations are sufficient. 20
1.5 Circuit to illustrate the incidence matrix and Tellegen's theorem. 23
1.6 (a) A circuit and (b) its dual. 27

2.1 A circuit with resistors, dependent and independent sources, used as an illustration for tableau analysis. 40
2.2 The network $\mathcal{N}$ contains linear circuit elements, dependent and independent sources. $\mathcal{N}'$ includes I_T that has been applied at the port of $\mathcal{N}$. 45
2.3 $\mathcal{N}$ contains linear circuit elements, independent sources, and linear dependent sources. $\mathcal{N}'$ includes V_N applied across the port. 46
2.4 Circuit simplification with the help of source transformations. 49
2.5 Compensation theorem 52
2.6 If $\mathcal{N}$ is passive, the reciprocity theorem (Theorem 2.6) states that I_{2_A} will be the same as I_{1_B}. 53
2.7 If $\mathcal{N}$ is passive, the reciprocity theorem (Theorem 2.7) states that V_{2_A} will be the same as V_{1_B}. 54

2.8 The maximum power transfer theorem (Theorem 2.8) states that maximum power is transferred from V_S to R_L if $R_L = R_S$. 57
2.9 We have plotted i_L as a function of v_L. In (a), the circuit corresponds to a voltage source V_S with a series resistance R_S. In (b), we have shown the corresponding characteristics of a photovoltaic cell. The available power is the area of the largest rectangle that can be drawn within the characteristics. 60

3.1 Series R-C circuit 75
3.2 Voltage across the capacitor for a series RC circuit, for different RC time-constants. The plot is from Prog. 3.1. 77
3.3 A series-RL circuit. The switch turns ON at $t = 0$. 78
3.4 Series RLC circuit. 85
3.5 A comparison of over-damped, critically damped and under-damped responses. The critically damped response has no overshoot, but a very low rise time. The under-damped response is very sharp, but is accompanied with ringing. The over-damped response is slow but has no ringing. 91
3.6 Metrics of the second-order under-damped response. t_p is the time at which the peak is attained; M_p is the peak overshoot. t_s is the settling time, t_f is the first time at which the response rises to 90% of the target. 92
3.7 Parallel RLC circuit 92

4.1 The phasor is a rotating complex vector. The actual voltage is the projection of the phasor on the real axis. 114
4.2 Mutual inductance. (a) The coupled terminals are both placed at the positive terminals, and (b) the coupling is between the positive and negative terminals. 122
4.3 Two mutually coupled inductors are driven by a voltage $\bar{V}_1$. The second inductor drives a load Z. 124
4.4 The power triangle for an inductive impedance. 133
4.5 The power triangle for a capacitive impedance. 134
4.6 Source trying to deliver power to a load 135
4.7 General three-phase configuration with a star-connected load. 139
4.8 General three-phase configuration with a delta-connected load. 140
4.9 Line and phase voltages and currents. (a) The line current and phase current are the same in a star-connected load. (b) The line voltage and phase voltage are the same in a delta-connected load. 141

4.10 (a) Phasor diagram for the line and phase voltages in a star-connected three-phase system. (b) Phasor diagram for the line and phase currents in a delta-connected three-phase system. . 142

5.1 Laplace domain model for a capacitor 174
5.2 Laplace domain model for an inductor 175
5.3 A circuit as a system. An independent voltage or current source is applied at the input. The output is any node voltage, or branch voltage, or branch current. 180
5.4 A system processes an input by multiplying its transfer function in the Laplace domain. 182
5.5 The frequency response of $H(s)$ is its value as we sweep s along the positive $j\omega$ axis. 188
5.6 A circuit with series LC resonance 189
5.7 $|H(j\omega)|$ as a function of ω for the circuit in Fig. 5.6. R, L, C were chosen as 1 Ω, 1 H, and 1 F, respectively. 190
5.8 Frequency response of the series RLC network as Q is increased from 1 to 32. 191
5.9 A circuit with shunt LC resonance 191
5.10 $|H(j\omega)|$ as a function of ω for the shunt-LC resonant circuit in Fig. 5.9. R, L, C were chosen as 1 Ω, 1 H, and 1F, respectively. 192
5.11 The bandwidth of the shunt-LC resonating circuit increases as Q is decreased. 193

6.1 Functions with only even symmetry. The axis of symmetry corresponding to the y-axis or $t = 0$ axis is implied. 212
6.2 Functions with only odd symmetry. The axis of symmetry corresponding to the y-axis or $t = 0$ is implied. 212
6.3 Functions with half-wave symmetry. When shifted by half of a period, the functions change in sign. The x-axis is implied. 214
6.4 A square wave between 0 and 1, with a period of 1 s, and a duty cycle of η. 217
6.5 sinc(x) 219
6.6 Spectrum of the square wave of Fig. 6.4. The duty cycle is 0.4. 220
6.7 Expansion or compression of the sinc function with duty cycle. 220
6.8 A full-wave rectifier with an R-L load. When $v(t)$ is positive, D_1 and D_3 conduct. When $v(t)$ is negative, D_4 and D_2 conduct. 224
6.9 Full-wave rectified voltage. In the circuit of Fig. 6.8, $v_{AB}(t)$ is shown assuming ideal diodes. 224
6.10 Full-wave rectified voltage and current through an RL load. . . 226

6.11 The voltage across the network is $v(t)$ and the current going into the network is $i(t)$. $i(t)$ and $v(t)$ have a period of T. . . . 228
6.12 Fourier coefficients as the period changes. 230

7.1 A two-port network in a circuit. The two-port network is an arbitrary connection of components. The voltage and current conventions are as shown. 241
7.2 A T-network. It is convenient to use the **Z**-parameters to abstract a T-network. 243
7.3 A Π-network. It is convenient to abstract a Π-network with the help of the **Y**-parameters. 244
7.4 Two experiments to evaluate the **H**-parameters. 252
7.5 The model for a BJT is often expressed with **H**-parameters. . 253
7.6 Circuit topology for driving point impedance evaluation. . . . 256
7.7 A two-port network used as a system. We have shown the input source as a voltage source in (a), as a current source in (b). The output quantity is either V_{out} or I_{out}. 258
7.8 Two-port network for example-7.9. 261
7.9 We need to perform two experiments to evaluate the transmission parameters. (a) open circuit at $\mathcal{P}_2$ and (b) short circuit at $\mathcal{P}_2$. 262
7.10 Two networks cascaded. Clearly $V_{2_a} = V_{1_b}$ and $-I_{2_a} = I_{1_b}$. . . 263
7.11 Two-port network for example-7.10. 264
7.12 Series-series configuration 266
7.13 Series-shunt configuration 267
7.14 Shunt-series configuration 267
7.15 Shunt-shunt configuration 268
7.16 Common interconnections that preserve port properties. . . . 269

8.1 Ideal filter characteristics. (a) Low-pass, (b) high-pass, (c) band-pass, and (d) band-stop filter characteristics. 281
8.2 Basic RC filter 281
8.3 Frequency response of RC filter. $\omega_0 = 1/(RC)$. The dashed line represents the desired ideal characteristics. 282
8.4 The response of the Butterworth filter for different orders, from $n = 1$ to $n = 7$. 282
8.5 (a) A loss-less two-port ladder network is configured as a filter. Source and load impedances of 1 Ω are chosen. Two possible LC-ladder networks as filters are shown in (b) and (c). 283
8.6 Simulated magnitude of the frequency response of the scaled Butterworth low-pass filter of Example 8.1. 286

8.7 Illustration of the frequency mapping plan for a low-pass to high-pass transformation. 287
8.8 Simulated transfer function of the circuit in Example 8.2. . . 288
8.9 $2 \mapsto 1$ map of a low-pass to band-pass transformation. Two points on the frequency axis of the final filter map to one point on the frequency axis of the prototype. 289
8.10 Simulated transfer function of the Butterworth band-pass filter of Example 8.4. The zoomed-in view on the right shows the flat nature of the pass-band. 291
8.11 Frequency mapping scheme for a low-pass to band-stop transformation. 292

List of Tables

5.1 Laplace transform pairs 164

6.1 Fourier transform pairs. These are in addition to the Laplace transform pairs of Tab. 5.1, where a replacement of s with $j\omega$ will give us a Fourier transform pair. 233

7.1 All possible combinations of independent and dependent variables for two-port networks, with their names and definitions. 242

7.2 Conversion between star (T or Y) and delta (Δ or Π) 249

7.3 Conversion between different two-port parameters. 265

Unit 1

Fundamentals of circuit analysis

Unit specifics

In this unit we have discussed the following:

- Device properties, resistors, capacitors, inductors, mutual inductance, independent and dependent voltage and current sources.
- The node-voltage method to analyze circuits.
- The mesh-current or loop-current method to analyze circuits.
- Tellegen's theorem.
- The dual of a circuit.

Rationale

This unit will help students enhance their circuit-analysis knowledge from basic KCL-KVL analysis of a circuit to a optimal circuit analysis. The node-voltage and mesh-current methods are optimal techniques to analyze circuits. The student will also obtain a gentle introduction to Octave (like MATLAB®), and will be nudged towards hands-on computation techniques. The unit will present a theoretical understanding of circuit networks, and develop the powerful Tellegen's theorem.

Circuits form the bedrock of Electrical Engineering. The study of electric circuits has two main parts, namely circuit analysis and circuit synthesis.

Circuit analysis is covered extensively in this text. This chapter forms the ground work for circuit analysis.

Pre-requisites

- Physics: Circuit analysis, KCL and KVL (Class XII)
- Mathematics: Basic matrix algebra (Class XII)
- Software: We will be using an open-source numerical package, Octave. Octave is compatible with the commercially available MATLAB®.

Unit outcomes

The list of outcomes of this unit are as follows.

U1-O1: Learn the node-voltage method, mesh-current method.

U1-O2: Learn and apply Tellegen's theorem.

U1-O3: Learn to handle dependent voltage and current sources.

U1-O4: Start using a numerical tool.

U1-O5: Be comfortable with circuit analysis at DC.

U1-06: Understand the orthogonality between voltages and currents.

Unit-1 outcomes	Expected mapping with course outcomes *(1: Weak, 2: medium, and 3: strong correlation)*					
	CO-1	CO-2	CO-3	CO-4	CO-5	CO-6
U1-O1	2	1	-	-	-	-
U1-O2	-	2	-	-	-	-
U1-O3	3	-	-	-	-	-
U1-O4	1	-	-	-	-	3
U1-O5	3	-	-	-	-	-
U1-O6	-	3	-	1	-	-

1.1 KCL and KVL

Kirchhoff's current law (KCL) and Kirchhoff's voltage law (KVL) form the very basis of circuit theory. KCL states that the sum of outward moving currents equals zero at any point in a circuit. KVL states that in any loop in a circuit, the sum of potential drops equals zero. In this chapter, we will look at how to efficiently analyze circuits with a minimal set of independent KCLs and KVLs.

The sketches in Fig. 1.1 pictorially depict KVL and KCL. The voltage **drops** around any loop of elements in a circuit add up to zero. In the Fig. 1.1, $v_a + v_b + v_c + v_d = 0$. In case of KCL, the sum of currents leaving any node is zero. In the Fig. 1.1, $i_a + i_b + i_c + i_d = 0$.

It is imperative to note that in any given circuit, both KVL and KCL are valid at all times and under all circumstances.

1.1.1 Origins of KVL and KCL

The origin of KCL is from conservation of charge. Charge can not be created or destroyed at any point in space, at any given time. In other words, the sum of all charges flowing out of a node at any point in time is zero; the sum of the rate of flow of charges flowing out of a node at any point in time is zero. This implies that the sum of currents flowing out of any node at any time is zero.

The origin of KVL is from conservation of energy. In electrostatics, the electric potential, e_1, at a point x_1 is defined as the energy required to move a unit charge from a point at zero potential to x_1. The energies required to move the unit charge from zero potential to x_2, x_3, x_4 are similarly defined as e_2, e_3, e_4 respectively. The energy required to return the charge back from x_1 to the point at zero potential is $-e_1$. The energy required to move the unit charge from x_1 to x_2, therefore, must be $e_2 - e_1$. Similarly, the energy required to move the unit charge from x_2 to x_3 will be $e_3 - e_2$; the energy required to move the unit charge from $x_3 to x_4$ will be $e_4 - e_3$; the energy required to move the unit charge from x_4 to x_1 is $e_1 - e_4$. The energy needed to move the unit charge around a loop, from x_1 to x_2 to x_3 to x_4 and then back to x_1 will therefore be $(e_2 - e_1) + (e_3 - e_2) + (e_4 - e_3) + (e_1 - e_4) = 0$.

1.2 Device properties

We will initially look at the different devices that we will use in this text.

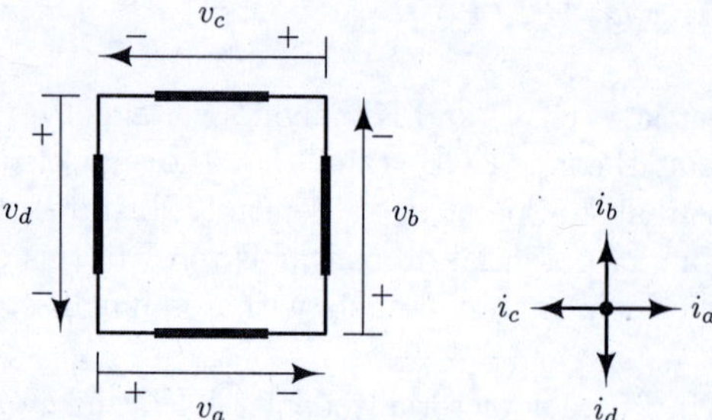

Figure 1.1: Kirchhoff's voltage law (KVL) and Kirchhoff's current law (KCL). The sum of voltage drops around a loop is zero. The sum of currents leaving any node is zero.

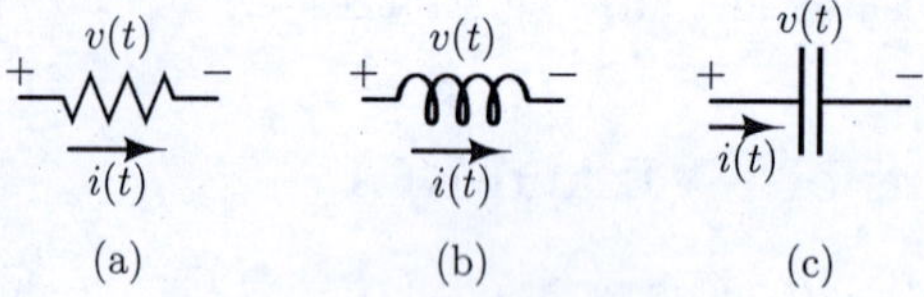

Figure 1.2: (a) A resistor, (b) an inductor, and (c) a capacitor, with definitions of currents through them and voltage drops across them.

Resistors: Resistors follow Ohm's law, i.e., the voltage across a resistor is proportional to the current through the resistor. R, the resistance in ohms, relates voltage and current through the well-known relationship:

$$v(t) = i(t)R$$

where $v(t)$ and $i(t)$ are the voltage across and current through the resistor, as shown in Fig. 1.2(a).

Inductors: Inductors relate the derivative of the current with the voltage across. The relationship for an inductor is:

$$v(t) = L\frac{di(t)}{dt}$$

where the voltage $v(t)$ and the current $i(t)$ are as shown in Fig. 1.2(b). *The voltage* **drop** *is L times the derivative of the current through the inductor.*[1]

[1]Many Physics texts refer to this as back-emf and create needless confusion.

Capacitors: Capacitors relate the derivative of the voltage across them with the current. The relationship for a capacitor is:

$$i(t) = C\frac{dv(t)}{dt}$$

where $v(t)$ is the voltage drop across the capacitor, and $i(t)$ is the current through the capacitor, as shown in Fig. 1.2(c).

Mutual inductance: A pair of magnetically coupled inductors form a mutual inductance. The element is a four-terminal network with three parameters, L_1, L_2, and M, where L_1 is the inductance of one coil, L_2 is the inductance of the other coil, and M is the mutual inductance. We have summarized the relationships in the pair of equations (1.1), (1.2).

$$v_1(t) = L_1\frac{di_1(t)}{dt} + M\frac{di_2(t)}{dt} \quad (1.1)$$

$$v_2(t) = M\frac{di_1(t)}{dt} + L_2\frac{di_2(t)}{dt} \quad (1.2)$$

where v_1, v_2, i_1, and i_2 are as shown in Fig. 1.3. The dots in a mutual inductance define the positive terminals of the two terminal-pairs.

The mutual inductance parameter, M, is related to the coupling between the two coils. A coupling coefficient, k, is the fraction of the magnetic flux generated by one coil coupled to the second coil. If the coupling coefficient, k, is 1, all the magnetic flux that goes through the first coil also goes through the second coil. If k is 0.5, 50% of the magnetic flux in the first coil goes through the second coil. The reverse is also implied and can be proven through electromagnetics [1], i.e., if k fraction of the flux generated by the first coil goes through the second, then k fraction of the flux generated by the second coil goes through the first.

k is always a positive quantity between 0 and 1. M is $k\sqrt{L_1 L_2}$. As such, the maximum value of M is $\sqrt{L_1 L_2}$, when $k = 1$. In such a case, one can prove that the mutual inductance simplifies to be an ideal transformer.

Scan the QR-code, and learn more about a variety of devices and their properties. Resistors, capacitors, inductors may not always be ideal.

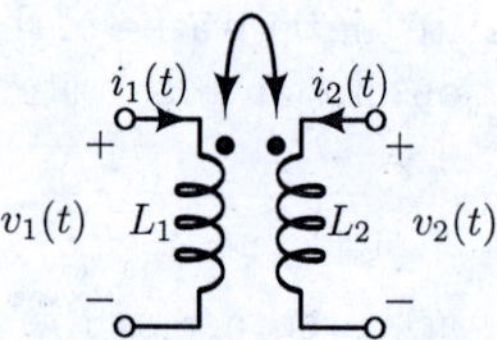

Figure 1.3: Definitions of the voltages and currents for a mutual inductance, as related to its equations given in (1.1) and (1.2).

Voltage sources: A voltage source defines the voltage across a branch, while the current through the branch is arbitrary.

Corollary 1.1. An ideal voltage source, V_0, in parallel with other components can be simply reduced to the voltage source, V_0.

Corollary 1.2. Two ideal voltage sources cannot be placed in parallel. Even if the values of the voltage sources are equal, the currents through them cannot be defined.

Placing voltage sources in parallel is generally bad practice. With a slight mismatch in values of the voltage sources, the internal source resistances will mitigate the mismatch. There will be large currents looping through the voltage sources, depending on the mismatch and the internal source resistances. For example, placing two Li-ion cells of the same voltage in parallel will create large circulating currents based on the mismatch of voltages, and rapidly degrade the capacity of the cells. Large capacity battery banks are assembled by placing **carefully matched, identical** cells in parallel.

Example 1.1. The internal resistances of two Li-ion cells were 0.02 Ω and 0.01 Ω, while the voltages of the two cells were 3.2 V and 3.3 V, respectively. The cells were placed in parallel with each other. What is the circulating current through the cells?

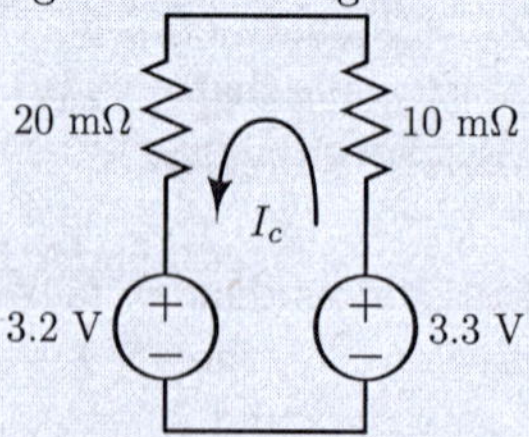

The voltage across the two resistors is $3.3 - 3.2 = 0.1$ V. The series combination of the two resistors is 0.03 Ω. The circulating current is therefore $0.1/0.03 = 3.33$ A.

Current sources: A current source defines the current through a branch, while the voltage across the branch is arbitrary.

Corollary 1.3. An ideal current source, I_0, in series with any other component can be simply reduced to the current source I_0.

Corollary 1.4. Two ideal current sources cannot be placed in series. Even if the values of the current sources are equal, the voltage drops across them cannot be defined.

If one needs to increase the power in a circuit, it is generally better to place multiple current sources in parallel. Alternatively, one could place multiple voltage sources in series to augment the power in a circuit.

Controlled voltage and current sources: Controlled sources (both voltage and current) define the voltage across or the current through a branch. The defined quantity is related to the voltage across or current through some other branch in the circuit. For instance, a voltage-controlled current source defines the current through a branch. This defined current is related to the voltage across some other specified branch. The controlled sources can be (1) voltage-controlled voltage sources (VCVS), (2) voltage-controlled current sources (VCCS), (3) current-controlled voltage sources (CCVS), and (4) current-controlled current sources (CCCS).

For instance, one may come across a circuit element that generates a voltage $H \cdot v_x$, where H is a dimensionless constant and v_x is the potential across two given terminals in the circuit. The circuit element is a voltage-controlled voltage source (VCVS). Similarly, one may find a circuit element that generates a current $G \cdot v_x$, where G is a constant with dimensions of conductance (S). In such a case, the circuit element is a VCCS. The proportionality constants required in the CCVS has dimensions of resistance while that required in the CCCS is dimensionless. Example 1.2 illustrates the usage of dependent controlled voltage and current sources.

There are no physical devices that directly emulate dependent voltage and current sources. However, these constructs are useful in modeling complex higher order devices and in circuit analysis.

Example 1.2. In the circuit schematic shown below, evaluate i_e.

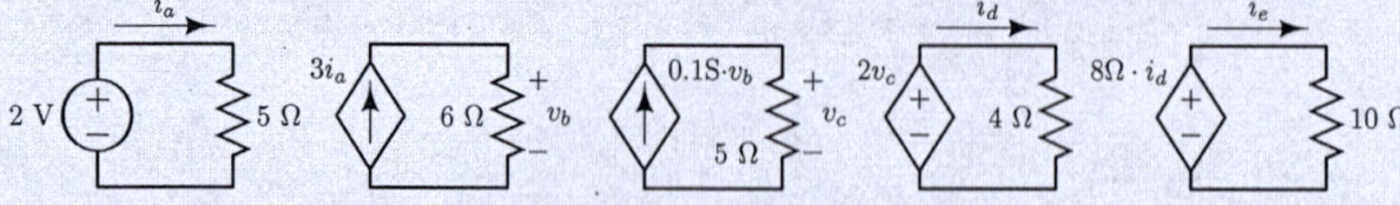

1. i_a is $2/5 = 0.4$ A.
2. The CCCS generates a current of $3 \times 0.4 = 1.2$ A.
3. v_b is $1.2 \times 6 = 7.2$ V.
4. The VCCS generates a current of $0.1 \times 7.2 = 0.72$ A.
5. v_c is $0.72 \times 5 = 3.6$ V.
6. The VCVS generates a voltage of $2 \times 3.6 = 7.2$ V.
7. i_d is $7.2/4 = 1.8$ A.
8. The CCVS generates a voltage of $1.8 \times 8 = 14.4$ V.
9. i_e is $14.4/10 = 1.44$ A.

1.3 The node-voltage method

The node-voltage method is based on forming a linearly independent but complete set of KCLs. The currents in the KCLs are evaluated from device properties.

1.3.1 The datum node

One of the nodes in the graph is chosen as a datum node. In common usage, this is called the "ground" or "earth." The datum node is nothing but a reference node, and we define potentials at the other nodes in the circuit with respect to the datum. In practice, the datum is often physically and electrically connected to the Earth or the ground. The Earth is deemed to be a good conductor at a constant potential of zero.

KCL is not written for the datum node.

Potentials at all other nodes are considered as variables. The potentials at the other nodes are all with respect to the datum node. Any node can be chosen as the datum.

1.3.2 Circuit examples for node-voltage analysis

Node-voltage analysis requires KCL equations for each node. Further KVLs are built into the system of equations and related to the branch currents through Ohm's law or device equations. KCL is not written at the datum; a KCL at the datum is linearly dependent on all other KCLs, as will be proved in theorem 1.1.

Our first example of node-voltage analysis is for a circuit with five nodes, resistors, and current sources as circuit elements, given below.

Example 1.3. Solve the circuit below. Assume I_1 is 1 A, I_2 is 2 A. Assume R_1 is 1 Ω, R_2 is 2 Ω, and likewise, R_k is k Ω. Evaluate the current through R_8 from left to right.

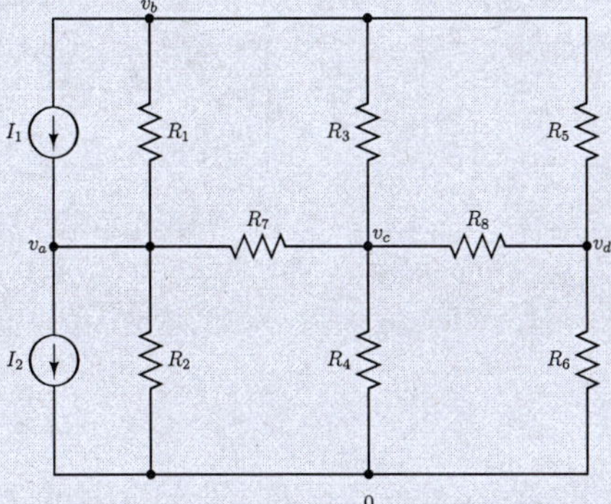

The node at the bottom, marked as '0', is the datum. KCLs are written at the nodes v_a, v_b, v_c and v_d. These four KCLs are sufficient. All other KCLs are linearly dependent and are not required. The current in each branch of the circuit is deduced from the corresponding device equation. For R_1, one end of R_1 is at potential v_b, the other end at potential v_a. The potential difference across R_1 is $v_b - v_a$. The current coming out of node b is therefore $(v_b - v_a)/R_1$. *Currents through all the branches of the circuit are evaluated in this manner, mentally, while writing out the KCLs.*

Now we write out the KCLs at the four nodes. The convention for KCL at a node is to equate the total current coming out of the node to zero.

$$-I_1 + I_2 + (v_a - v_b)/R_1 + v_a/R_2 + (v_a - v_c)/R_7 = 0$$
$$I_1 + (v_b - v_a)/R_1 + (v_b - v_c)/R_3 + (v_b - v_d)/R_5 = 0$$
$$(v_c - v_a)/R_7 + v_c/R_4 + (v_c - v_b)/R_3 + (v_c - v_d)/R_8 = 0$$
$$(v_d - v_c)/R_8 + (v_d - v_b)/R_5 + v_d/R_6 = 0$$

The above set of four equations can be re-written as:

$$\mathbf{G} \cdot \begin{bmatrix} v_a \\ v_b \\ v_c \\ v_d \end{bmatrix} = \begin{bmatrix} I_1 - I_2 \\ -I_1 \\ 0 \\ 0 \end{bmatrix}$$

where $\mathbf{G}$ is

$$\begin{bmatrix} \frac{1}{R_1} + \frac{1}{R_2} + \frac{1}{R_7} & -\frac{1}{R_1} & -\frac{1}{R_7} & 0 \\ -\frac{1}{R_1} & \frac{1}{R_1} + \frac{1}{R_3} + \frac{1}{R_5} & -\frac{1}{R_3} & -\frac{1}{R_5} \\ -\frac{1}{R_7} & -\frac{1}{R_3} & \frac{1}{R_7} + \frac{1}{R_3} + \frac{1}{R_4} + \frac{1}{R_8} & -\frac{1}{R_8} \\ 0 & -\frac{1}{R_5} & -\frac{1}{R_8} & \frac{1}{R_5} + \frac{1}{R_8} + \frac{1}{R_6} \end{bmatrix}$$

Inverting matrix $\mathbf{G}$ one can obtain all the node voltages.

$$\begin{bmatrix} v_a \\ v_b \\ v_c \\ v_d \end{bmatrix} = \mathbf{G}^{-1} \cdot \begin{bmatrix} I_1 - I_2 \\ -I_1 \\ 0 \\ 0 \end{bmatrix}$$

It is not easy to invert a matrix by hand. However, one can make use of Octave to quickly invert the matrix. The code-snippet below illustrates the manner in which the question may be solved using Octave.

```
octave:1> [r1, r2, r3, r4, r5, r6, r7, r8] = deal(1,2,3,4,5,6,7,8);
octave:2> [i1, i2] = deal(1, 2);
octave:3> G = [1/r1+1/r2+1/r7, -1/r1, -1/r7, 0; ...
-1/r1, 1/r1+1/r3+1/r5, -1/r3, -1/r5; ...
-1/r7, -1/r3, 1/r7+1/r3+1/r4+1/r8, -1/r8; ...
0, -1/r5, -1/r8, 1/r5+1/r8+1/r6];
octave:4> V = inv(G)*[i1-i2; -i1; 0; 0]
V =
   -2.5440
   -2.9204
   -1.8128
   -1.6489
```

Finally, the current through R_8 from left to right is nothing but $(v_c - v_d)/R_8$, and works out to -20.5 mA.

In a second example, we will demonstrate node voltage analysis for a circuit with voltage sources and resistors. The current through a voltage source is indeterminate, i.e., any current can flow through an ideal voltage source. As such, it is not possible to write KCLs at nodes attached to voltage sources. How do we perform node voltage analysis then?

Example 1.4. Assume V_1 is 1 V, V_2 is 2 V, R_1 is 1 Ω, R_2 is 2 Ω, and likewise R_k is k Ω. What is the current through R_7 from left to right?

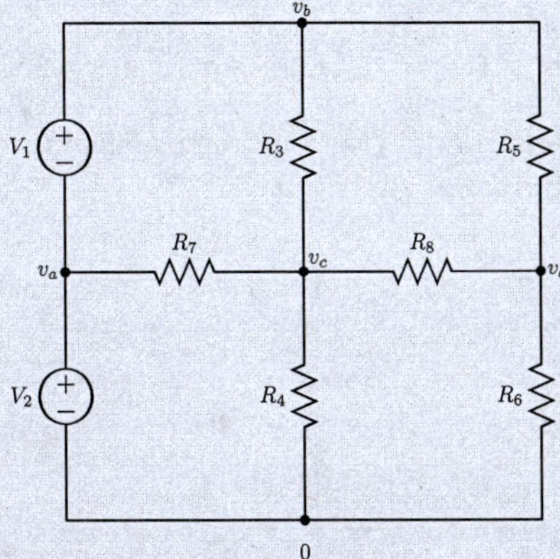

It is not possible to write KCLs at nodes a and b. However, observe that it is not necessary to write the KCL at these nodes at all. The voltage sources already define the potentials at nodes a and b.

$$v_a = V_2, \quad \text{and, } v_b = V_2 + V_1$$

We will only write KCLs at the remaining two nodes:

$$v_c/R_4 + (v_c - V_2)/R_7 + (v_c - V_1 - V_2)/R_3 + (v_c - v_d)/R_8 = 0$$
$$v_d/R_6 + (v_d - v_c)/R_8 + (v_d - V_1 - V_2)/R_5 = 0$$

or in matrix form:

$$\begin{bmatrix} \frac{1}{R_4} + \frac{1}{R_7} + \frac{1}{R_3} + \frac{1}{R_8} & -\frac{1}{R_8} \\ -\frac{1}{R_8} & \frac{1}{R_6} + \frac{1}{R_8} + \frac{1}{R_5} \end{bmatrix} \begin{bmatrix} v_c \\ v_d \end{bmatrix} = \begin{bmatrix} \frac{V_2}{R_7} + \frac{V_1}{R_3} + \frac{V_2}{R_3} \\ \frac{V_1}{R_5} + \frac{V_2}{R_5} \end{bmatrix}$$

The Octave code snippet below evaluates the question numerically.

```
octave:1> [r1, r2, r3, r4, r5, r6, r7, r8] = deal(1,2,3,4,5,6,7,8);
octave:2> [v1, v2] = deal(1, 2);
octave:3> G = [1/r4+1/r7+1/r3+1/r8, -1/r8; -1/r8, 1/r6+1/r8+1/r5];
octave:4> V = inv(G)*[v2/r7+v1/r3+v2/r3; v1/r5+v2/r5];
octave:5> I = (v2 - V(1))/r7
I = 0.034967
```

As such, the current through R_7 from left to right is 35 mA.

Performing node analysis with voltage sources is not always so straightforward. In the previous example, the two voltage sources were directly establishing the potentials at a and b with respect to the datum. In some situations, the voltage source may not directly establish a potential at a node. An example is shown next.

Example 1.5. In the circuit below, V_1 is 1 V, V_2 is 2 V, R_1 is 1 Ω, R_2 is 2 Ω, and likewise, R_k is k Ω. Use node voltage analysis and evaluate the current through R_3 from top to bottom.

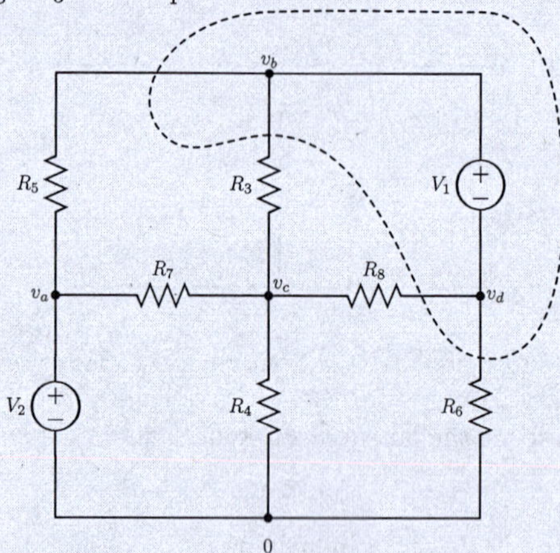

The branches with voltage sources are a challenge to the node voltage method, where we are to write out KCLs at all the nodes. The potential v_a is nothing but V_2, and as such, the node voltage equation at a is not required. v_b is $v_d + V_1$. However, we need to write out at least two KCLs to solve the two unknowns, v_c, and v_d. The voltage source V_1 can be tackled using a super-node around the two nodes connected to it, combining b and d as shown with the dashed curve. We will write out one KCL for c and one KCL for the super-node combining b and d. The node voltage equations are then:

$$(v_c - V_2)/R_7 + v_c/R_4 + (v_c - v_d)/R_8 + (v_c - v_d - V_1)/R_3 = 0$$
$$(v_d + V_1 - V_2)/R_5 + (v_d + V_1 - v_c)/R_3 + (v_d - v_c)/R_8 + v_d/R_6 = 0$$

or,

$$\begin{bmatrix} \frac{1}{R_7} + \frac{1}{R_4} + \frac{1}{R_8} + \frac{1}{R_3} & -\frac{1}{R_8} - \frac{1}{R_3} \\ -\frac{1}{R_3} - \frac{1}{R_8} & \frac{1}{R_5} + \frac{1}{R_3} + \frac{1}{R_8} + \frac{1}{R_6} \end{bmatrix} \begin{bmatrix} v_c \\ v_d \end{bmatrix} = \begin{bmatrix} \frac{V_2}{R_7} + \frac{V_1}{R_3} \\ \frac{V_2}{R_5} - \frac{V_1}{R_5} - \frac{V_1}{R_3} \end{bmatrix}$$

Once again, it is instructive to solve for the current through R_3 using Octave.

```
octave:1> [v1,v2]=deal(1,2);
octave:2> [r1,r2,r3,r4,r5,r6,r7,r8]=deal(1,2,3,4,5,6,7,8);
octave:3> G = [1/r7+1/r4+1/r8+1/r3, -1/r8-1/r3; ...
-1/r8-1/r3, 1/r5+1/r3+1/r8+1/r6];
octave:4> V = inv(G)*[v2/r7+v1/r3; v2/r5-v1/r5-v1/r3];
octave:5> I = ((V(2)+v1)-V(1))/r3
I = 0.1441
```

The current through R_3 from v_b to v_c is 144 mA.

Observation 1.1. The node-voltage analyses circuit examples show that the matrix **G** is symmetric in all three cases. **G** is always symmetric as long as there are no dependent sources in the circuit. Symmetry in **G** can frequently be used as a sanity check while solving circuits. A path from node m to n through a resistor R_{mn} will create an entry $-1/R_{mn}$ for G_{mn} and G_{nm}. As such, a circuit with independent sources and resistors will result in a symmetric **G**.

1.3.3 Circuit examples with dependent sources

So far we have restricted our circuits to resistors and independent current and voltage sources. Node voltage analysis for circuits with dependent voltage and current sources is not much different. As before, one cannot write out a KCL equation at nodes connected to dependent voltage sources. These nodes will be eliminated. Dependent current sources will have to be handled just like regular current sources. It is straightforward to handle any current source, independent or dependent, in the node-voltage method.

The next example illustrates a circuit with an independent voltage source, a dependent voltage-controlled current source, and a dependent current-controlled voltage source. Two nodes, one corresponding to the independent voltage source, and a second corresponding to the dependent voltage source, will need to be eliminated.

Example 1.6. In the circuit below, V_1 is 1 V, G_2 is 1/2 S, R_1 is 1 Ω, R_2 is 2 Ω, and likewise R_k is k Ω. Evaluate the current i_1.

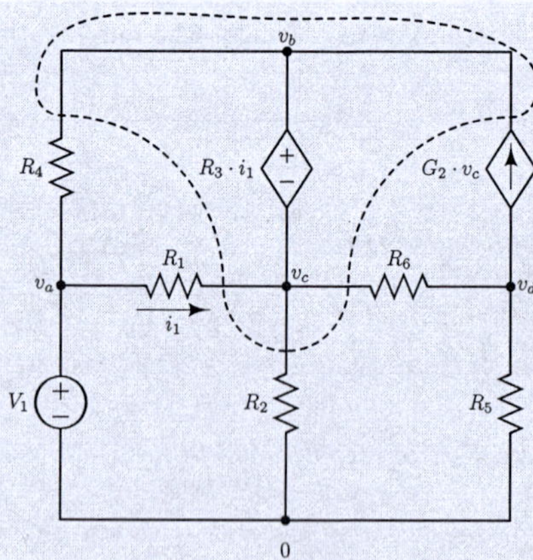

Similar to our previous example 1.5, in this case, a super-node is constructed around the CCVS, as shown in the dashed region. Before we start writing out KCLs, we observe that v_a is V_1, i_1 is $(V_1 - v_c)/R_1$, and v_b is $v_c + (V_1 - v_c) \cdot R_3/R_1$. There are only two unknown node voltages, namely v_c and v_d. Now we will write out KCLs for the node d and the super-node combining b and c.

$$\begin{aligned} (v_c - v_a)/R_1 + v_c/R_2 + (v_c - v_d)/R_6 + (v_b - v_a)/R_4 - G_2 v_c &= 0 \\ (v_d - v_c)/R_6 + v_d/R_5 + G_2 v_c &= 0 \end{aligned}$$

In the two equations, we can substitute the values of v_a and v_b as expressed earlier. We now combine the two equations in a matrix form, with only two variables, namely v_c and v_d.

$$\begin{bmatrix} \frac{1}{R_1} + \frac{1}{R_2} + \frac{1}{R_6} + \frac{1}{R_4} - \frac{R_3}{R_1 R_4} - G_2 & -\frac{1}{R_6} \\ G_2 - \frac{1}{R_6} & \frac{1}{R_6} + \frac{1}{R_5} \end{bmatrix} \cdot \begin{bmatrix} v_c \\ v_d \end{bmatrix} = \begin{bmatrix} \frac{V_1}{R_1} + \frac{V_1}{R_4} - \frac{V_1 R_3}{R_1 R_4} \\ 0 \end{bmatrix}$$

The final solution using Octave is given below.

```
octave:1> [v1,g2]=deal(1,1/2);
octave:2> [r1,r2,r3,r4,r5,r6]=deal(1,2,3,4,5,6);
octave:3> G = [1/r1+1/r2+1/r6+1/r4-r3/r1/r4-g2, -1/r6;...
g2-1/r6, 1/r6+1/r5];
octave:4> V = inv(G)*[v1/r1+v1/r4-v1*r3/r1/r4; 0];
octave:5> I = (v1-V(1))/r1
I = 0.3889
```

A more difficult situation may arise when the controlled source is dependent on itself. The next example illustrates a circuit. The general principle remains the same, i.e., if the number of nodes in a circuit is N, and the number of dependent and independent voltage sources is K, then the number of KCL equations to be solved is $N - 1 - K$.

Example 1.7. In the circuit shown below, the resistors R_1, R_2, R_3, R_4 are 1 Ω, 2 Ω, 3 Ω and 4 Ω respectively. Z is 5 Ω. I_0 is 1 A. Evaluate i_3.

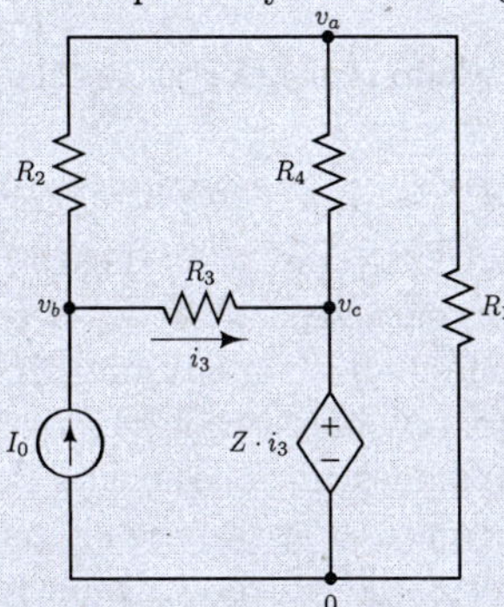

v_a and v_b are unknown. v_c is Zi_3, where i_3 is $(v_b - v_c)/R_3$. In other words,

$$v_c = Z \cdot \frac{v_b - v_c}{R_3}$$

The expression can be simplified to arrive at v_c in terms of v_b. v_c is, therefore, $v_b Z/(R_3 + Z)$.

Now we will write out KCLs at nodes a and b.

$$\begin{aligned} v_a\left(\frac{1}{R_1} + \frac{1}{R_2} + \frac{1}{R_4}\right) - v_b\left(\frac{1}{R_2} + \frac{Z/R_4}{R_3 + Z}\right) &= 0 \\ -v_a\frac{1}{R_2} + v_b\left(\frac{1}{R_2} + \frac{1}{R_3} - \frac{Z/R_3}{R_3 + Z}\right) &= I_0 \end{aligned}$$

As before, the remainder of the problem can be solved with the help of Octave.

```
octave:1> [i0,z]=deal(1,5);
octave:2> [r1,r2,r3,r4]=deal(1,2,3,4);
octave:3> G = [1/r1+1/r2+1/r4, -(1/r2+z/r4/(r3+z));...
-1/r2, 1/r2+1/r3-z/r3/(r3+z)];
octave:4> V = inv(G)*[0; i0];
octave:5> I = (V(2) - V(2)*z/(r3+z))/r3
I = 0.2857
```

Observation 1.2. With dependent sources, the **G** matrix is no longer symmetric. In example 1.6, applying $G_2 = 0$ will introduce symmetry in the **G** matrix. In example 1.7, applying $Z = 0$ will make the **G** matrix symmetric. In other words, dependent sources create asymmetry in the **G** matrix during node-voltage analyses.

1.4 The mesh or loop-current method

Definition 1.1 (Loop): A loop is a set of circuit elements that form a closed path.

Definition 1.2 (Mesh): A mesh is a loop which does not have a loop within.

KVL equations can be written out around each mesh of a planar circuit. These KVL equations are guaranteed to be linearly independent and sufficient. The definitions of mesh and loop are not very strict; it is possible to redraw a circuit in a different manner to convert a loop into a mesh and vice-versa. For a circuit with N nodes and B elements (branches), the number of mesh equations that are required to solve the circuit is $B - (N - 1)$. These mesh equations need to cover all elements of the circuit.

1.4.1 Circuit examples for mesh or loop analysis

The mesh-current method is easiest to work with when there are voltage sources, and no current sources. The first example will demonstrate the mesh-current method in a circuit with independent voltage sources and resistors.

Example 1.8. In the circuit shown below, V_1 and V_2 are 1 V and 2 V respectively. R_3 is 3 Ω, R_4 is 4 Ω, and likewise, R_k is k Ω. Evaluate the voltage across R_3, from top to bottom.

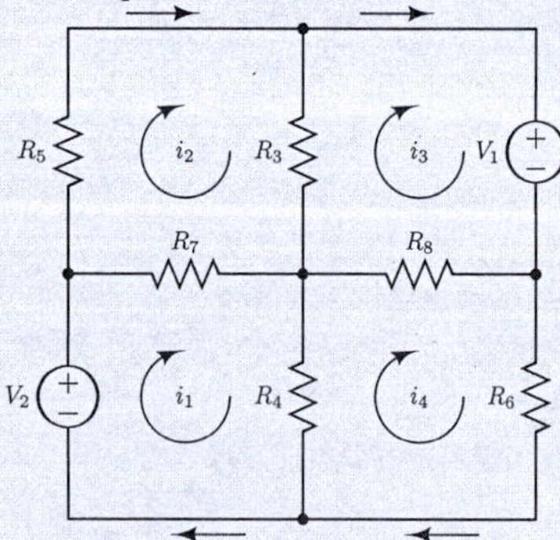

There are four meshes in the circuit. Let us visualize currents circulating in these meshes, as indicated. The currents in each element are the superposition of the mesh currents. For example, the current through R_4 is $i_1 - i_4$ from top to bottom, the current through R_8 is $i_4 - i_3$ left to right, the current through R_5 is just i_2 going up, the current through R_3 is $i_2 - i_3$ from top to bottom.

Now we will write out KVL equations around each mesh.

$$\begin{aligned}(i_1 - i_2)R_7 + (i_1 - i_4)R_4 &= V_2 \\ (i_2 - i_1)R_7 + i_2R_5 + (i_2 - i_3)R_3 &= 0 \\ (i_3 - i_4)R_8 + (i_3 - i_2)R_3 &= -V_1 \\ (i_4 - i_1)R_4 + (i_4 - i_3)R_8 + i_4R_6 &= 0\end{aligned}$$

or,

$$\mathbf{Z} \cdot \begin{bmatrix} i_1 \\ i_2 \\ i_3 \\ i_4 \end{bmatrix} = \begin{bmatrix} V_2 \\ 0 \\ -V_1 \\ 0 \end{bmatrix}$$

where

$$\mathbf{Z} = \begin{bmatrix} R_7 + R_4 & -R_7 & 0 & -R_4 \\ -R_7 & R_3 + R_5 + R_7 & -R_3 & 0 \\ 0 & -R_3 & R_8 + R_3 & -R_8 \\ -R_4 & 0 & -R_8 & R_4 + R_6 + R_8 \end{bmatrix}$$

We can invert the matrix $\mathbf{Z}$ and compute all the mesh currents. The voltage across R_3 is $(i_2 - i_3)R_3$. All the branch voltages, all the branch currents can be computed from the mesh currents. As before, Octave can be used for rapid computation.

```
octave:1> [v1,v2]=deal(1,2);
octave:2> [r3,r4,r5,r6,r7,r8]=deal(3,4,5,6,7,8);
octave:3> Z = [r7+r4, -r7, 0, -r4;...
-r7, r3+r5+r7, -r3, 0;...
0, -r3, r3+r8, -r8;...
-r4, 0, -r8, r4+r6+r8];
octave:4> I = inv(Z)*[v2; 0; -v1; 0];
octave:5> V = (I(2)-I(3))*r3
V = 0.4324
```

Introducing current sources in the circuit introduces complexity in some situations. If the current sources are on the outer branches of the circuit, the corresponding mesh currents are directly decided by these current sources. In the next example, there are current sources along the outer branches of the circuit.

Example 1.9. In the circuit below I_1 and I_2 are 1 A and 2 A respectively. R_k is k Ω. What is the voltage across R_3 from top to bottom?

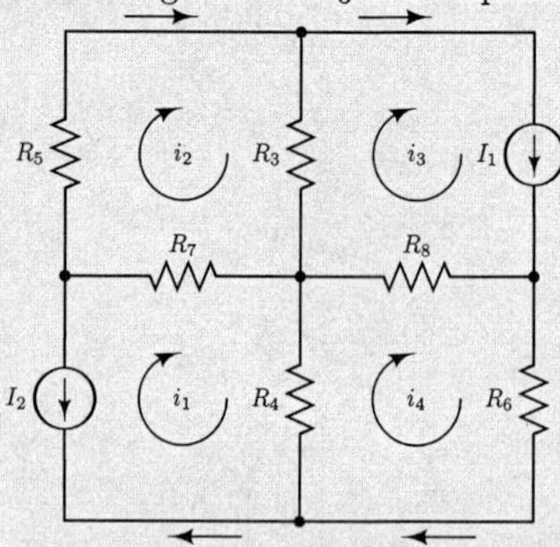

The voltage across a current source can be anything, and as such, one cannot write a KVL through current sources. However, if the mesh current is the same as the current through the current source, the corresponding KVL equation is no longer required. In this case, $i_1 = -I_2$ and $i_3 = I_1$. We have written KVLs for the other loops.

$$\begin{aligned}(i_2 + I_2)R_7 + i_2R_5 + (i_2 - I_1)R_3 &= 0 \\ (i_4 + I_2)R_4 + (i_4 - I_1)R_8 + i_4R_6 &= 0\end{aligned}$$

or,

$$\begin{bmatrix} R_7 + R_5 + R_3 & 0 \\ 0 & R_4 + R_8 + R_6 \end{bmatrix} \begin{bmatrix} i_2 \\ i_4 \end{bmatrix} = \begin{bmatrix} I_1R_3 - I_2R_7 \\ I_1R_8 - I_2R_4 \end{bmatrix}$$

As before, Octave may be used for calculations.

```
octave:1> [i1,i2]=deal(1,2);
octave:2> [r3,r4,r5,r6,r7,r8]=deal(3,4,5,6,7,8);
octave:3> Z = [r3+r5+r7, 0; 0, r4+r6+r8];
octave:4> I = inv(Z)*[i1*r3-i2*r7; i1*r8-i2*r4];
octave:5> V = (I(1)-i1)*r3
V = -5.2000
```

Then, how do we analyze using the mesh-current method when there are current sources inside the circuit? The method is now adjusted to create a super-mesh; a combination of multiple meshes. Instead of the regular mesh currents, we introduce the current of a super-mesh, such that every current source directly decides the current in a mesh. The next example is illustrates the technique.

Example 1.10. In the circuit below I_1 and I_2 are 1 A and 2 A respectively. R_k is k Ω. What is the voltage across R_3 from top to bottom?

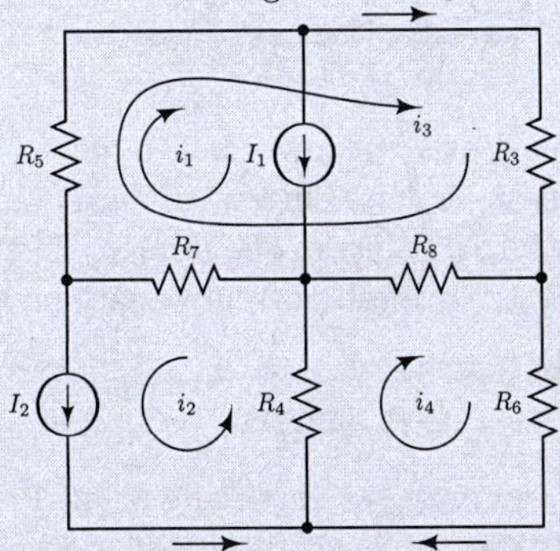

The current sources need to decide mesh currents. As such, we have drawn i_1; the current source I_1 decides the value of i_1. We draw the mesh current i_2 aligned with the current source I_2. For I_1 to uniquely decide the value of the mesh current i_1, we should not declare the adjacent mesh current as a variable. Instead, we can choose a loop (a super-mesh) that combines two or more meshes and passes through R_3, but avoids I_1. i_3 is a super-mesh that has accordingly been chosen. The mesh current i_4 is chosen as usual.

Now we have the current going through R_5 as $i_1 + i_3$ from bottom to top, the current through R_7 as $i_1+i_2+i_3$ from right to left, the current through R_8 as i_3-i_4 from right to left, the current through R_4 as i_2+i_4 from bottom to top. Note that the current i_1 is just I_1, and the current i_2 is just I_2.

The KVLs for the remaining two loops corresponding to i_3 and i_4 are:

$$\begin{aligned} i_3R_3 + (i_3 - i_4)R_8 + (i_3 + I_2 + I_1)R_7 + (i_3 + I_1)R_5 &= 0 \\ i_4R_6 + (i_4 + I_2)R_4 + (i_4 - i_3)R_8 &= 0 \end{aligned}$$

or,

$$\begin{bmatrix} R_3 + R_5 + R_7 + R_8 & -R_8 \\ -R_8 & R_4 + R_6 + R_8 \end{bmatrix} \begin{bmatrix} i_3 \\ i_4 \end{bmatrix} = \begin{bmatrix} -I_1(R_5 + R_7) - I_2R_7 \\ -I_2R_4 \end{bmatrix}$$

The voltage across R_3 from top to bottom is nothing but i_3R_3. We can solve the remainder of the problem with the help of Octave, or any other tool. The voltage across R_3 is evaluated as -4.56 V.

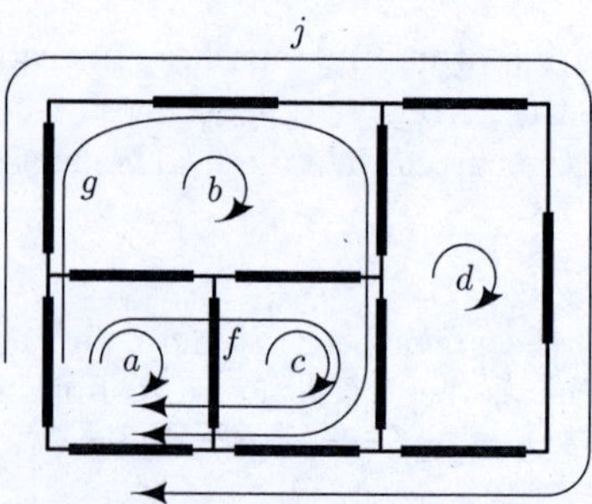

Figure 1.4: Several loops exist in any given circuit. The number of loops required to perform mesh or loop-current analysis is $B - (N - 1)$, where there are B elements and N nodes in the circuit. The circuit shown has 13 elements, 10 nodes. 4 loop or mesh-current equations are sufficient.

1.4.2 Choose the right loops

There are many loops and meshes in a circuit. How do we pick a set of loops and meshes that allow us to analyze the circuit?

Fig. 1.4 shows a circuit with 13 elements, 10 nodes. As such 4 mesh current equations are sufficient to completely analyze the circuit. However, if these 4 mesh equations are not available, which 4 loops do we choose? a, b, c, d are the defined meshes. Let us define the other loops as follows:

$$e = a \cup b, \quad f = a \cup c, \quad g = a \cup b \cup c, \quad h = a \cup b \cup d, \quad i = a \cup c \cup d$$
$$j = a \cup b \cup c \cup d, \quad k = b \cup c, \quad l = b \cup d, \quad m = b \cup c \cup d, \quad n = c \cup d$$

In addition to meshes a, b, c, d, the loops f, g and j are also illustrated in Fig. 1.4.

Loop current analysis can be performed in many different ways. Meshes $\{a, b, c, d\}$ will work well, unless there is a current source in between the meshes. If the element shared by a and c is a current source, we can choose meshes $\{a, b, d, f\}$, or $\{b, c, d, f\}$. If the element shared by a and b is a current source, we can choose meshes $\{a, c, d, e\}$, or $\{b, c, d, e\}$. If the elements shared between $b - d$ and $c - d$ are current sources, we can choose $\{a, b, c, j\}$.

Scan the QR-code, and learn more about how to pick an optimal set of loop-current equations by studying fundamental loop analysis.

1.4.3 Circuit examples with dependent sources

Working with dependent sources is straightforward with mesh or loop-current analysis. As before, we need to make sure that if there are current sources, we pick the set of loops such that KVLs through current sources are not required. We will repeat the examples done with node-voltage analysis, example 1.6 and example 1.7.

Example 1.11. In the circuit below, V_1 is 1 V, G_2 is 1/2 S, R_1 is 1 Ω, R_2 is 2 Ω, and likewise R_k is k Ω. Evaluate the current i_1.

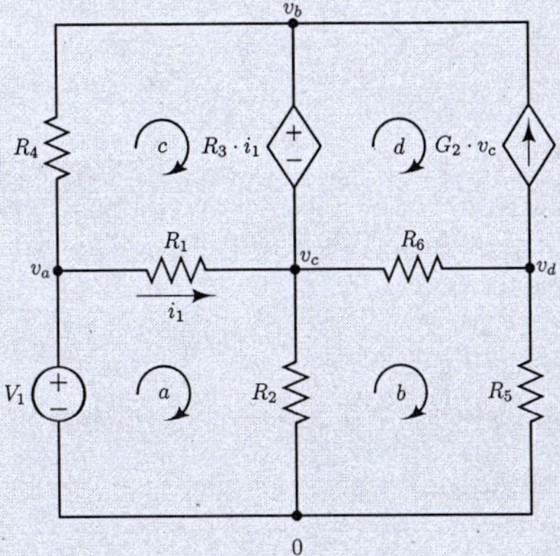

Four mesh currents are identified. Clearly $i_1 = i_a - i_c$. Further, $v_c = (i_a - i_b)R_2$. It is also apparent that $i_d = -G_2 v_c = -G_2 R_2 (i_a - i_b)$. The KVLs for the meshes a, b and c are as follows.

$$\begin{aligned} i_a(R_1 + R_2) - i_b R_2 - i_c R_1 &= V_1 \\ -i_a R_2 + i_b(R_2 + R_6 + R_5) - i_d R_6 &= 0 \\ -i_a R_1 + i_c(R_1 + R_4) + R_3 i_1 &= 0 \end{aligned}$$

Substituting the values for i_1 and i_d, we can simplify the mesh equations and rearrange in the form of a single matrix equation.

$$\begin{bmatrix} R_1 + R_2 & -R_2 & -R_1 \\ G_2 R_2 R_6 - R_2 & R_2 + R_5 + R_6 - G_2 R_2 R_6 & 0 \\ -R_1 + R_3 & 0 & R_1 + R_4 - R_3 \end{bmatrix} \begin{bmatrix} i_a \\ i_b \\ i_c \end{bmatrix} = \begin{bmatrix} V_1 \\ 0 \\ 0 \end{bmatrix}$$

Computation of $i_1 = i_a - i_c$ gives the same result as before, 389 mA.

Example 1.12. In the circuit shown below, the resistors R_1, R_2, R_3, R_4 are 1 Ω, 2 Ω, 3 Ω and 4 Ω respectively. Z is 5 Ω. I_0 is 1 A. Evaluate i_3.

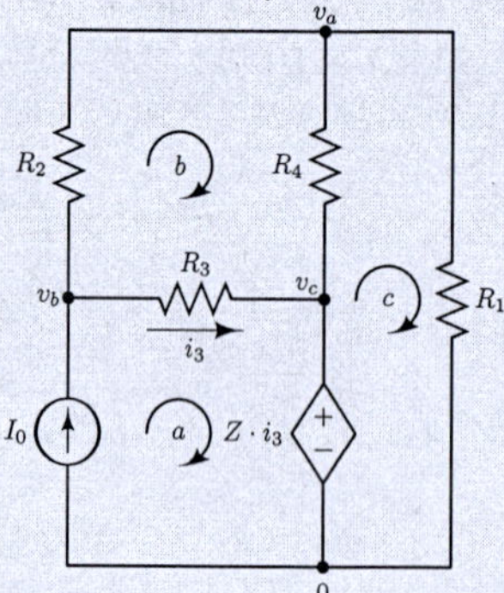

Three mesh currents are identified, of which $i_a = I_0$. Further, i_3 is $I_0 - i_b$, and the potential across the CCVS is $Z(I_0 - i_b)$. We can now write out the two mesh equations for i_b and i_c.

$$\begin{aligned}(R_2 + R_3 + R_4)i_b - R_4 i_c &= I_0 R_3 \\ -R_4 i_b + (R_1 + R_4)i_c &= Z(I_0 - i_b)\end{aligned}$$

A matrix representation of the equation-pair is given below.

$$\begin{bmatrix} R_2 + R_3 + R_4 & -R_4 \\ Z - R_4 & R_1 + R_4 \end{bmatrix} \begin{bmatrix} i_b \\ i_c \end{bmatrix} = \begin{bmatrix} R_3 I_0 \\ Z I_0 \end{bmatrix}$$

We can compute the answer using Octave.

```
octave:1> [i0,z,r1,r2,r3,r4]=deal(1,5,1,2,3,4);
octave:2> Z = [r2+r3+r4, -r4; z-r4, r1+r4];
octave:3> I = inv(Z)*[i0*r3; i0*z];
octave:4> i3 = i0-I(1)
i3 = 0.2857
```

Observation 1.3. The matrix **Z** obtained through mesh analysis is symmetric when there are no dependent sources. Let us consider a resistor R shared between the mth and nth loops. If the directions of currents in the mth and nth loop through this resistor are the same, then this resistor will create an entry $+R$ for Z_{mn} and Z_{nm}. If the currents in the loops through this resistor are opposite, the resistor will create an entry $-R$ for Z_{mn} and Z_{nm}.

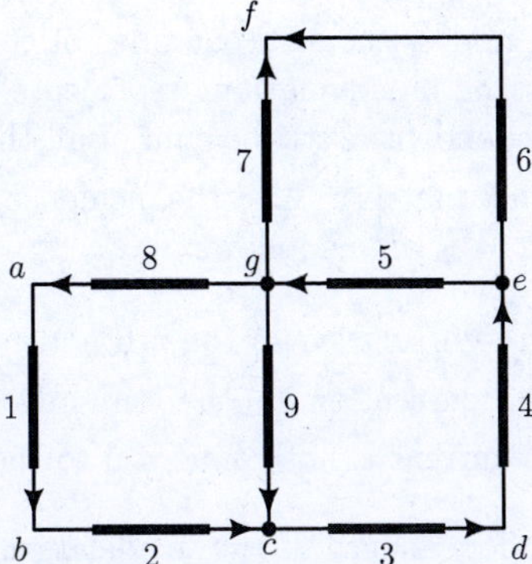

Figure 1.5: Circuit to illustrate the incidence matrix and Tellegen's theorem.

1.5 Tellegen's theorem

1.5.1 The incidence matrix

Consider a circuit as shown in Fig. 1.5. The nodes are marked a through g, while the circuit components are labeled 1 through 9. The directions of currents are also marked. Although arbitrary, the directions are required for reference in the following discussion. The properties of the components are irrelevant for this discussion.

For each node in the circuit we can write a KCL. We can form 7 KCL equations for the nodes a through g. Our convention will be to use a positive sign when a current flows out of the node. Let $\{i_1, i_2, \ldots, i_9\}$ be the currents going through the components 1 through 9, in the directions marked in Fig. 1.5.

$$\begin{bmatrix} 1 & 0 & 0 & 0 & 0 & 0 & 0 & -1 & 0 \\ -1 & 1 & 0 & 0 & 0 & 0 & 0 & 0 & 0 \\ 0 & -1 & 1 & 0 & 0 & 0 & 0 & 0 & -1 \\ 0 & 0 & -1 & 1 & 0 & 0 & 0 & 0 & 0 \\ 0 & 0 & 0 & -1 & 1 & 1 & 0 & 0 & 0 \\ 0 & 0 & 0 & 0 & 0 & -1 & -1 & 0 & 0 \\ 0 & 0 & 0 & 0 & -1 & 0 & 1 & 1 & 1 \end{bmatrix} \begin{bmatrix} i_1 \\ i_2 \\ i_3 \\ \vdots \\ i_9 \end{bmatrix} = 0$$

or,

$$\mathbf{A}_i\mathbf{I} = 0$$

$\mathbf{A}_i$ has 7 rows and 9 columns. In general, it will have the same number of rows as the number of nodes, the same number of columns as the number of elements. Note that the nine equations represented by the given matrix equation are linearly dependent; e.g., the sum of all the rows of $\mathbf{A}_i$ is 0.

Definition 1.3 (Incidence matrix): $\mathbf{A}_i$ is the incidence matrix. The elements of the incidence matrix, A_{pq}, are as follows:

$$A_{pq} = \begin{cases} +1 \text{ if the } q\text{th current leaves the } p\text{th node,} \\ -1 \text{ if the } q\text{th current enters the } p\text{th node,} \\ 0 \text{ if the } q\text{th current is not connected to the } p\text{th node.} \end{cases} \tag{1.3}$$

Theorem 1.1 (Linear dependence of the incidence matrix): The incidence matrix is linearly dependent.

Proof. For every current that leaves a node it also has to enter another node. As such, for that column, there will be two non-zero entries, one with a +1 and the other with a −1. Therefore the sum of every column is 0. ■

Definition 1.4 (The reduced incidence matrix): The incidence matrix with the equation for the datum node removed is the reduced incidence matrix.

Example 1.13. In our example, if we arbitrarily choose node g as the datum, the reduced incidence matrix equation becomes:

$$\begin{bmatrix} 1 & 0 & 0 & 0 & 0 & 0 & 0 & -1 & 0 \\ -1 & 1 & 0 & 0 & 0 & 0 & 0 & 0 & 0 \\ 0 & -1 & 1 & 0 & 0 & 0 & 0 & 0 & -1 \\ 0 & 0 & -1 & 1 & 0 & 0 & 0 & 0 & 0 \\ 0 & 0 & 0 & -1 & 1 & 1 & 0 & 0 & 0 \\ 0 & 0 & 0 & 0 & 0 & -1 & -1 & 0 & 0 \end{bmatrix} \begin{bmatrix} i_1 \\ i_2 \\ i_3 \\ \vdots \\ i_9 \end{bmatrix} = 0$$

Theorem 1.2 (The reduced incidence matrix is linearly independent): For any circuit with n nodes, the KCL equations for any $n-1$ nodes form a set of $n-1$ linearly independent equations.

Proof. The reduced incidence matrix, $\mathbf{A}$, is constructed by removing the row corresponding to the datum node from $\mathbf{A}_i$. The elements connected to the datum node now appear only once in $\mathbf{A}$, i.e., the qth branch going out of the datum node to the pth node has an entry of −1 in the pth row and qth column of the reduced incidence matrix. The corresponding +1 would appear in the row corresponding to the datum node, which has now been removed. There is no way to linearly combine the rows to obtain 0, and **A** is linearly independent.

$$\mathbf{AI} = 0 \tag{1.4}$$

■

1.5.2 Independent KVLs

We define the potential at each node with respect to the datum node as e_i.

Example 1.14. In the circuit of Fig. 1.5, the potential differences, $\{v_k\}$, across each circuit element, is represented in the matrix form as follows. The potential difference is taken in the same direction as the current.

$$\begin{bmatrix} 1 & -1 & 0 & 0 & 0 & 0 \\ 0 & 1 & -1 & 0 & 0 & 0 \\ 0 & 0 & 1 & -1 & 0 & 0 \\ 0 & 0 & 0 & 1 & -1 & 0 \\ 0 & 0 & 0 & 0 & 1 & 0 \\ 0 & 0 & 0 & 0 & 1 & -1 \\ 0 & 0 & 0 & 0 & 0 & -1 \\ -1 & 0 & 0 & 0 & 0 & 0 \\ 0 & 0 & -1 & 0 & 0 & 0 \end{bmatrix} \begin{bmatrix} e_a \\ e_b \\ e_c \\ e_d \\ e_e \\ e_f \end{bmatrix} = \begin{bmatrix} v_1 \\ v_2 \\ v_3 \\ \vdots \\ v_9 \end{bmatrix}$$

Let us assume there are b components in the circuit and n nodes including the datum. Then, in general, $\mathbf{ME} = \mathbf{V}$, where $\mathbf{M}$ has b rows and $n-1$ columns, $\mathbf{E}$ has $n-1$ rows and 1 column, and $\mathbf{V}$ has b rows and 1 column.

$$M_{pq} = \begin{cases} +1 \text{ if the } p\text{th current leaves the } q\text{th node,} \\ -1 \text{ if the } p\text{th current enters the } q\text{th node,} \\ 0 \text{ if the } p\text{th current is not connected to the } q\text{th node.} \end{cases} \tag{1.5}$$

From (1.3) and (1.5), we observe that $M_{pq} = A_{qp}$, or in other words,

$$\mathbf{M} = \mathbf{A}^T, \quad \mathbf{A}^T\mathbf{E} = \mathbf{V} \tag{1.6}$$

1.5.3 Tellegen's theorem

Theorem 1.3 (Tellegen's theorem): Consider two circuits that have the same topology (i.e., arrangement of nodes and elements, without considering the component properties.) $\{v_k\}$ are the potential differences across the b elements of the first circuit, and $\{i_k\}$ are the currents through the b elements of the second circuit. The potential differences, $\{v_k\}$, and the currents, $\{i_k\}$, are specified in the same direction. Then

$$\sum_{k=1}^{b}(v_k i_k) = 0 \tag{1.7}$$

The two circuits can also be the same circuit, as a special case.

Proof. (1.7) can be rephrased as $\mathbf{V}^T \cdot \mathbf{I} = 0$. Since both circuits have the same topology, the incidence matrix is the same for both. From (1.6)

$$\mathbf{V} = \mathbf{A}^T\mathbf{E}$$

or,

$$\mathbf{V}^T = \mathbf{E}^T\mathbf{A}$$

Introducing (1.4),

$$\mathbf{V}^T \cdot \mathbf{I} = \mathbf{E}^T\mathbf{A}\mathbf{I} = 0$$

■

Tellegen's theorem is extremely powerful and can be applied to two different circuits with the same topology, a circuit at two different points in time, a circuit at the same point in time. With observations at the same point of time for the same circuit, Tellegen's theorem reduces to the law of conservation of energy (instantaneous power). However, the law of conservation of energy is only a particular case of Tellegen's theorem. Wide circuit theoretical conclusions can be made with judicious application of Tellegen's theorem.

Corollary 1.5 (Conservation of power)**.** In a circuit, the total instantaneous power absorbed is equal to zero, or $\mathbf{V}^T \cdot \mathbf{I} = 0$. (Or, the sum of all power generated equals the sum of all power absorbed.)

1.6 Duality

Tellegen's theorem tells us that $\mathbf{V}^T \cdot \mathbf{I} = 0$, or in other words, the dot product of the set of branch voltages with the set of branch currents is 0. This could also be viewed as orthogonality between the voltage vector and the current vector. If we interchange all the voltages and all the currents, Tellegen's theorem will still be satisfied. For a given circuit $\mathcal{C}_1$, therefore, can we find another circuit, $\mathcal{C}_2$, for which the branch currents are equal to the branch voltages of $\mathcal{C}_1$, and vice-versa?

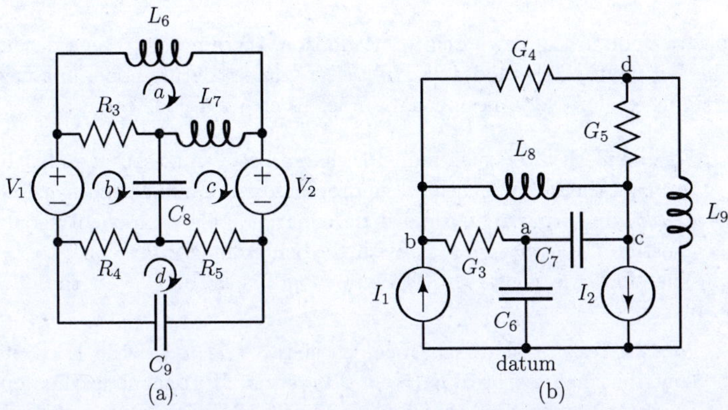

Figure 1.6: (a) A circuit and (b) its dual.

Definition 1.5 (Dual of a network): If the voltages and currents of a network $\mathcal{C}_1$ are $\mathbf{V}_1$ and $\mathbf{I}_1$ respectively, then the dual of $\mathcal{C}_1$ is such a network, $\mathcal{C}_2$, for which the set of branch voltages are numerically $\mathbf{I}_1$, and the set of branch currents are numerically $\mathbf{V}_1$.

For every KVL in the original network, we will have a KCL in the dual network. For every node in the original network, we will have a mesh in the dual network. For every mesh in the original network, we will have a node set in the dual network. Further, we need to replace every resistance with a conductance of the same value, every capacitance with an inductance of the same value, and vice-versa, every voltage source with a current source, and vice-versa, to satisfy device properties.

It turns out that for every **planar** circuit, it is possible to construct its dual. The mesh equations for the dual circuit will be the same as the node equations for the original and vice-versa. A planar circuit can be drawn on paper, with no crossings. A non-planar circuit cannot be drawn on paper without branch crossings, even after any amount of re-organization. It is not easy to find out if a given circuit is indeed planar or not. A quick test for planarity is to check if any element in the circuit is shared between more than 2 meshes. If more than two meshes share an element, the network is not planar. It will not be possible to find a dual for such a non-planar network.

Example 1.15. Let us try to construct the dual of the circuit shown in Fig. 1.6(a). There are four meshes in the circuit, labeled as a, b, c and d.

Now we will attempt to develop a dual for this graph. For every mesh, we expect a node in the new circuit. In addition, there needs to be a datum node. Let us mark the nodes, 'a', 'b', 'c', 'd', and a datum node, in a new circuit. Next, for every mesh equation, let us create a node equation.

1. For the mesh a, L_7 is an inductor shared between meshes a and c. In terms of the mesh equation, the drop across the inductor is $L_7d(i_a - i_c)/dt$. In terms of its dual node equation, this component should contribute $C_7d(v_a - v_c)/dt$, with C_7 being numerically equal to L_7. Therefore, we construct C_7 between nodes 'a' and 'c' in the dual network.
2. R_3 is a resistor shared between meshes a and b. In the KVL, its contribution is a drop of $R_3(i_a - i_b)$. In terms of its dual node equation, the component should contribute a current of $G_3(v_a - v_b)$, where G_3 is a conductance numerically equal to R_3. Therefore in the dual network we have constructed G_3 between nodes 'a' and 'b'.
3. L_6 is an inductor that creates a drop of L_6di_a/dt. In the dual circuit, it should create a current outflow of C_6dv_a/dt, where C_6 is numerically equal to L_6. Therefore we have constructed C_6 between node 'a' and the datum.
4. Next let us look at mesh b. The component C_8 is shared between meshes b and c. In the mesh equation, it creates a voltage drop of $1/C_8 \int(i_b - i_c)dt$. For the dual node equation, there should be a current outflow of $1/L_8 \int(v_b - v_c)dt$, with L_8 numerically equal to C_8. This works out to an inductor L_8 between nodes 'b' and 'c'.
5. R_4 is a resistor shared between meshes b and d. In the mesh equation it contributes a voltage drop of $R_4(i_b - i_d)$. In the corresponding node equation for the dual, there should be a current outflow of $G_4(v_b - v_d)$, where G_4 is numerically equal to R_4. This works out to a conductance of G_4 between nodes 'b' and 'd' in the dual circuit.
6. V_1 is a voltage source that creates a drop of $-V_1$ in the mesh equation. The corresponding node equation in the dual circuit should have a contribution of $-I_1$ as the current outflow, with I_1 numerically equal to V_1. This works out to a current source I_1 from the datum to node 'b'.
7. Next we look at mesh c. The voltage source V_2 creates a drop of V_2 in the mesh equation. The corresponding node equation for the dual circuit should have a current outflow of I_2, with I_2 numerically equal to V_2. This evaluates to a current source I_2 from node 'c' to the datum.
8. The resistor R_5 is shared between meshes c and d. As such, for the dual circuit, it will contribute a conductor G_5 between nodes 'c' and

'd', with G_5 numerically equal to R_5.

9. Finally we will look at mesh d, which has one remaining component. The capacitor C_9 contributes a voltage drop of $1/C_9 \int i_d dt$, in the mesh equation. For the dual circuit, the corresponding node equation requires a term $1/L_9 \int v_d dt$ as current outflow. This evaluates to an inductor L_9 between 'd' and the datum.

Every KCL in Fig. 1.6(a) has an identical KVL in Fig. 1.6(b), and vice-versa. Every mesh current in Fig. 1.6(a) has an identical node voltage, and vice-versa.

1.7 Unit summary

- The voltage drop across a resistor is iR. The voltage drop across an inductor is $L\dot{i}$. The current through a capacitor is $C\dot{v}$.

- It is illegal to set up a loop of independent and dependent voltage sources. It is illegal to have all elements attached to a node to be current sources.

- Dependent voltage and current sources can be controlled by voltages across or currents through other branches. Such dependent sources are mathematical constructs, useful for circuit analysis, understanding and simplification.

- The datum node (or ground, or earth) is an arbitrarily chosen node in the circuit, designated to be at 0 potential.

- KCLs written for all nodes other than the datum node are linearly independent and sufficient to solve a circuit. This method is the node voltage method.

- Voltage sources may create difficulty in applying the node voltage method. Every voltage sources reduces the number of node equations by one. A super-node around a voltage source may be constructed if the voltage source does not clearly define a node voltage.

- If KVLs are written for each mesh, the set of equations obtained are linearly independent and sufficient to solve a circuit. This method is the mesh current method.

- Current sources may create difficulty in applying the mesh-current method. Every current source reduces the number of mesh equations by one. A super-mesh may have to be constructed to isolate current sources shared between two or more meshes.

- The matrices obtained from node voltage analysis and mesh analysis are symmetric, if and only if there are no dependent sources in the network.

- Tellegen's theorem proves that a set of voltages that obey KVL, and a set of currents that obey KCL, for the same circuit topology, are orthogonal to each other. Guaranteed conservation of power is a particular case of Tellegen's theorem.

- It is possible to generate the dual of any planar network. Each KCL in the original network corresponds to a KVL in its dual. Each KVL in the original network corresponds to a KCL in the dual. Between a network and its dual, the number of branches remains the same. In the dual network, meshes map to nodes, and vice-versa. The dual of the dual of a network is the network itself. Circuit component-types have to be inverted, keeping the value the same.

Computation

In this text, we need to rely on a tool that allows us to compute quickly. Nevertheless, we do not want to bother ourselves with the intricacies of computer programming. A quick interactive tool that gives us this benefit is Octave. MATLAB® is the original computational tool; however, MATLAB® is proprietary and requires a license. Octave is an open-source tool that is similar to MATLAB®. Most commands or functions in Octave are compatible with the corresponding functions in MATLAB®; however, there may be minor variations. In this text, we will assume the availability of Octave. The detailed manual for Octave is available at [2].

The programming style in Octave is slightly different from regular computer programming. In Octave, it is more efficient to process an array (or a vector) in one shot than to cycle through each element of a vector and process them individually. For example, if we have two vectors X and Y, each of length n, and if we want to work out $\sum_1^n x_i y_i$, then in most computer programming languages (C, C++, Python, Perl, Java), we will write a `for` loop which adds up the product of each element pair product. In Octave, we will simply use `transpose(X)*Y` (or `X*transpose(Y)` in case `X` and `Y` are row-vectors) to obtain the dot product of the two vectors. A `for` loop in Octave takes more time to execute than a vector or matrix operation.

Installation instructions and other details regarding the usage of Octave in this text have been provided in Appendix A. Octave (like MATLAB®) has extensive documentation readily available. If you do not know what a specific command does, say `deal`, you just have to type `help deal` to obtain the relevant documentation and see examples. The advantage of a tool like Octave over others is its focus on mathematics and the availability of plotting tools. Like most computing languages, you will get a better grasp of the language as you use it. Till you use the tool yourself hands-on, you will not be at ease with it. Electrical Engineers use MATLAB®/Octave extensively. Circuit theory is a convenient study to get familiar and comfortable with Octave. Before starting the exercises, obtain a copy of Octave and start tinkering. To start with, make sure you can solve the worked-out examples.

1.8 Exercises

Multiple choice type questions

1.1 Two inductors of 0.1 H and 0.2 H are coupled to each other through a coupling coefficient of 0.5. What is the mutual inductance?
(a) 0.05 H (b) 0.0707 H (c) 0.1 H (d) 0.1414 H

1.2 Two inductors of 10 mH and 2.5 mH are mutually coupled. What is the maximum mutual inductance?
(a) 2.5 mH (b) 5 mH (c) 10 mH (d) 20 mH

1.3 A mutual inductance of 2 mH is designed using inductors of 1 mH and 9 mH. What should be the coupling coefficient?
(a) 1/8 (b) 1/6 (c) 1/3 (d) 2/3

1.4 Which of the following is true for the coupling coefficient, k, of a mutual inductance?
(a) $k \in (-\infty, \infty)$ (b) $k \in [0, \infty)$ (c) $k \in [-1, 1]$ (d) $k \in [0, 1]$

1.5 Which of the following is the outcome of Tellegen's theorem?
(a) $\mathbf{VI} = 0$ (b) $\mathbf{V}^T\mathbf{I} = 0$ (c) $\mathbf{VI}^T = 0$ (d) $\mathbf{V}^T\mathbf{I}^T = 0$

Short answer type questions

1.6 The incidence matrix is guaranteed to be linearly ______.

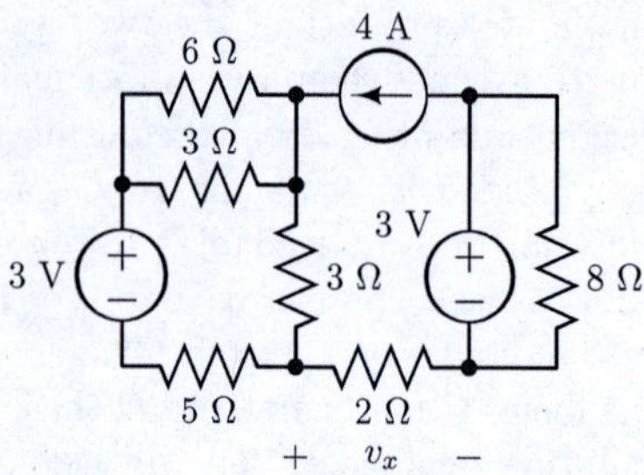

Figure 1.7: Circuit for question 1.16

1.7 Removal of any one row of the incidence matrix gives the ______, which is linearly ______.

1.8 A circuit does not have any dependent sources. While analyzing using the ______ method, the **G** matrix will be ______.

1.9 One should not place two or more voltage sources in ______.

1.10 One should not place two or more current sources in ______.

1.11 If a circuit has 8 nodes and 12 elements, what is the number of equations in node-voltage analysis?

1.12 In an 8-node 12-element circuit, what is the number of mesh equations in mesh-current analysis?

1.13 In the 8-node 12-element circuit, what is the maximum number of independent and dependent current sources? (Answer: 5)

1.14 In the 8-node 12-element circuit, what is the maximum number of independent and dependent voltage sources? (Answer: 7)

1.15 In the 8-node 12-element circuit, there are 4 voltage sources and 3 current sources. Which analysis method will have lesser equations - node voltage method or mesh current method? Why? (Answer: mesh method will have 2 equations, while node voltage method will need 3 equations.)

Numericals

1.16 In the circuit shown in Fig. 1.7, find the value of v_x.

1.17 Solve the circuits in Fig. 1.8 using the node voltage method. Arrive at the matrix equation $\mathbf{GV} = \mathbf{U}$ for each of the circuits. Invert the matrix **G** using Octave.

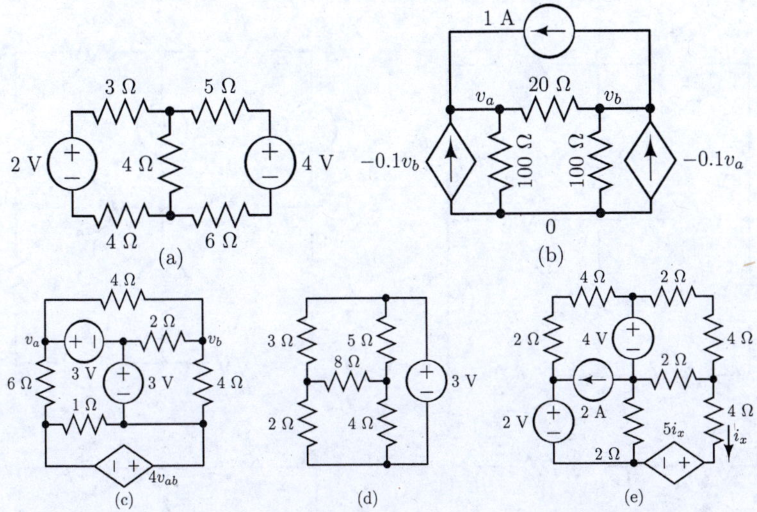

Figure 1.8: Circuits for questions 1.17, 1.18, and 1.19.

1.18 Solve the circuits in Fig. 1.8 using the mesh method. Arrive at the matrix equation $\mathbf{ZI} = \mathbf{U}$ for each of the circuits. Invert the matrix $\mathbf{A}$ using Octave. Cross-check your answers with those obtained earlier in 1.17.

1.19 For each of the circuits in Fig. 1.8 (let us call the circuit as $\mathcal{C}_a$), come up with a new circuit $\mathcal{C}_b$, such that (i) the mesh equations of $\mathcal{C}_a$ are the same as the node equations of $\mathcal{C}_b$, and (ii) the node equations of $\mathcal{C}_a$ are the same as the mesh equations of $\mathcal{C}_b$.

1.20 Schematics are shown in Fig. 1.9. The component values are to be taken as 1, 2, 3,...12, numerically. For example, a component R_5 will have a value of 5 Ω, V_6 will have a value of 6 V, I_7 a value of 7 A, K_8 a value of 8, Z_9 a value of 9 Ω, G_{10} a value of 10 S. Solve the four circuits using the node-voltage method first and the mesh method next.

1.21 Consider a current-controlled voltage source (CCVS). Suppose the voltage across this CCVS is defined as $10i_x$, where i_x is the current through the CCVS itself. Can you simplify the representation of this CCVS?

1.22 Consider a voltage-controlled current source (VCCS). Suppose the current through this VCCS is defined as $0.1v_x$, where v_x is the voltage across the VCCS itself. Can you simplify the representation of this VCCS?

1.23 Consider a voltage-controlled voltage source (VCVS). Is it possible to have the value of the voltage source to be proportional to the voltage across the

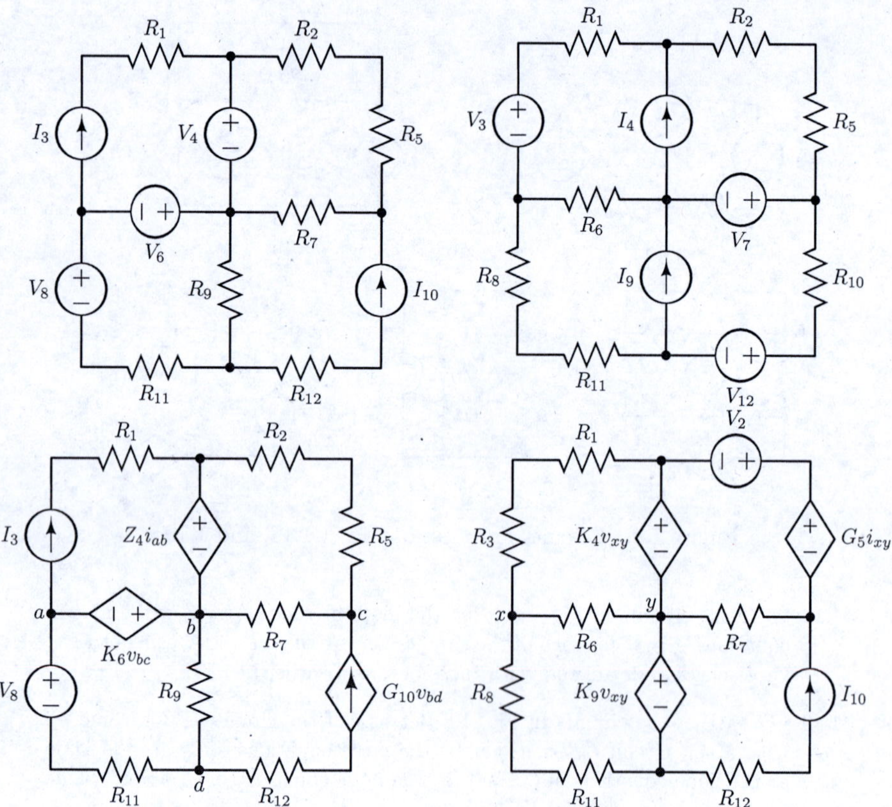

Figure 1.9: Circuits for questions 1.20.

VCVS itself? Think and decide as to what happens if a proportionality constant $\neq 1$ is used. What happens if a proportionality constant $= 1$ is used? Repeat the exercise for a current-controlled current source (CCCS).

1.24 A resistive network with unknown resistances was used in two experiments, as shown in Fig. 1.10. First R_L was set to 2 Ω, v_1 was 8 V, i_i was measured as -2 A, v_L was measured as 2 V. Next R_L was changed to 4 Ω, and the applied v_1 was changed to 12 V. The measured i_1 was -2.4 A. What would be v_L in the second experiment? (*Hint: Use Tellegen's theorem.*)

1.25 In the schematic of Fig. 1.11, two experiments are shown with a two-port network. In experiment-A, a 1-A 50-Ω source is applied on the left side, and a 50-Ω resistor is connected on the right. The current measured is 0.1 A.

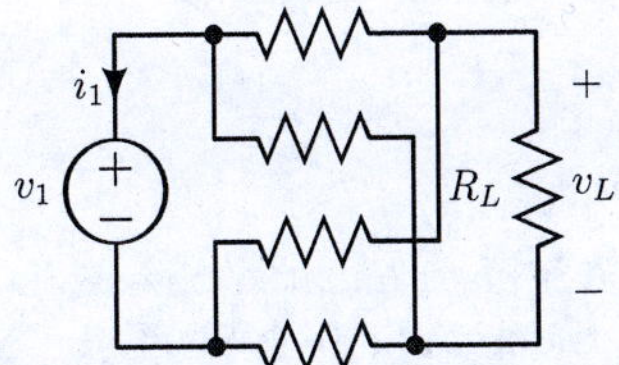

Figure 1.10: Circuit for question 1.24

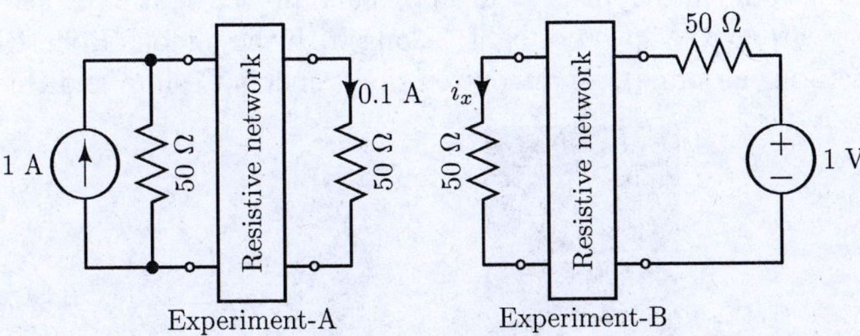

Figure 1.11: Schematics for question 1.25

In experiment-B, a 1-V 50-Ω source is applied on the right side, and a 50-Ω resistor is connected on the left. What will be the current i_X? (*Hint: Use Tellegen's theorem. Any other method will lead to significant wastage of time.*)

1.26 If you apply mesh-method to solve the circuit of Fig. 1.12, what is the least number of equations you will obtain? What is the least number of equations you will obtain if you apply node-analysis? Use the more straightforward method and solve for i_x.

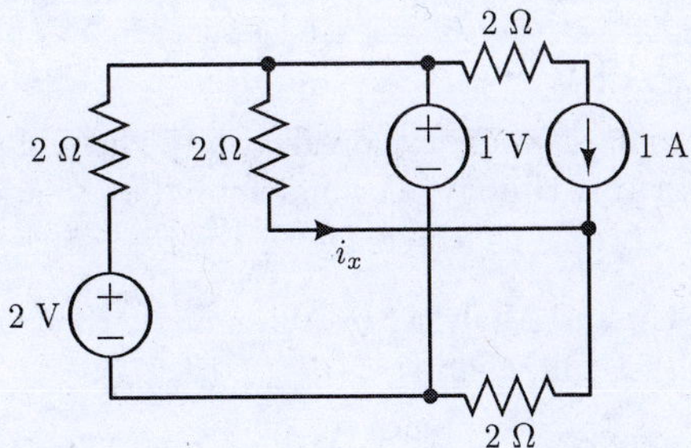

Figure 1.12: Circuit for question 1.26

Know more

Historical profiles

Gustav Kirchhoff (1824-1887) formulated the KVL and KCL as a course project when he was an undergraduate student. Later, this became his doctoral thesis. A German-Russian scientist, Kirchhoff's work encompassed electromagnetics, black-body radiation, and spectroscopy. Even before Maxwell and his laws came into being, he calculated and proved that electricity moves through a wire at the speed of light. Along with the famous Robert Bunsen (of Bunsen-burner fame), he discovered the elements Cesium and Rubidium.

Gustav Kirchhoff‡ Bernard Tellegen†

Bernard Tellegen (born in 1900, died 1990) was a Dutch electrical engineer who formulated the Tellegen's theorem in 1952. A large part of Tellegen's career was at Philips Research. His other notable contributions included the gyrator and the pent-ode. The gyrator was an important contribution in electronic circuits. Tellegen won the IEEE Edison medal in 1973 for his invention of the gyrator.

‡: Photograph taken from the public domain.

†: Photograph re-used with permission from Royal Philips, Philips Company Archives.

Understand in depth

The topics in this chapter were developed with the help of a branch of Mathematics known as **graph theory**. For understanding this topic from ground-up, and for practice problems, the reader is referred to the following:

1. Engineering Circuit Analysis, by William H. Hayt and Jack E. Kemmerley, McGraw-Hill, chapter 3.

2. Basic Circuit Theory, by C.A. Desoer and E.S. Kuh, McGraw-Hill, chapters 9, 10, 11.

Unit 2

Network theorems

Unit specifics

In this unit we present the following topics:

- Tableau analysis
- The superposition theorem
- The Thévenin theorem
- The Norton theorem
- Source transformations
- The compensation theorem
- The reciprocity theorem
- The maximum power transfer theorem

Rationale

In the previous unit the student had learned to analyze a circuit systematically, but at the same time, optimally. This unit will give the reader intuition about the circuit, without a detailed analysis. Network theorems are often used as short-cuts to make predictions regarding circuits without performing a full analysis.

In this unit we will restrict ourselves to resistive circuits with strictly linear components, namely, R, linear dependent sources, and independent sources. The network theorems, with minor modifications, are also valid for circuits with L, C and M. However, these will be discussed later, in Unit-4, after we have started working fluently with inductors and capacitors.

Most of the network theorems are valid only for linear circuits. Non-linear circuits are often linearized for suitable application of the network theorems.

Pre-requisites

- Physics: Circuit analysis, an understanding of Unit-1.
- Mathematics: Basic matrix algebra, differentiation (Class XII)
- Software: We will be using an open-source numerical package, Octave. Octave is compatible with the commercially available MATLAB®.

Unit outcomes

The list of outcomes of this unit are as follows.

U2-O1: Learn to apply various network theorems to simplify circuits.

U2-O2: Understand the concept of a power source.

U2-O3: Critically appreciate maximum power transfer.

U2-O4: Understand the main internal working of computer programs that simulate circuits.

Unit-2 outcomes	Expected mapping with course outcomes *(1: Weak, 2: medium, and 3: strong correlation)*					
	CO-1	**CO-2**	**CO-3**	**CO-4**	**CO-5**	**CO-6**
U2-O1	-	3	-	-	1	-
U2-O2	-	3	-	-	1	-
U2-O3	-	2	-	-	2	-
U2-O4	3	-	-	-	-	3

2.1 Tableau analysis

Nodal analyses of a circuit (KCLs at all nodes except the datum) led us to

$$\mathbf{A} \cdot \mathbf{I} = \mathbf{0} \tag{2.1}$$

where **A** is the reduced incidence matrix, **I** is the vector of all the branch currents. Further, if we assign a potential to every node (the potential of the datum is 0), all the branch voltages can be expressed in terms of these potentials (KVLs). The above statement worked out to

$$\mathbf{A}^T \cdot \mathbf{E} = \mathbf{V} \tag{2.2}$$

where **E** is the vector of all the node (not datum) potentials, **V** is the vector of all the branch voltages. Further, each two-terminal device in the circuit will define a relationship for the corresponding branch current and voltage. Let us assume that the allowed devices in the circuit are independent current and voltage sources, dependent current and voltage sources, and resistors. We can express the device characteristics as:

$$\mathbf{M} \cdot \mathbf{V} + \mathbf{N} \cdot \mathbf{I} = \mathbf{U} \tag{2.3}$$

where **M** and **N** are $b \times b$ matrices that relate the branch voltages and currents, **U** is a vector with the independent voltage and current sources only. Now it is possible to assemble (2.1), (2.2), and (2.3) into a single compact form as follows:

$$\begin{bmatrix} \mathbf{0} & \mathbf{0} & \mathbf{A} \\ -\mathbf{A}^T & \mathbf{1}_b & \mathbf{0} \\ \mathbf{0} & \mathbf{M} & \mathbf{N} \end{bmatrix} \cdot \begin{bmatrix} \mathbf{E} \\ \mathbf{V} \\ \mathbf{I} \end{bmatrix} = \begin{bmatrix} \mathbf{0} \\ \mathbf{0} \\ \mathbf{U} \end{bmatrix} \tag{2.4}$$

(2.4) can be written succinctly as:

$$\mathbf{T} \cdot \mathbf{W} = \hat{\mathbf{U}} \tag{2.5}$$

where **T** is the square matrix in (2.4), **W** is the column vector of all the unknowns: node potentials, the branch voltages, the branch currents. $\hat{\mathbf{U}}$ is a column vector with the independent voltage and current sources. (2.5) may be solved to obtain **W**, i.e., all the node potentials, branch voltages, and branch currents by inverting **T** and multiplying with $\hat{\mathbf{U}}$.

As an illustration, we will analyze a circuit with resistors, dependent and independent sources, shown in Fig. 2.1. The circuit of Fig. 2.1 has six

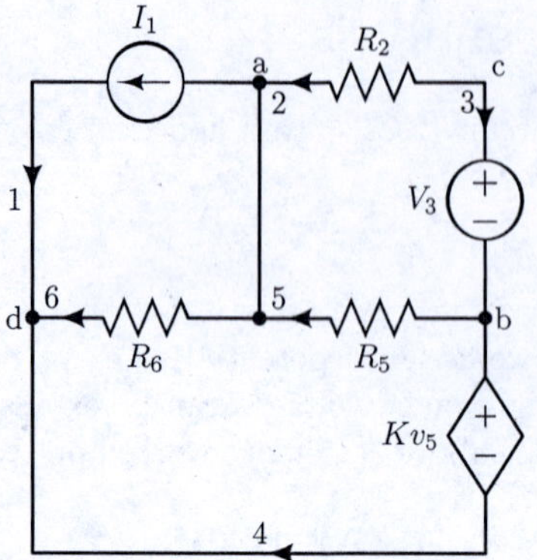

Figure 2.1: A circuit with resistors, dependent and independent sources, used as an illustration for tableau analysis.

elements and four nodes. Let us assume d is the datum. We have indicated the directions of branch voltages and currents. KCLs at a, b, c give:

$$\begin{bmatrix} 1 & -1 & 0 & 0 & -1 & 1 \\ 0 & 0 & -1 & 1 & 1 & 0 \\ 0 & 1 & 1 & 0 & 0 & 0 \end{bmatrix} \cdot \begin{bmatrix} i_1 \\ i_2 \\ i_3 \\ i_4 \\ i_5 \\ i_6 \end{bmatrix} = \begin{bmatrix} 0 \\ 0 \\ 0 \end{bmatrix} \text{, or, } \mathbf{A} \cdot \mathbf{I} = \mathbf{0}$$

We have written KVLs to express every branch voltage in terms of the node potentials. We obtain the following:

$$\begin{bmatrix} v_1 \\ v_2 \\ v_3 \\ v_4 \\ v_5 \\ v_6 \end{bmatrix} - \begin{bmatrix} 1 & 0 & 0 \\ -1 & 0 & 1 \\ 0 & -1 & 1 \\ 0 & 1 & 0 \\ -1 & 1 & 0 \\ 1 & 0 & 0 \end{bmatrix} \cdot \begin{bmatrix} e_a \\ e_b \\ e_c \end{bmatrix} = \begin{bmatrix} 0 \\ 0 \\ 0 \\ 0 \\ 0 \\ 0 \end{bmatrix} \text{, or, } \mathbf{V} - \mathbf{A}^T \cdot \mathbf{E} = \mathbf{0}$$

Thirdly, we can express the device characteristics of every circuit element.

$$\begin{bmatrix} 0 & 0 & 0 & 0 & 0 & 0 \\ 0 & 1 & 0 & 0 & 0 & 0 \\ 0 & 0 & 1 & 0 & 0 & 0 \\ 0 & 0 & 0 & 1 & -K & 0 \\ 0 & 0 & 0 & 0 & 1 & 0 \\ 0 & 0 & 0 & 0 & 0 & 1 \end{bmatrix} \begin{bmatrix} v_1 \\ v_2 \\ v_3 \\ v_4 \\ v_5 \\ v_6 \end{bmatrix} + \begin{bmatrix} 1 & 0 & 0 & 0 & 0 & 0 \\ 0 & -R_2 & 0 & 0 & 0 & 0 \\ 0 & 0 & 0 & 0 & 0 & 0 \\ 0 & 0 & 0 & 0 & 0 & 0 \\ 0 & 0 & 0 & 0 & -R_5 & 0 \\ 0 & 0 & 0 & 0 & 0 & -R_6 \end{bmatrix} \begin{bmatrix} i_1 \\ i_2 \\ i_3 \\ i_4 \\ i_5 \\ i_6 \end{bmatrix} = \begin{bmatrix} I_1 \\ 0 \\ V_3 \\ 0 \\ 0 \\ 0 \end{bmatrix}$$

which is essentially, $\mathbf{M} \cdot \mathbf{V} + \mathbf{N} \cdot \mathbf{I} = \mathbf{U}$.

For a circuit with 6 branches and 4 nodes there are 6 branch voltages, 6 branch currents, and 3 node potentials. The three matrix equations can be combined into a single matrix equation with $2 \times 6 + 3 = 15$ rows, as $\mathbf{T} \cdot \mathbf{W} = \hat{\mathbf{U}}$. In Octave, after defining `A`, `M`, `N`, we can describe `T` as:

```
k = rows(A);
b = columns(A);
T = [zeros(k,k),    zeros(k,b), A; ...
   -transpose(A), eye(b),       zeros(b, b);...
     zeros(b,k),    M,          N];
```

In the above code snippet, note that `k` is the number of nodes independent of the datum, i.e., $n-1$. For a circuit with 4 nodes, one will write out only 3 node voltage equations. The node voltage equation at the datum is linearly dependent with the other node voltage equations.

Tableau analysis is often how a computer simulation program will solve a circuit and is purely mechanical. (2.5) can be solved by inverting $\mathbf{T}$, and is given by:

$$\mathbf{W} = \mathbf{T}^{-1} \cdot \hat{\mathbf{U}} \tag{2.6}$$

Inverting $\mathbf{T}$ and multiplying with $\hat{\mathbf{U}}$ is how all the node potentials, branch voltages, and branch currents can be obtained.

2.2 Superposition theorem

Theorem 2.1 (Superposition theorem): A network $\mathcal{N}$ has a set of k **independent** sources, $\{u_1, u_2, \cdots u_k\}$, resistors, and linear dependent sources. The output quantity of $\mathcal{N}$ is a certain node potential, or a certain branch voltage or a certain branch current, w. w_i is the measured output quantity when only u_i is ON, and the remaining independent sources are OFF (value of zero). Then the output quantity w when all the independent sources are ON together will be the sum of all w_i, that is,

$$w = \sum_{i=1}^{k} w_i$$

Proof. The output quantity, w, is an element of $\mathbf{W}$, the vector comprising all node potentials, branch voltages, and branch currents. From (2.6), $\mathbf{W}$ is a linear combination of all the elements of $\hat{\mathbf{U}}$. In other words,

$$w = a_1 u_1 + a_2 u_2 + \cdots a_k u_k$$

If we switch ON only u_i and zero the remaining independent sources, the output is $a_i u_i$. So, w_i is $a_i u_i$. Hence, $w = \sum_{i=1}^{k} w_i$. ■

The message of the superposition theorem is quite simple: the effect of all the independent sources together is the sum of their individual effects. Suppose a circuit has an independent current source I_1 and an independent voltage source V_2. In that case, we can analyze the circuit by first solving the circuit applying only I_1 and with $V_2 = 0$, then solving the circuit applying only V_2 and with $I_1 = 0$. We add the two solutions to obtain the final result.

To null a voltage source, the voltage source must be replaced with a short circuit ($V = 0$). To null a current source, the current source must be replaced with an open circuit ($I = 0$).

The superposition theorem is valid when the circuit contains only linear circuit elements, i.e., independent voltage and current sources, linearly dependent voltage and current sources, and resistors. Later, we will show that superposition is valid even with inductors, capacitors, and mutual inductance.

Example 2.1. In the circuit shown below, V_1 is 1 V, I_2 is 2 A, R_3, R_4, R_5 are 3 Ω, 4 Ω and 5 Ω respectively. Evaluate the voltage across R_4.

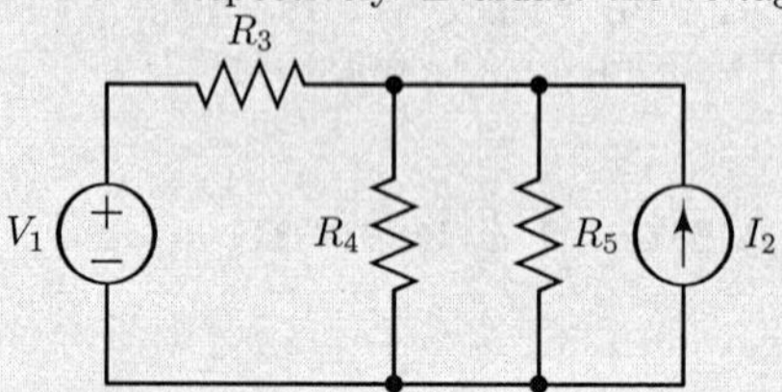

It is certainly possible to solve the circuit using the node voltage method or the mesh current method. However, the analysis is simpler using superposition.

First let us apply V_1 and turn I_2 OFF. Turning a current source OFF implies its value becoming 0 A, or the current source becoming an open circuit. The circuit is now a simple voltage divider and the voltage across R_4 can be evaluated orally as $V_1 \cdot (R_4 \parallel R_5)/(R_3 + R_4 \parallel R_5)$ or 0.43 V.

Now let us apply I_2 and turn V_1 OFF. Turning a voltage source OFF implies its value becoming 0 V, or the voltage source becoming a short circuit. Now we have the current I_2 being pushed through a parallel combination of three resistors. The voltage across R_4 is simply $I_2 \cdot (R_3 \parallel R_4 \parallel R_5)$ or 2.55 V.

Overall, the voltage across R_4 is 2.98 V.

Example 2.2. Let us now analyze the circuit in Fig. 2.1 and see if superposition offers any advantage.

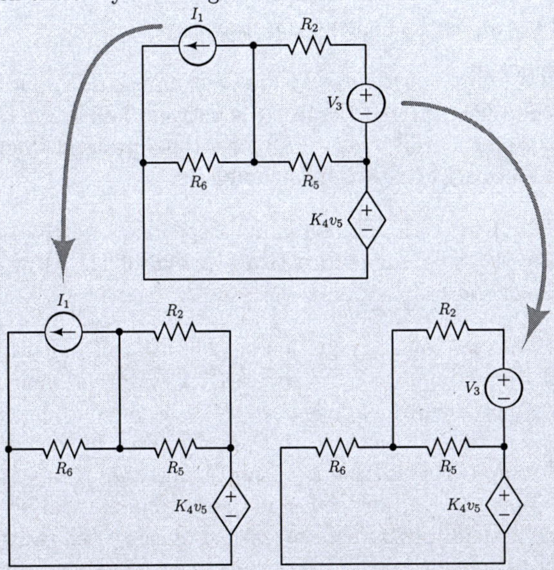

The circuit has two independent sources, I_1 and V_2. The diagram above shows the two-step technique for analysis using superposition. First, we switch ON I_1 and null V_2, i.e., replace V_2 with a short circuit. The circuit is analyzed. Next, we switch ON V_2 and null I_1, i.e., replace I_1 with an open circuit. The circuit is analyzed again. The analysis of the original circuit (any branch voltage, any branch current, any node voltage) is the sum of these two analyses. Superposition is a valuable technique in simplifying the circuit. With the short-circuit and the open-circuit, it is conceivable that the complexity of the circuit can be substantially reduced.

Definition 2.1 (System): A system is a relationship between one or more input excitations and one or more outputs.

Definition 2.2 (Linear system): A linear system has the following two properties:

1. When an input u_1 is applied to a system, its output is v_1. When an input u_2 is applied to a system, its output is v_2. The system is linear with respect to its input *if and only if* the output is $v_1 + v_2$ when the

input is $u_1 + u_2$. This is known as the superposition property.

2. When an input u is applied to a system, its output is v. When an input Au is applied to the system, its output is Av, where A is any constant. This is known as the homogeneity property.

Definition 2.3 (Time invariant system): When an input $u(t)$ is applied to a system, its output is $v(t)$. If the input is delayed, i.e., if the applied input is $u(t - t_0)$, the output will be $v(t - t_0)$ for a time-invariant system, i.e., the output will be delayed by the same amount.

Corollary 2.1. A network $\mathcal{N}$, with only independent sources, linear dependent sources, resistors, is **linear and time-invariant (LTI)** with respect to each independent source.

Proof. We can use the tableau equation of (2.6). We may identify any single element of $\hat{\mathbf{U}}$ as the input, u. Let us reset all the other elements of $\hat{\mathbf{U}}$ to zero. Also, we may identify any element of $\mathbf{W}$ as the output, w. Now w is proportional to u. So $\mathcal{N}$ is linear with respect to each independent source.

With one source at a time, w is proportional to u. If $u(t)$ is delayed by t_0, i.e., if we apply $u(t - t_0)$ instead of $u(t)$, the response will be $w(t - t_0)$ as opposed to $w(t)$. If all the sources are delayed by t_0, so will the output. This proves that $\mathcal{N}$ is time invariant with respect to each independent source. ■

Corollary 2.2. Inductors, capacitors, mutual inductance are linear time-invariant circuit elements, as long as there are no (zero) initial conditions.

Proof. Let us assume that for each of these components, the input quantity is current, and the output quantity is voltage.

For inductors, $v(t) = L\frac{di}{dt}$. If the current is i_1, $v_1 = L\frac{di_1}{dt}$. If the current is i_2, $v_2 = L\frac{di_2}{dt}$. If the current is $i_1 + i_2$ the voltage is $L\frac{d(i_1+i_2)}{dt} = v_1 + v_2$.

For capacitors, $v(t) = 1/C \int_0^t i(t)dt + v(0)$. Given zero initial conditions, this becomes $v(t) = 1/C \int_0^t i(t)dt$. For input current $i_1(t)$, $v_1(t) = 1/C \int_0^t i_1(t)dt$. For input current $i_2(t)$, $v_2(t) = 1/C \int_0^t i_2(t)dt$. For input current $i_1(t) + i_2(t)$, the output voltage is $1/C \int_0^t (i_1(t) + i_2(t))dt = v_1(t) + v_2(t)$.

If the inputs are delayed, the corresponding outputs will be equally delayed. The reader can easily prove this. The components are time-invariant.

Similarly, for inductors and capacitors, if the input quantity is voltage and the output quantity is current, the behavior is linear and time-invariant. Also, similar arguments will hold for mutual inductance. ■

Definition 2.4 (Port): A port $\mathcal{P}$ is a pair of wires coming out of a network.

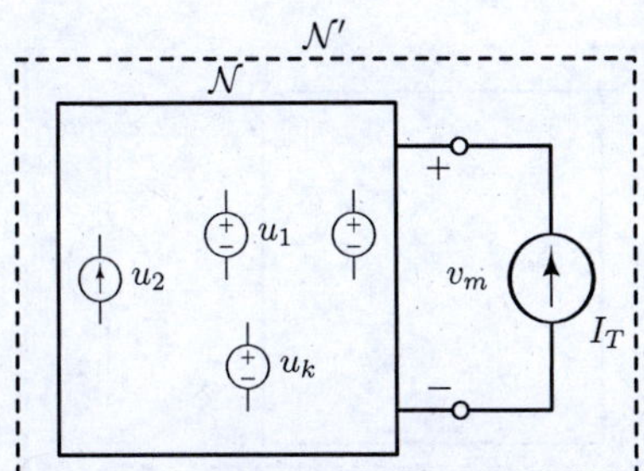

Figure 2.2: The network $\mathcal{N}$ contains linear circuit elements, dependent and independent sources. $\mathcal{N}'$ includes I_T that has been applied at the port of $\mathcal{N}$.

Definition 2.5 (Driving point impedance): The driving point impedance of a network $\mathcal{N}$, looking into a port $\mathcal{P}$, is the resistance measured across $\mathcal{P}$ when all the independent sources in $\mathcal{N}$ are nulled (i.e., voltage sources short-circuited, current sources open-circuited.)

2.3 Thévenin and Norton theorems

Theorem 2.2 (Thévenin's theorem): A network $\mathcal{N}$ comprises linear circuit elements (independent sources, linear dependent sources, resistors) and $\mathcal{N}$ has one port. Looking at $\mathcal{N}$ from outside the port, the behavior of $\mathcal{N}$ is identical to that of a single voltage source V_T in series with a resistor R_T, where (i) V_T is the open-circuit voltage of $\mathcal{N}$, and (ii) R_T is the driving point resistance of the network.

Proof. Consider the one-port network $\mathcal{N}$ as shown in Fig. 2.2. A current source I_T is applied from outside the network, and the voltage v_m is measured. $\mathcal{N}$ contains a set of independent sources $\{u_1, u_2, \cdots u_k\}$. The network, including I_T, is labeled as $\mathcal{N}'$, as shown in Fig. 2.2. $\mathcal{N}'$ may be solved using the tableau method. Therefore, from (2.6):

$$v_m = a_1 u_1 + a_2 u_2 + \cdots a_k u_k + a_T I_T,$$

because there are k independent sources within $\mathcal{N}$ and I_T applied from outside $\mathcal{N}$. We combine the first k terms into a single term, V_T, and rewrite the last term as $R_T I_T$.

$$v_m = V_T + R_T I_T \tag{2.7}$$

(i) As per (2.7), when I_T is nulled (open circuited), the measured v_m is V_T. V_T is also the open-circuit voltage, measured when the port is open.

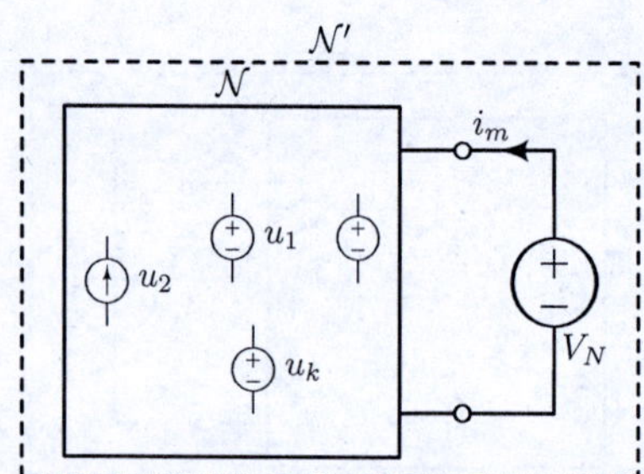

Figure 2.3: $\mathcal{N}$ contains linear circuit elements, independent sources, and linear dependent sources. $\mathcal{N}'$ includes V_N applied across the port.

(ii) When we null the k independent sources (short-circuit voltage sources, open-circuit current sources), from (2.7), v_m is only $R_T I_T$. R_T is also the driving point resistance of the network.

As such, $\mathcal{N}$ may be replaced with a voltage source, V_T, in series with a resistor, R_T. ■

Theorem 2.3 (Norton's theorem): A network $\mathcal{N}$ is comprised of linear circuit elements and has one port $\mathcal{P}$. Looking at $\mathcal{N}$ from outside $\mathcal{P}$, the behavior of $\mathcal{N}$ is identical to that of a single current source I_N in shunt with a resistor R_N, where (i) I_N is the short-circuit current of $\mathcal{N}$, and (ii) R_N is the driving point resistance of the network.

Proof. Consider the one-port network $\mathcal{N}$ as shown in Fig. 2.3. A voltage source V_N is applied from outside the network, and the current i_m is measured. $\mathcal{N}$ contains a set of independent sources $\{u_1, u_2, \cdots u_k\}$. The network, including V_N, is labeled as $\mathcal{N}'$. $\mathcal{N}'$ may be solved using the tableau method. Therefore, from (2.6):

$$i_m = b_1 u_1 + b_2 u_2 + \cdots b_k u_k + b_N V_N$$

We arrive at the above equation because there are k independent sources within $\mathcal{N}$ and V_N applied from outside $\mathcal{P}$. We combine the first k terms into a single term, I_N, and rewrite the last term as V_N/R_N.

$$i_m = I_N + V_N/R_N \tag{2.8}$$

(i) As per (2.8), when V_N is nulled (short-circuited), the measured i_m is I_N. I_N is also the short-circuit current, measured when $\mathcal{P}$ is short-circuited.

(ii) When the k independent sources are nulled (voltage sources short-circuited, current sources open-circuited), from (2.8) i_m is only V_N/R_N. R_N is also the driving point resistance of the circuit. Nulling a source implies making its value zero.

As such, $\mathcal{N}$ may be conveniently replaced with I_N in shunt with R_N. ■

The Thévenin and Norton theorems are duals of each other. Any linear circuit can be modeled as its Thévenin or Norton equivalent.

Example 2.3. In the circuit shown below, R_1, R_2, R_3, R_4 are 1 Ω, 2 Ω, 3 Ω and 4 Ω respectively. R_L is 5 Ω and V_0 is 6 V. Evaluate the current through the resistor R_L in the direction indicated.

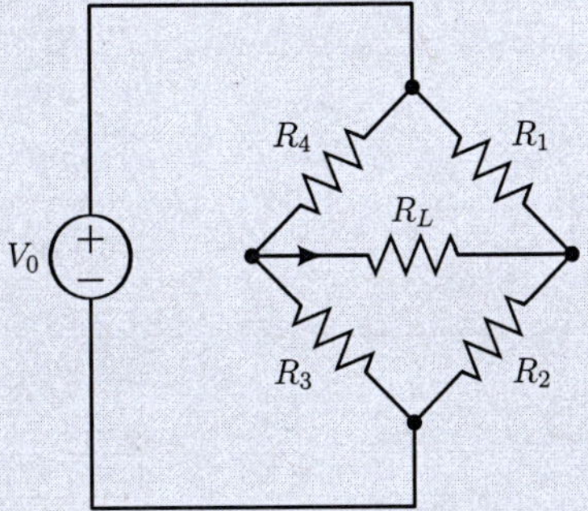

Notice that the component values are such that the familiar Wheatstone-bridge is *not set up*. Let us evaluate the Thévenin equivalent of the network, looking back from R_L. We will now pull R_L out of the network. For clarity, this is shown in the diagram below.

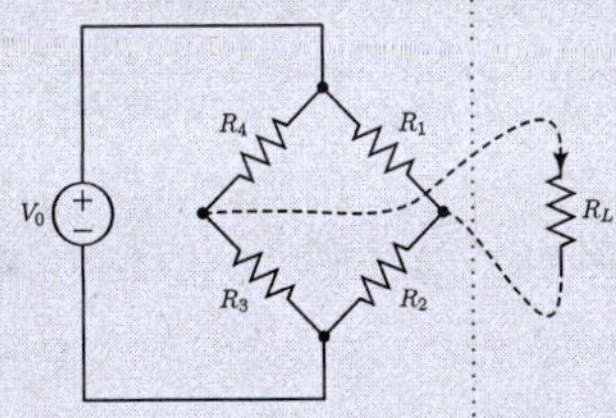

We can now disconnect R_L, and work out the open-circuit voltage across the two dangling terminals. If we consider the negative terminal of V_0 to be the datum, the upper dangling terminal is at $V_0R_3/(R_3+R_4) = 2.57$ V. The lower dangling terminal is at $V_0R_2/(R_1+R_2) = 4$ V. The open circuit voltage is -1.43 V.

Next we can work out the driving point resistance looking into these two dangling terminals. We will short-circuit V_0. This works out to $(R_3 \parallel R_4) + (R_1 \parallel R_2) = 2.38\ \Omega$.

Now the model of the dotted region is simply a voltage source of -1.43 V in series with a resistance of 2.38 Ω. We now return R_L back into the circuit. The current through R_L works out to -0.19 A.

Example 2.4. In the circuit shown in the schematic below, evaluate the voltage across the 6 Ω load resistor.

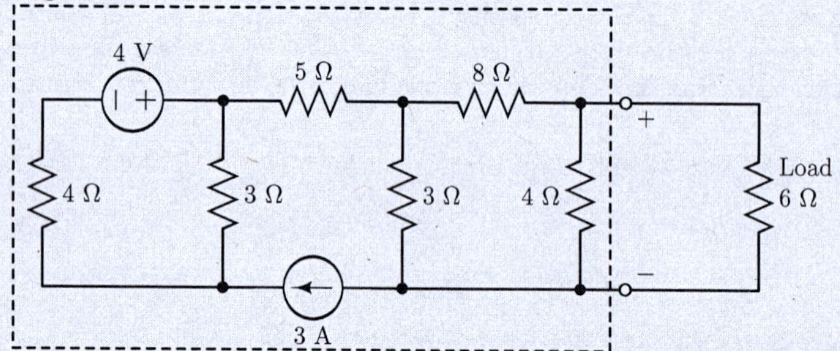

First, we will remove the load and measure the open-circuit voltage. Then, we will null all the internal voltage and current sources and find the driving point resistance. The open-circuit voltage, V_T, is $3 \times 3/15 \times 4 = 2.4$ V. The driving point resistance, R_T, is $4 \parallel 11 = 2.933\Omega$. The circuit can now be re-drawn as in the schematic (a) below, where the voltage across the load can be found using potential division, as 1.612 V.

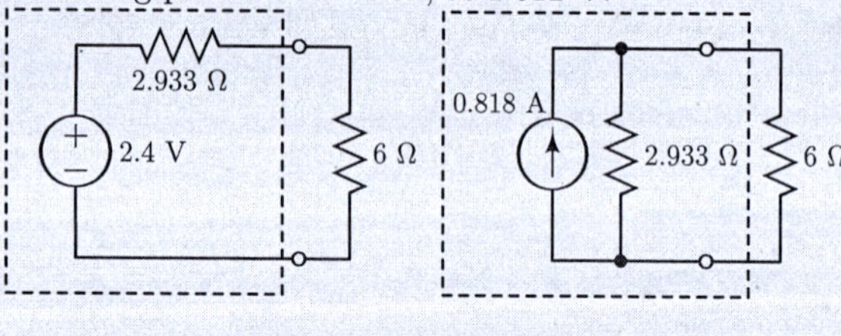

(a) Thévenin equivalent (b) Norton equivalent

The same circuit can also be analyzed using Norton's theorem. The short-circuit current, I_N, is $3 \times 3/11 = 0.818$ A. The driving point resistance, $R_N = R_T = 2.933\Omega$. The circuit can be re-drawn as in the schematic (b) above, and the voltage across the load can be worked out using parallel combining of resistors, as 1.612 V.

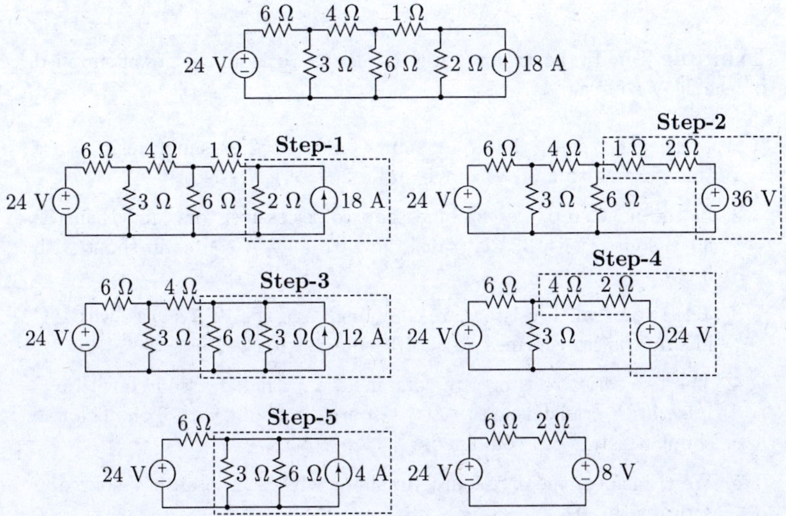

Figure 2.4: Circuit simplification with the help of source transformations.

2.4 Source transformations

Theorem 2.4 (Transforming a voltage source into a current source and vice-versa): A linear circuit has a Thévenin equivalent of V_T and R_T. Then its Norton equivalent is $I_N = V_T/R_T$ and $R_N = R_T$. If the linear circuit has a Norton equivalent of I_N and R_N, its Thévenin equivalent will be $V_T = I_N R_N$ and $R_T = R_N$.

Proof. The Thévenin model has a short-circuit current of V_T/R_T. Therefore the Norton equivalent current will be $I_N = V_T/R_T$. Further, R_T and R_N are the same driving point resistance of the circuit.

Similarly, the Norton model will have an open-circuit voltage of $I_N R_N$. Therefore the Thévenin equivalent voltage will be $V_T = I_N R_N$. ■

We can transform a voltage source in series with a resistor to a current source in shunt with a resistor. For example, a 10 V battery with an internal resistance of 2 Ω can be transformed to a 5 A current source with a shunt resistance of 2 Ω. The transformation is both ways, and a current source in shunt with a resistor can be transformed to a voltage source in series with a resistor.

Example 2.5. In the circuit of Fig. 2.4, let us work out the power supplied by the 24 V source.

1. In the first step, the 18 A source and the 2 Ω shunt resistor are transformed to a 36 V source and a 2 Ω series resistor.
2. In the next step, we combine the two series resistors, and then the 36 V source with 3 Ω is transformed to a 12 A source in shunt with 3 Ω.
3. The two parallel resistors are combined, and the 12 A source with 2 Ω in shunt is transformed to a 24 V source with 2 Ω in series.
4. The two series resistors are combined, and the 24 V source with 6 Ω in series is transformed to a 4 A source in shunt with 6 Ω. The two shunt resistors are combined.
5. We transform the 4 A source in shunt with 2 Ω to an 8 V source in series with 2 Ω.
6. Now the circuit reduces in complexity and the current coming out of the 24 V source works out to 2 A. The power supplied by the 24 V source is therefore 48 W.

Observation 2.1. A practical voltage source always has an internal series resistance. Therefore it can be transformed into a current source in shunt with the internal resistance. Every source of **finite** power (voltage or current source) can be transformed one way or the other. There is nothing sacrosanct about a voltage source; in some cases, it may be easier to treat it as a current source.

2.5 Compensation theorem

The compensation theorem has been stated in the literature in several different forms [3], [4]. All the statements are equivalent in terms of network theory. The statement in this text follows that of [4].

Theorem 2.5 (Compensation theorem): The compensation theorem is illustrated with the help of the schematics in Fig. 2.5. An LTI network $\mathcal{N}$ is connected to a resistor R. A current I flows through R as shown in Fig. 2.5(a). When the resistor R is changed to $R + \Delta R$, the current changes to $I + \Delta I$, as shown in Fig. 2.5(b). $R_{\mathcal{N}}$ is the driving point impedance of

$\mathcal{N}$. ΔI is the same as the current driven by a voltage source $-I\Delta R$ driving $R_{\mathcal{N}} + R + \Delta R$, as illustrated in Fig. 2.5(c).

Proof. Let us mentally replace the network $\mathcal{N}$ with its Thévenin equivalent model, V_T in series with R_T. R_T is nothing but the driving point impedance of the network, $R_{\mathcal{N}}$.

In the first scenario, V_T is driving $R_{\mathcal{N}} + R$ and a current I flows.

$$\therefore\ V_T = IR + IR_{\mathcal{N}}$$

In the second scenario, V_T is driving $R_{\mathcal{N}} + R + \Delta R$ and a current $I + \Delta I$ flows.

$$\begin{aligned} V_T &= (I + \Delta I)(R_{\mathcal{N}} + R + \Delta R) \\ \text{or, } V_T &= \underbrace{IR + IR_{\mathcal{N}}}_{V_T} + I\Delta R + \Delta I(R_{\mathcal{N}} + R + \Delta R) \\ \text{or, } V_T &= V_T + I\Delta R + \Delta I(R_{\mathcal{N}} + R + \Delta R) \\ \therefore\ -I\Delta R &= \Delta I(R_{\mathcal{N}} + R + \Delta R) \end{aligned}$$

■

Example 2.6. In the circuit below the strength of the voltage and current sources are unknown. The kth resistor, R_k, is of k Ω. When R_L is 10 Ω, I_L is 1 A. What will be I_L if R_L is 15 Ω?

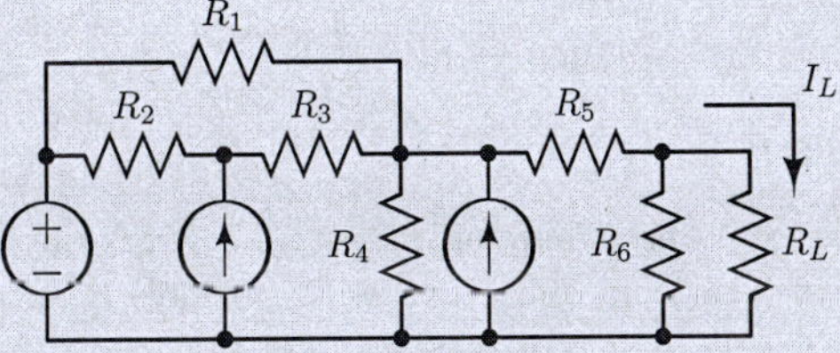

First let us work out the driving point impedance of the circuit looking backwards from R_L. To evaluate the driving point impedance, all independent sources need to be switched OFF (or zeroed), i.e., all voltage sources are to be short circuited and all current sources are to be open circuited. The remaining circuit, looking back from R_L, not including R_L, has an impedance of:

$$R_{\mathcal{N}} = R_6 \parallel [R_5 + \{R_4 \parallel R_1 \parallel (R_2 + R_3)\}] = 2.92\ \Omega$$

We do not know the strengths of the voltage and current sources; as such let us not use the Thévenin or Norton theorems, but apply the compensation

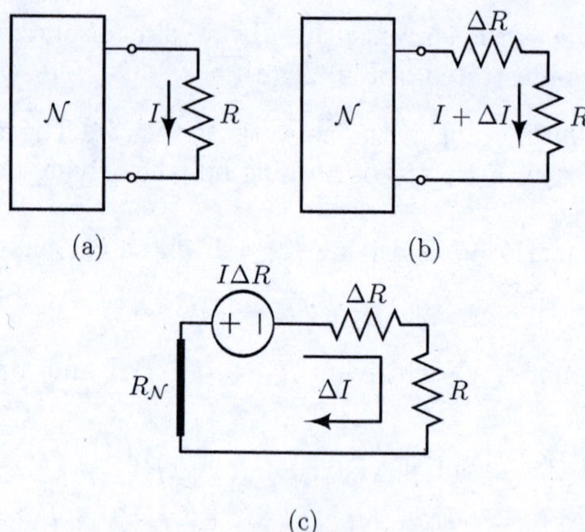

Figure 2.5: [Compensation theorem] (a) An LTI network $\mathcal{N}$ is connected to a resistor R. A current I flows through R. (b) R is changed to $R + \Delta R$. The current I will no longer flow, and change to $I + \Delta I$. (c) Compensation theorem says that ΔI will be the current driven by a voltage source $I\Delta R$ driving $R + \Delta R$ and the driving point impedance of $\mathcal{N}$. The polarity of the voltage source is against the direction of current.

theorem. The resistor R_L is changed by $\Delta R = 5\ \Omega$. The extra current, ΔI_L will be that caused by a voltage source of -1 A$\times$5 Ω driving $R_{\mathcal{N}}+R_L+\Delta R$.

$$\therefore \quad \Delta I_L = -5/(2.92 + 15) = -0.279$$

So the new I_L is 0.721 A.

Alternately, we could use Thévenin's theorem to work out this problem. Imagine the network looking back from R_L to have a certain Thévenin equivalent voltage V_T and resistance $R_{\mathcal{N}}$. In that case, $V_T = I_L(R_{\mathcal{N}} + R_L) = 12.92$ V, for the original circuit. Now when we change R_L to 15 Ω, $12.92 = I_L(2.92 + 15)$, or I_L is 0.721 A.

2.6 Reciprocity theorem

Definition 2.6 (Passive network): A passive network is a network in which no individual component generates power. A typical passive network will not contain controlled voltage and current sources, independent voltage and

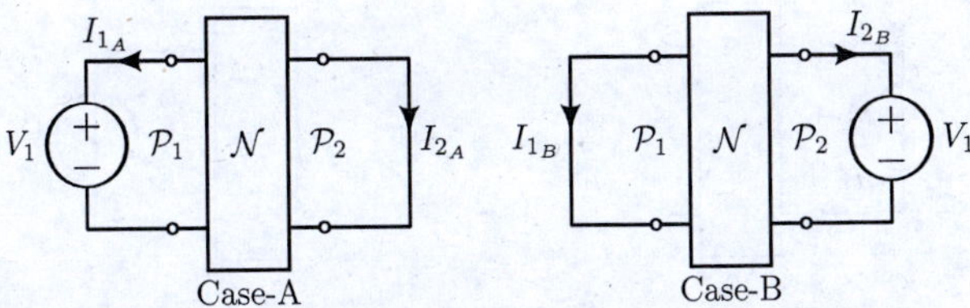

Figure 2.6: If $\mathcal{N}$ is passive, the reciprocity theorem (Theorem 2.6) states that I_{2_A} will be the same as I_{1_B}.

current sources.

Theorem 2.6 (Reciprocity theorem with a voltage source, short circuit current): The passive (resistive) network $\mathcal{N}$ has two ports $\mathcal{P}_1$ and $\mathcal{P}_2$. An excitation voltage V_1 is applied at $\mathcal{P}_1$, and the short circuit current is measured at $\mathcal{P}_2$ as I_2. Then, if we apply the same excitation V_1 at $\mathcal{P}_2$, the short circuit current measured at $\mathcal{P}_1$ will be I_2.

Proof. Let us assume that the short circuit in both cases is a 0-Ω resistor. The network di-graph in both cases is identical. We will apply Tellegen's theorem in the following manner: first, the currents in Case-A, voltages in Case-B, and next, the currents in Case-B, voltages in Case-A.

Let us also assume there are k resistors in $\mathcal{N}$, of values $\{R_1, R_2, \cdots R_k\}$, the currents through these k resistors being $\{i_{R1_A}, i_{R2_A}, \cdots i_{Rk_A}\}$ in Case-A, and $\{i_{R1_B}, i_{R2_B}, \cdots i_{Rk_B}\}$ in Case-B.

$$I_{1_A} \times 0 + I_{2_A} V_1 + \sum_{j=1}^{j=k} (i_{Rj_A} R_j i_{Rj_B}) = 0 = V_1 I_{2_B} + 0 \times I_{1_B} + \sum_{j=1}^{j=k} (i_{Rj_B} R_j i_{Rj_A})$$

The above results in $I_{2_A} = I_{2_B}$. ■

Theorem 2.7 (Reciprocity theorem with a current source, open circuit voltage): The passive (resistive) network $\mathcal{N}$ has two ports $\mathcal{P}_1$ and $\mathcal{P}_2$. An excitation current I_1 is applied at $\mathcal{P}_1$, and the open-circuit voltage is measured at $\mathcal{P}_2$ as V_2. Then, if we apply the same excitation I_1 at $\mathcal{P}_2$, the open-circuit voltage measured at $\mathcal{P}_1$ will be V_2.

Proof. Let us assume that the open circuit in both cases is a resistor of infinite value. The network di-graph in both cases is identical. We will apply Tellegen's theorem again: first, the currents in Case-A with voltages in Case-B, and next, the currents in Case-B with voltages in Case-A. Again,

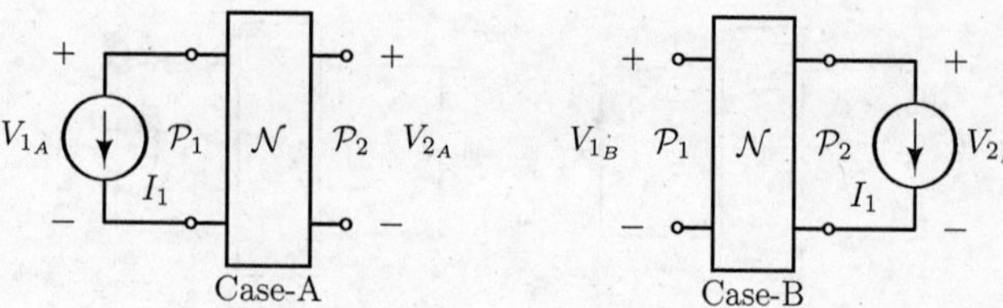

Figure 2.7: If $\mathcal{N}$ is passive, the reciprocity theorem (Theorem 2.7) states that V_{2_A} will be the same as V_{1_B}.

let us assume there are k resistors in $\mathcal{N}$ and have currents as in the proof of theorem-2.6.

$$I_1 V_{2_B} + 0 \times V_{1_B} + \sum_{j=0}^{k} (i_{Rj_A} R_j i_{Rj_B}) = 0 = 0 \times V_{1_A} + I_1 V_{2_A} + \sum_{j=0}^{k} (i_{Rj_B} R_j i_{Rj_A})$$

The statement can be simplified to show that $V_{2_A} = V_{2_B}$. ■

Reciprocity is valid as long as the network is linear and passive. Reciprocity is also valid in electro-magnetism; for example, a passive antenna is reciprocal because its characteristics while transmitting and receiving are identical. Theorems 2.6 and 2.7, which we have just proved, form the basis for the reciprocity of an antenna. A mutual inductor is also reciprocal, in the sense that the effect of a changing current in the primary coil on the secondary coil will be the same as the effect of a changing current in the secondary coil on the primary coil.

Example 2.7. In the circuit shown below, the values of the unmarked are unknown. When a current of 5 A is applied in the schematic on the left, a current of 1 A is measured through the 10 Ω resistor. What will be i_x, the current in the 3 Ω resistor when a current of 5 A is applied as in the schematic on the right?

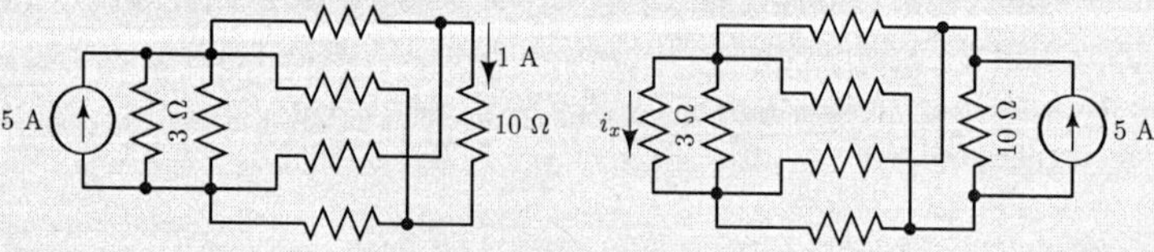

For the schematic on the left, a current of 1 A through the 10 Ω resistor creates a 10 V drop. Let us consider a passive two-port network which includes both the 3 Ω and the 10 Ω resistors, in addition to all the other resistors shown in the schematic on the left. Applying a 5 A current at the port on the left creates an open-circuit voltage of 10 V at the port on the right.

By the reciprocity theorem, applying a 5 A current at the port on the right should create a 10 V open-circuit voltage at the port on the left. For the schematic on the right, this is the case. Therefore, the current i_x through the 3 Ω resistor will be 3.33 A.

Reciprocity can also be applied to multi-port (more than 2-port) networks quite easily. One only needs to consider two ports at a time, while ignoring (either by short-circuiting, or by open-circuiting, or even by connecting an agreed-upon load at) the other ports.

Example 2.8. In the schematics shown below, $\mathcal{N}$ is a passive three-port network. When a 10 V voltage source with source resistance of 1 Ω is applied at $\mathcal{P}_1$, a voltage of 2 V is measured across a 2 Ω resistor at $\mathcal{P}_2$, a voltage of 3 V is measured across a 3 Ω resistor at $\mathcal{P}_3$ (illustrated in the schematic on the left). What is the voltage v_x measured across a 1 Ω resistor at $\mathcal{P}_1$ (illustrated in the schematic on the right) if 1 V in series with 2 Ω is applied at $\mathcal{P}_2$ and 1 V in series with 3 Ω is applied at $\mathcal{P}_3$ at the same time?

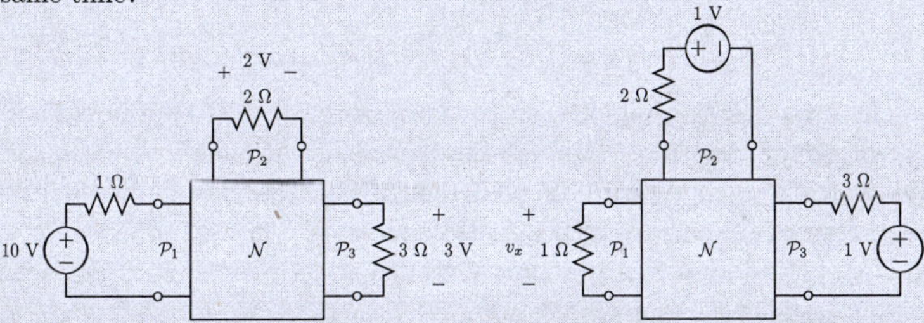

It is possible to solve this question with the help of Tellegen's theorem. Instead, we will apply the reciprocity theorem, with the aid of superposition and homogeneity. (Any question that can be solved with the help of the reciprocity theorem can also be directly solved with the help of Tellegen's theorem.)

First, let us consider the network $\mathcal{N}'$, as shown in the schematic below. The 1 Ω resistor at $\mathcal{P}_1$, the 2 Ω resistor at $\mathcal{P}_2$ and the 3 Ω resistor at $\mathcal{P}_3$ have been absorbed into $\mathcal{N}'$.

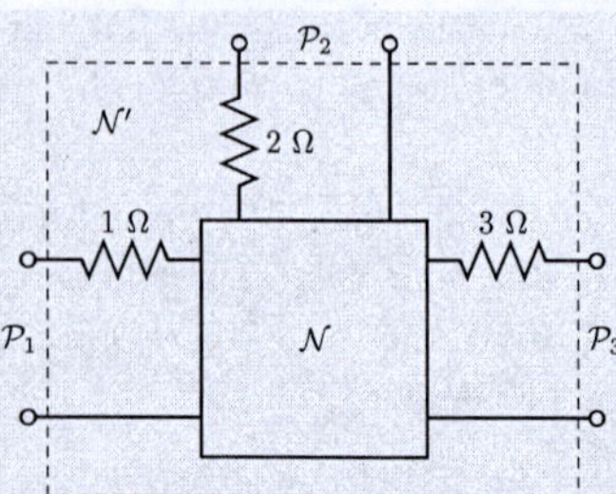

Now, as per problem statement, when 10 V is applied at the new $\mathcal{P}_1$, and the new $\mathcal{P}_2$ and $\mathcal{P}_3$ are short-circuited, the currents in $\mathcal{P}_2$ and $\mathcal{P}_3$ are 1 A and 1 A. If we apply reciprocity theorem now, and apply 10 V at $\mathcal{P}_2$, (with a short circuit at $\mathcal{P}_3$,) $\mathcal{P}_1$ should carry 1 A. Similarly if we apply 10 V at $\mathcal{P}_3$, (with a short circuit at $\mathcal{P}_2$,) $\mathcal{P}_1$ should carry 1 A. To summarize in a table:

Voltage at $\mathcal{P}_2$	Voltage at $\mathcal{P}_3$	Current at $\mathcal{P}_1$
10 V	0 V	1 A
1 V	0 V	0.1 A
0 V	10 V	1 A
0 V	1 V	0.1 A
1 V	1 V	0.2 A

Overall, with the help of reciprocity and superposition, we have evaluated the current at $\mathcal{P}_1$ to be 0.2 A, the voltage v_x to be 0.2 V.

In the previous example 2.8, observe the usage of the reciprocity theorem along with superposition. We also have used homogeneity. That is, if the applied voltage of 10 V results in a current of 1 A, then an applied voltage of 1 V will result in a current of 0.1 A. Homogeneity can be applied only when there is a single independent source ON at a time, or when all independent sources are scaled together.

2.7 Maximum power transfer theorem

The following discussion on the maximum power transfer theorem applies to circuits with resistors and linear dependent sources. We will study a re-statement of the theorem for general circuits with R, L, C, M, and dependent sources, later in Unit-4.

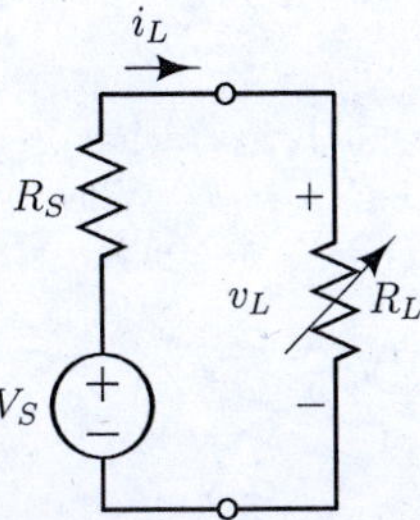

Figure 2.8: The maximum power transfer theorem (Theorem 2.8) states that maximum power is transferred from V_S to R_L if $R_L = R_S$.

Theorem 2.8 (Maximum power transfer theorem for a resistive network): When a non-ideal voltage source (voltage source in series with a non-zero resistor) drives a resistive network, maximum power is drawn from the source to the network when the driving point resistance of the network is equal to the source resistance.

Proof. i_L is $V_S/(R_S + R_L)$. v_L is $V_S R_L/(R_S + R_L)$. Therefore the power transferred to the load is $P_L = V_S^2 R_L/(R_S + R_L)^2$. P_L is maximum when dP_L/dR_L is zero.

$$\frac{dP_L}{dR_L} = V_S^2 \frac{\{(R_S + R_L)^2 - 2R_L(R_S + R_L)\}}{(R_S + R_L)^4} = 0$$

Simplifying, we obtain $R_L = R_S$. ■

We have stated the maximum power transfer theorem for a non-ideal voltage source. However, equivalently it could also be formulated for a non-ideal current source (current source in shunt with a resistor). We can transform a current source in shunt with a source resistance into a voltage source in series with the same source resistance, and then the theorem in its current wording may be applied.

Corollary 2.3. When a non-ideal current source (current source in shunt with a non-zero resistor) drives a resistive network, maximum power is drawn from the source to the network when the driving point resistance of the network is equal to the source resistance.

Proof. The current source in shunt with the source resistance can be transformed to a voltage source in series with the same source resistance. The proof follows. ■

Example 2.9. In the circuit below, what is the value of R_L such that maximum power is delivered to R_L? What is this maximum power in watts? Assume V_0 is 10 V and R_k is k Ω.

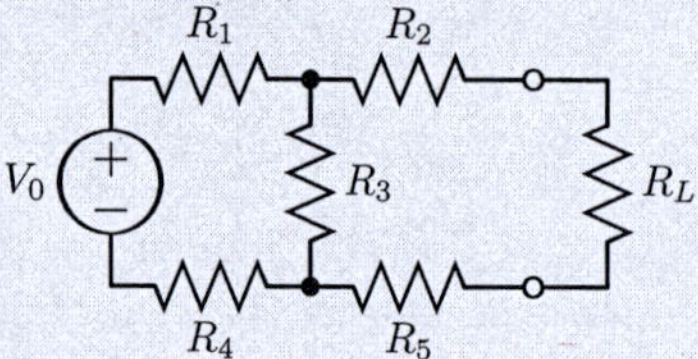

Let us first construct a Thévenin equivalent of the network, looking in from R_L. The driving point impedance can be calculated by short-circuiting V_0; the rest of the circuit can be resolved by series and parallel combinations. The driving point impedance, $R_{\mathcal{N}}$ is:

$$R_{\mathcal{N}} = \{(R_1 + R_4) \parallel R_3\} + R_2 + R_5 = 8.875\ \Omega$$

Maximum power will be delivered to R_L if R_L is 8.875 Ω.

The Thévenin equivalent voltage, V_T, can be evaluated by considering the open-circuit voltage across the terminals. This simply works out to:

$$V_T = V_0 \frac{R_3}{R_1 + R_3 + R_4} = 3.75\ \text{V}$$

The power in the R_L is the voltage across R_L times the current through R_L. When R_L is the optimal 8.875 Ω, the voltage across R_L is $V_T/2$, and the current is $V_T/(2R_{\mathcal{N}})$. Therefore, the maximum power that can be delivered to R_L, when R_L is 8.875 Ω, is $(V_T/2) \cdot (V_T/(2R_{\mathcal{N}})) = 0.396$ W.

Observation 2.2. Looking back from a load, R_L, if the Thévenin equivalent voltage is V_T, then the maximum power that can be delivered to the load resistor is $V_T^2/(4R_L)$.

Observation 2.3. The maximum power transfer theorem is misapplied in many cases. The theorem states that for a given source resistance, the load resistance must be equal to the source resistance to draw maximum power. It does not talk about what the source resistance should be. The premise of the maximum power transfer theorem is a situation where the source resistance is constant. What should be the source resistance to deliver maximum power from the source to the load? The answer is zero for a voltage source, ∞ for a current source. If you have the option of changing the source resistance of a power source, to maximize power transfer your objective should be to make the source as ideal as possible.

Example 2.10. The model of a BJT amplifier is shown in the two-port network below. The input is from a microphone of source resistance R_S. The output is delivered to a speaker of resistance R_L. i_C is a variable bias current that the amplifier designer can control. v_T, V_A, β are BJT device parameters that are fixed. Find the values of R_L and i_C such that maximum power is delivered to the load speaker.

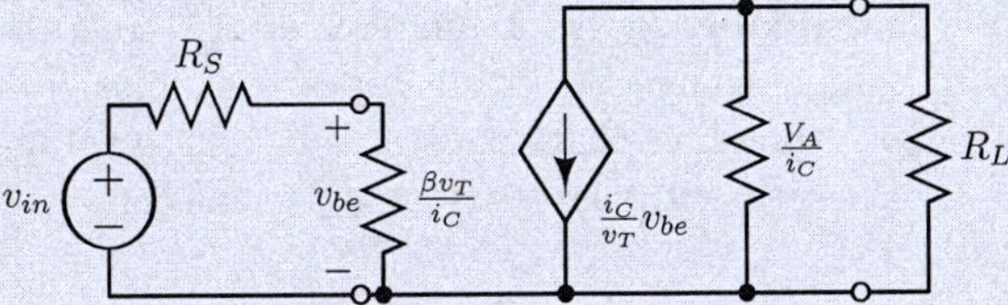

Let us first evaluate the driving point impedance of the circuit looking back from R_L. Only the independent voltage source, v_{in}, is to be nulled for this evaluation. If v_{in} is zero, the voltage v_{be} is also zero; the current through the controlled current source is therefore zero. The driving point impedance is therefore simply V_A/i_C. For maximum power transfer to R_L, R_L must be equal to V_A/i_C. We do not have a value for i_C yet.

Next, let us evaluate the actual power delivered to R_L. The voltage across the input port is $v_{in}\frac{\beta v_T/i_C}{\beta v_T/i_C+R_S}$. The current through the controlled source is the input port voltage times i_C/v_T. Only half of this current flows through R_L, assuming we have already made $R_L = V_A/i_C$. The power, P, in R_L is therefore:

$$P = \left(\frac{1}{2}\cdot\frac{i_C}{v_T}\cdot\frac{\beta v_T/i_C}{R_S+\beta v_T/i_C}\right)^2 (V_A/i_C) = \frac{i_C\beta^2 V_A}{4(i_C R_S+\beta v_T)^2}$$

To maximize the power in R_L, we need to evaluate dP/di_C and equate it to zero. This results in:

$$\begin{aligned} \beta^2 V_A(i_C R_S+\beta v_T)^2 &= 2i_C\beta^2 V_A(i_C R_S+\beta v_T)R_S \\ \text{or, } i_C R_S+\beta v_T &= 2i_C R_S \\ \text{or, } i_C &= \beta v_T/R_S \end{aligned}$$

We can rephrase the result to $\beta v_T/i_C = R_S$. This result is very interesting and satisfying, because the driving point impedance of the network looking rightwards into the input port is $\beta v_T/i_C$ and this needs to be equal to the source resistance. This means that maximum power needs to be delivered to the input port, and that defines the value of i_C.

Finally, we have $i_C = \beta v_T/R_S$, and $R_L = R_S V_A/(\beta v_T)$.

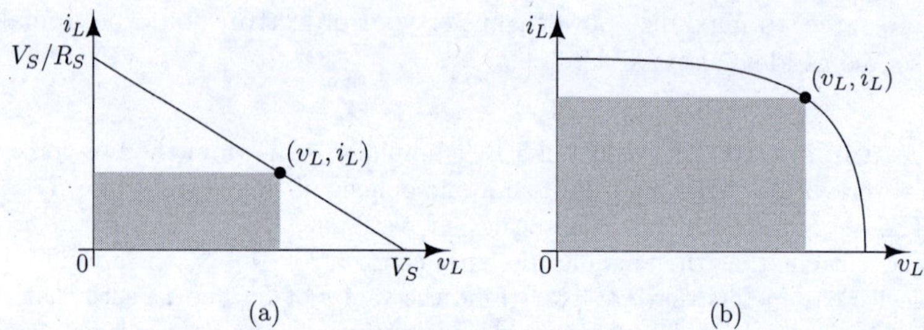

Figure 2.9: We have plotted i_L as a function of v_L. In (a), the circuit corresponds to a voltage source V_S with a series resistance R_S. In (b), we have shown the corresponding characteristics of a photovoltaic cell. The available power is the area of the largest rectangle that can be drawn within the characteristics.

Observation 2.4. When the load resistance is equal to the source resistance, P_L is $V_S^2/(4R_S)$. The source resistance defines the maximum power available from a given voltage source. The maximum power available from a given voltage source is the available power.

Definition 2.7 (Available power): The maximum power that a practical source can deliver is known as its available power.

Let us plot i_L as a function of v_L. (i_L and v_L are shown in Fig. 2.8.) The power delivered to the load is the area of a rectangle drawn from a point on the curve with the axes, as shown in Fig. 2.9(a).

The concept of available power is valid for all sources of power. A source of power is always an element where energy is converted from either mechanical, light, or chemical energy to electrical energy. The maximum power that one can draw from such a source is physically limited by the mechanical, chemical, or electromagnetic power given to the energy conversion system. There is no ideal voltage source in practice. In circuit theory, ideal voltage and current sources are convenient mathematical tools that help us analyze and design. In practice, every source is a source of power and has finite available power.

2.8 Unit summary

- If an input u_1 is applied to a system, the output is v_1. For an input u_2 the output is v_2. The system is linear if on application of $u_1 + u_2$ at the input, the output is $v_1 + v_2$.

- A circuit with R, L, C, M, independent sources, and linear dependent sources is an LTI system.

- In the superposition theorem, the response to multiple independent sources is the sum of the response to each independent source taken separately. When applying one of many independent sources, make sure to null the other sources. To null a voltage source, make it 0 volts, i.e., replace it with a short circuit. To null a current source, make it 0 amperes, i.e., replace it with an open circuit.

- A linear network between two terminals can be simplified into a single voltage source in series with a single resistance. (Thévenin's theorem.) Alternatively, a linear network between two terminals can be simplified into a single current source in shunt with a single resistance. (Norton's theorem.)

- The equivalent resistance obtained using either Thévenin's theorem or Norton's theorem is also known as the driving point impedance. The driving point impedance between two terminals is evaluated by nulling all *independent* sources in the circuit and then finding the equivalent resistance between the two terminals.

- The Thévenin equivalent voltage, V_T, of a network is the open-circuit voltage of the network. The Norton equivalent current, I_N, of a network is the short-circuit current of the network. V_T and I_N of a network are related to each other by the driving point impedance. However, one should not, in general, try to obtain the driving point impedance by evaluating V_T and I_N, as this may occasionally result in a division by zero.

- A voltage source, V_0, in series with a resistance, R can be transformed into a current source, V_0/R, in shunt with a resistance, R. A current source, I_0 in shunt with a resistance R can be transformed into a voltage source, $I_0 R$ in series with a resistance, R.

- A passive two-port network is reciprocal. A voltage source V_0 (or current source I_0) at port-1 creates a short-circuit current I_s at port-2 (or open-circuit voltage V_o at port-2.) Then the same voltage source V_0 at port-2 (or current source I_0 at port-1) will create the same short-circuit current I_s at port-1 (or the same open-circuit voltage V_o at port-1.)

- A non-ideal voltage source drives maximum power into a resistive load when the load resistance is the same as the source resistance. A non-ideal voltage source can be transformed into a non-ideal current source.

The non-ideal current source drives maximum power into a resistive load when the load resistance is the same as the source resistance.

2.9 Exercises

Multiple choice type questions

2.1 A network is excited by two sources. If the first is ON and the second is OFF, a current of 2 A is measured. If both are ON, a current of 1.5 A is measured. What is the current when the first is OFF but the second is ON?
(a) -0.5 A (b) +0.5 A (c) -3.5 A (d) 3.5 A

2.2 A battery applies a current of 13.6 mA across a load of 100 Ω, and 11.6 mA across a load of 120 Ω. What is the open circuit voltage of the battery?
(a) 1.20 V (b) 1.36 V (c) 1.39 V (d) 1.58 V

2.3 A voltage source of source resistance 10 Ω energizes a load through a wire of resistance 1 Ω. What should be the load resistance for maximum power transferred to the load?
(a) 1 Ω (b) 9 Ω (c) 10 Ω (d) 11 Ω

2.4 For maximum power transfer from a voltage source to the parallel combination of two resistors of 10 Ω and 20 Ω, the source resistance should be:
(a) 0 Ω (b) 10 Ω (c) 20 Ω (d) 30 Ω

2.5 A 1 A current source has a source resistance of 10 Ω. What is the available power from the current source?
(a) 2.5 W (b) 5 W (c) 7.5 W (d) 10 W

2.6 An LTI passive two-port network, $\mathcal{N}$, is excited at port-1 by a 1 V voltage source and it provides a short-circuit current of 2 A at port-2. What is the short-circuit current at port-1 when $\mathcal{N}$ is excited by a 5 V voltage source at port-2?
(a) 0.4 A (b) 2 A (c) 4 A (d) 10 A

Short answer type questions

2.7 A good voltage source should have a ______ (large/small) series resistance. A good current source should have a ______ (large/small) shunt resistance.

2.8 If the driving point impedance of a network is $R_{\mathcal{N}}$ and the Thévenin equivalent voltage is V_T, then the Norton equivalent current is ______.

2.9 A 10 V voltage source in series with 10 Ω can be transformed into a ______ A current source in ______ (series/shunt) with a ______ Ω resistor.

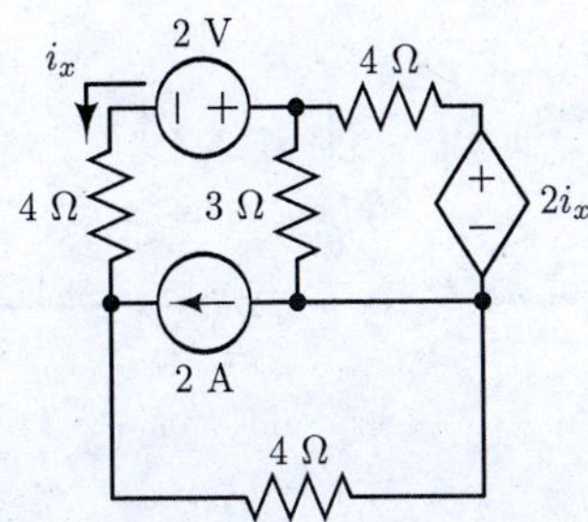

Figure 2.10: Circuit for questions 2.13, 2.14

2.10 The superposition, Thévenin's, and Norton's theorem apply to ______ circuits only.

2.11 The reciprocity theorem applies to a multi-port network that is strictly a ______ and ______ network.

2.12 An LTI network has a single independent source. If this source is 1 V, a measured current is i_x. It is easy to prove (using the superposition theorem) that if the source is 10 V, the measured current will be $10i_x$. It is not so easy to prove that if the source is 0.1 V, the measured current will be $0.1i_x$. How will you prove this second statement?

Numericals and long answer type questions

2.13 Use superposition or Thévenin's or Norton's theorems to determine the value of i_x in Fig. 2.10.

2.14 Is the tableau matrix **T** guaranteed to be invertible? Set up the tableau matrix for the circuit in Fig. 2.10 and invert it using Octave.

2.15 In the circuit of Fig. 2.11, all voltages ($V_0 \ldots V_3$, V_{out}) are referred to the datum node with the ground/earth symbol. The node labeled V_0 is connected to a voltage source V_0 with respect to ground; the node labeled V_1 is connected to a voltage source V_1 with respect to ground etc. Use Thévenin's theorem and superposition to arrive at an expression for V_{out} in terms of V_0, V_1, V_2, and V_3. The circuit is commonly used as a digital to analog converter. If V_0 is $b_0 \cdot 1$, V_1 is $b_1 \cdot 1$, etc., then show that V_{out} is the digital word $b_3b_2b_1b_0$ times 1/16 volts.

2.16 In the circuit of Fig. 2.12, use superposition and Thévenin, Norton's theorems to find the power supplied by the 1-volt voltage source and the power supplied by the 2-volt voltage source.

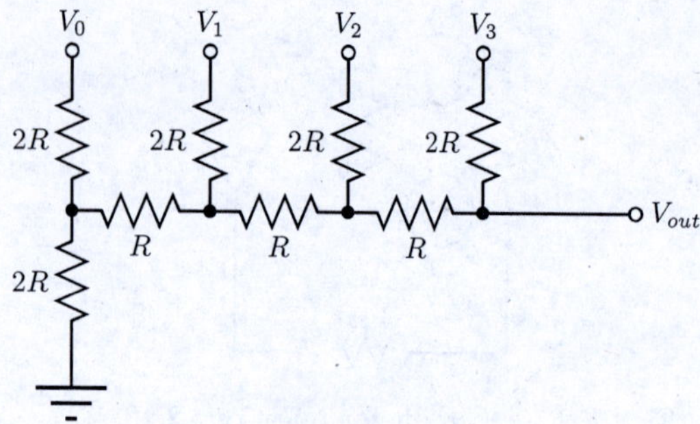

Figure 2.11: Circuit for question 2.15

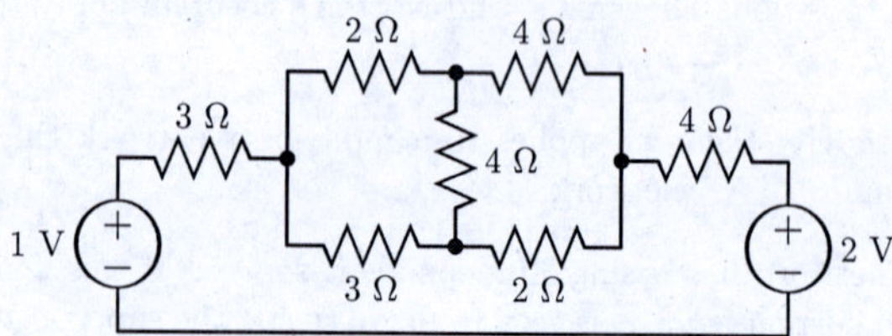

Figure 2.12: Circuit for question 2.16.

2.17 A two-port network shown in Fig. 2.13 has only linear circuit elements. When only the 2-A current source is applied, v_1 is 0.3 V, and i_2 is 0.4 A. When only the 3-V voltage source is applied, v_1 is 1.5 V, and i_2 is -0.1 A. What powers are delivered by the 2-A source and the 3-V source when both are switched on together?

2.18 When connected to a load of 10 Ω, a car battery provides a current of 2 A. When connected to a load of 20 Ω, the same battery provides a current of 1.05 A. What is the open-circuit voltage of the battery? What is the internal source resistance of the battery?

2.19 A circuit as configured in Fig. 2.14 responds to different values of R_L as

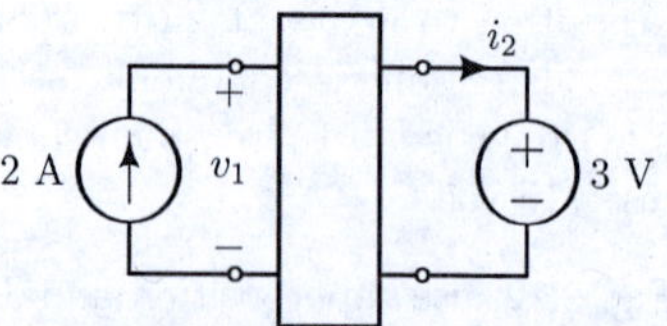

Figure 2.13: Circuit for question 2.17

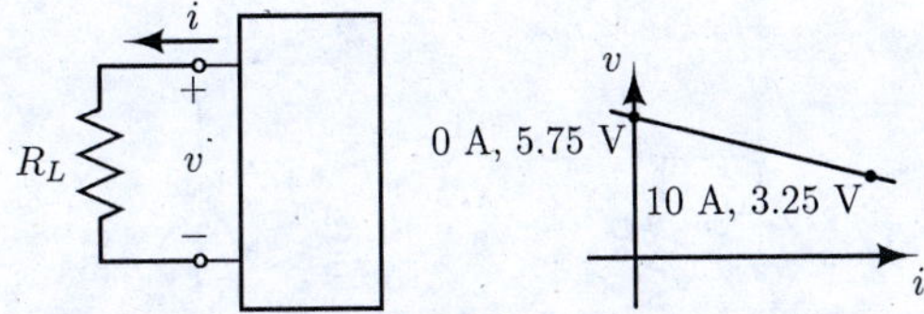

Figure 2.14: Circuit and graph for question 2.19

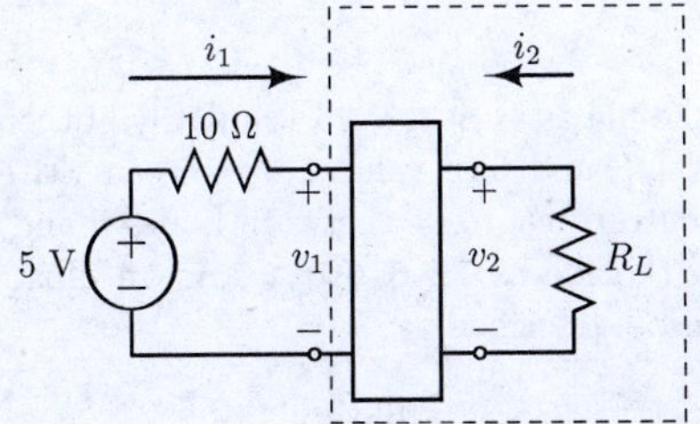

Figure 2.15: Circuit for question 2.20

shown in the corresponding graph. What is the value of R_L for which the temperature of R_L rises the fastest?

2.20 The two-port network shown in Fig. 2.15 has linear circuit elements and does not contain independent voltage or current sources. The voltages v_1, v_2, and the currents i_1, i_2 are related by the following equations:

$$\begin{aligned} v_1 &= 5i_1 + 6i_2 \\ v_2 &= 6i_1 + 12i_2 \end{aligned}$$

What should be R_L for maximum power transfer from the input voltage source to the dashed box? What should be R_L for maximum power transfer to R_L? Explain why the two answers are different.

2.21 In the circuit shown in Fig. 2.16, find the driving point impedance looking in from the terminal-pair labeled with positive and negative symbols.

Project activity

2.22 A simplified SPICE circuit description format is as follows. A circuit is described in a file, with each line describing a single circuit component. If there are n nodes in a circuit, the nodes can be labeled as 0, 1, $\ldots n-1$. If there is a resistor, R_k, between nodes 2 and 5 of value 26 Ω, we can

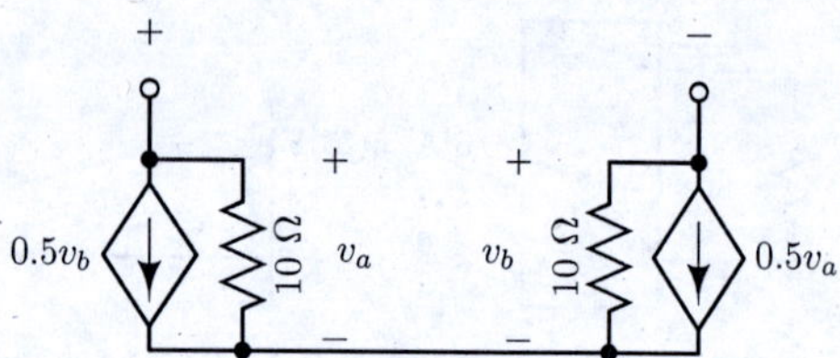

Figure 2.16: Circuit for question 2.21

write a line in the file as `Rk 2 5 26`. Similarly, if there is a voltage source, V_0 between nodes 3 and 0 of value 2 V, we can write a line in the file as `V0 3 0 2`. A current source I_x between nodes 6 and 4 of value 0.1 A can be described in the file as `Ix 6 4 0.1`. As an illustration, the circuit in example 2.4 can be described as:

file.sp

```
Ra 1 0 4
Vb 2 1 4
Rc 2 0 3
Rd 2 3 5
Re 3 4 3
Rf 3 5 8
Rg 5 4 4
Ih 4 0 3
RL 5 4 6
```

Write a tableau solving program. The program will be in Octave. The program will take in a file with a circuit description using the SPICE syntax and output the tableau matrix $\mathbf{T}$, and other components of the tableau equation, $\mathbf{W}$, and $\hat{\mathbf{U}}$. Restrict the input circuit to have only resistors and independent voltage and current sources. The program can flow as follows. First, read the entire file. An example is given in the code in Prog. 2.1. Be aware that the code will not work for elements that have more than two nodes.

Program 2.1. spicereadin.m: Octave script to read in a SPICE netlist and store it as instances, nodes, and values. The code will work only for two terminal elements.

```
fp = fopen('file.sp', 'r'); % Open the file

nodeas = []; nodebs = []; values = []; % Blank matrices.
instances = fscanf(fp, "%s", 1); % First element name
while (~(feof(fp)))
  nodeas = [nodeas; fscanf(fp, "%d", 1)]; % First terminal
  nodebs = [nodebs; fscanf(fp, "%d", 1)]; % Second terminal
  values = [values; fscanf(fp, "%f", 1)]; % Value
  instances = [instances; fscanf(fp, "%s", 1)]; % Next element name
```

```
    if(feof(fp)) instances = instances(1:end-1,:); end; % Termination
    % Note, the above line is needed only in Octave. Remove it for MATLAB.
end
fclose(fp); % Close the file
```

Estimate the number of branches (the number of instances) and the number of nodes. Then construct an empty (filled with zeros) tableau matrix $\mathbf{T}$ and an empty (filled with zeros) column vector $\hat{\mathbf{U}}$. Now process each branch and keep filling up $\mathbf{T}$ and $\hat{\mathbf{U}}$. Finally, work out $\mathbf{W}$ by computing $\mathbf{T}^{-1}\hat{\mathbf{U}}$.

2.23 As an extension of the previous exercise, if you can incorporate voltage and current-controlled voltage and current sources, you will have a full-fledged circuit simulator for DC circuits. You will need to examine the syntax of SPICE [5].

2.24 Based on your experience of question 2.15, design a 6-bit digital to analog converter. Demonstrate its operation in the laboratory. Use 1% accurate resistors (brown for the tolerance color code.) Make a table of the applied digital code and the observed output voltage of your D/A converter.

Know more

Historical profiles

Léon Charles Thévenin (1857 to 1926) was a French engineer who worked at the corps of telegraph engineers, which later became the French PTT (which have now split into French Telecom and La Poste). Thévenin worked on long-distance underground telegraph cables. A graduate of École Polytechnique de Paris, he developed the Thévenin theorem in 1882. He was appointed as a teaching inspector for the École Supérieure. Later he became the Director of the Telegraph Engineering School and the Chief Engineer of the telegraph workshops.

Léon Thévenin‡ Edward Norton†

‡: Photograph in the public domain.

†: Courtesy of AT&T Archives and History Center.

Edward Lawry Norton (1898 to 1983) was an American engineer. A graduate of MIT (B.S. 1922) and Columbia University (M.A. 1925), he spent his career in Bell Labs. During his days at Bell Labs, Norton was a legendary figure, single-handedly driving network theory work. Norton's theorem was independently derived in 1926 by Edward Norton and Hans Mayer.

Hans Ferdinand Mayer worked in the Siemens research laboratories in Berlin. In 1943 he was arrested by the Nazi regime for political reasons. Hans Mayer was the author of the famous 1939 Oslo report, a severe breach of German security during World War II. At Siemens, he had broad access to ongoing research in electronic weapons systems and radar. He survived the war through various Nazi concentration camps. After the war, he was the head of Siemens' research department for communications till 1962.

Understand in depth

The network theorems, such as reciprocity theorem and compensation theorem, have numerous applications in the world of electromagnetics. Refer to the seminal work of G.D. Monteath to explore the possibilities.

- G.D. Monteath, "Application of the compensation theorem to certain radiation and propagation problems," Proceedings of the IEE-Part IV: Institution Monographs, vol. 98, no. 1, pp. 23–30, 1951.

An excellent resource for laboratory verification of the network theorems is at the Virtual-Labs website by IIT Kharagpur. Scan the QR-code and visit the analog signals, network and measurement virtual laboratory.

Unit 3

First and second order networks

Unit specifics

In this unit the following topics are discussed:

- First order differential equations
- Series and parallel RL and RC circuits
- Initial and final conditions
- Second order differential equations
- Series and parallel RLC circuits
- Forced and free responses
- Time constants

Rationale

The student has learned circuit analysis at DC over the previous two units. In this unit the student will be introduced to dynamical circuits with first and second order circuits involving resistors, capacitors and inductors. We will form differential equations and solve them directly.

The circuits in this unit are a small set of possible first and second order circuits. However, the intuitive techniques shown will help the student analyze other complex circuit problems.

Pre-requisites

- Mathematics: Differential equations and basic calculus. Complex numbers.
- Physics: Inductors and capacitors. Charge sharing in capacitors.
- Software: We will continue using Octave as a numerical analysis package. Octave will be used for both programming as well as for calculations.

Unit outcomes

The list of outcomes of this unit are as follows.

U3-O1: Be able to frame first and second order linear differential equations for a circuit.

U3-O2: Be able to solve first and second order linear differential equations.

U3-O3: Be able to write-out solutions to first and second order circuits by inspection.

U3-O4: Be able to sketch and graph the time-domain responses of first and second order circuits.

Unit-3 outcomes	**Expected mapping with course outcomes** *(1: Weak, 2: medium, and 3: strong correlation)*					
	CO-1	**CO-2**	**CO-3**	**CO-4**	**CO-5**	**CO-6**
U3-O1	3	1	-	-	-	-
U3-O2	3	1	-	-	-	2
U3-O3	3	3	-	-	-	-
U3-O4	2	3	-	-	1	3

3.1 First order differential equations

Linear first order differential equations are of the form:

$$\frac{dx(t)}{dt} = ax(t) + p(t)$$

where $x(t)$ is the unknown quantity which is a function of time t, a is a known constant, and $p(t)$ is a known function of time, also known as the input.

3.1.1 Forms of the input

In circuit theory, we like to consider inputs i.e., $p(t)$, which are of the form Be^{st}, where B and s are arbitrary constants. It turns out that a simple form like the above can cover large classes of possible inputs. Typically, the input function is considered to be zero for $t < 0$, unless otherwise stated.

$s = 0$: This will make the input a constant for $t \geq 0$. In signal theory, such an input is known as a step function, symbolized by $u(t)$.

$s < 0$, $s \in \mathbb{R}$: In other words, s is a negative real number. This will result in a damped exponential, an input exponentially decreasing from B towards 0, with time.

$s > 0$, $s \in \mathbb{R}$: In other words, s is a positive real number. This will result in a rising exponential, an input exponentially increasing towards ∞ from B, with time.

$s \in \mathbb{C}$, $\text{Re}(s) < 0$: s is a complex number, but with a negative real part. Such an input is complex. Usually this input comes along with its complex conjugate; the pair of inputs (with s and its complex conjugate $\text{Conj}[s]$) together form a damped sinusoid. The amplitude of the damped sine wave exponentially decreases from B at $t = 0$ to 0.

$s \in \mathbb{C}$, $\text{Re}(s) > 0$: s is a complex number with a positive real part. Such an input comes along with its complex conjugate as a pair. The pair of inputs together form an exponentially rising sinusoid. The amplitude of the sine wave exponentially increases from B at $t = 0$ to ∞.

$s \in \mathbb{C}$, $\text{Re}(s) = 0$: s is a purely imaginary number. Such an input comes along with its complex conjugate as a pair. The pair of inputs together form a pure cosine (or sine) wave.

We should also realize that once we turn to complex numbers for s, there should not be any restriction on B. B should also be allowed to be complex.

A complex input, as such, is physically meaningless. However, a complex input along with its conjugate, in superposition, form a real input. A large class of real inputs can be synthesized by tweaking the properties of s; real, positive, negative, zero, imaginary, complex with positive and negative real parts.

3.1.2 Form of the solution

In mathematics, differential equations are usually solved by looking up templates. **The standard template for all linear differential equations that we will encounter in circuit analysis is Ce^{st}.** This includes first order, second order, and all higher order circuits, as long as they are built out of linear circuit elements (R, L, C, M, and linear dependent sources). The template for the solution, is in fact, the same as the template for the input.

The differential equation will have two parts to the solution. The first part is known as the homogeneous, or free response, or zero-input solution. The second part is known as the forced response, or steady-state, or zero-state solution. The complete solution is the sum of the zero-input and zero-state solutions.

3.1.3 Zero-input solution, free response

The first part of the solution is the zero-input solution, or free response, or homogeneous solution. This is the response of the circuit without any input applied to the differential equation. The free response of the circuit is primarily due to initial conditions, i.e., charge stored on a capacitor, or magnetic flux stored in an inductor, at $t = 0$. In most practical circuits with R, L, C and M, the free response decays to zero as $t \to \infty$. The template solution for the free response is Ce^{st}, where C is a constant that depends on the initial conditions of the circuit and s is a constant that depends on the component values and topology of the circuit.

Multiple free responses (values of s) could be possible; a second order differential equation will have two possible solutions; an nth order differential equation will have n possible solutions. The free response is the sum of all the possible solutions, i.e., $\sum C_k e^{s_k t}$. A distinct scaling constant, C_k, is associated with every possible s_k in the free response.

3.1.4 Zero-state solution, forced response

The second part of the solution is the zero-state solution, or forced response, or the particular solution. The forced response of the differential equation is the response to the input, without considering any initial conditions. For an input of the form $Be^{s_0 t}$, the forced response will be of the form $Ce^{s_0 t}$. In other words, the form of the solution is the same as that of the input, other than a scaling constant. (Note, we have allowed complex numbers in the discussion, and therefore, the scaling constant could also be complex!)

3.1.5 Complete solution

Let us now try to solve the following first order linear differential equation:

$$\frac{dx(t)}{dt} = ax(t) + Be^{s_0 t} \tag{3.1}$$

a, B, s_0 are known constants.

To start with, *let us assume the input is zero.*

$$\frac{dx(t)}{dt} = ax(t) \tag{3.2}$$

In (3.2), let us plug in a template solution, $x(t) = Ae^{st}$. This will give

$$Ase^{st} = aAe^{st}$$

In other words, $s = a$ will satisfy (3.2). This solution is known as the zero-input solution, or the homogeneous solution, or the free response.

However, $x(t) = Ae^{at}$ will not satisfy the complete differential equation of (3.1). To solve (3.1), let us pick a different template solution, $x(t) = Ce^{s_0 t}$. If we plug this in, we obtain:

$$Cs_0 e^{s_0 t} = aCe^{s_0 t} + Be^{s_0 t}$$

This gives us $C = B/(s_0 - a)$. This second solution is known as the zero-state solution, or the particular solution, or the forced response.

Our complete solution is the sum of the zero-input solution and the zero-state solution.

$$x(t) = Ae^{at} + \frac{B}{s_0 - a}e^{s_0 t}$$

Notice that A is still an unknown. The value of A is to be deduced from initial conditions, or from conditions at a particular point in time.

If the value of $x(t)$ at $t = 0$ is given as x_0, then A can be deduced. At $t = 0$, the value of our solution is:

$$x(0) = A + B/(s_0 - a) = x_0$$

This gives $A = x_0 - B/(s_0 - a)$.

Finally, the solution to the complete linear differential equation is ready. Between first, second, and higher order differential equations, the only differences are:

1. For an nth order differential equation, there will be n possible solutions to the homogeneous equation. These will have to be combined with n different A coefficients.

2. To solve for the n different values of A corresponding to the n solutions to the homogeneous equations, we will need n different conditions (as opposed to just one initial condition.)

The reader is encouraged to plug the solution back into the original differential equation of (3.1) and check that the solution is satisfactory. Differential equation theory in mathematics guarantees that this is the only solution to the differential equation.

3.1.6 Summary of steps

The following is a summary of the steps to solve the differential equation.

1. Find the solution to the homogeneous equation. The unknown constant(s) are to be left alone. The value of s in the template is to be solved for. This is the free response.

2. Find the particular solution to the given input. This is the forced response.

3. Add the forced and free responses. Now solve for the unknowns in the free response part by plugging-in at given time-points.

3.2 Series RC and RL circuits

3.2.1 Series RC circuit

In the schematic shown in Fig. 3.1, the switch is turned ON at $t = 0$. The voltage across the capacitor before the switch closes is V_0. Let us solve the circuit for $v_C(t)$ and $i(t)$, as indicated in the schematic.

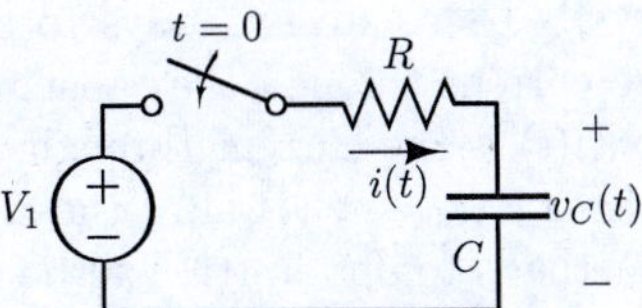

Figure 3.1: Series R-C circuit

The circuit has a single mesh and for $t \geq 0$ the mesh equation can be written out as below:

$$V_1 = i(t)R + v_C(t)$$

The current through the capacitor is the same as the current through the resistor. As such,

$$V_1 = RC\frac{dv_C(t)}{dt} + v_C(t) \tag{3.3}$$

Let us now proceed with our three-step solution technique, as discussed in section 3.1.6.

1. The homogeneous equation is as given below.

   $$\frac{dv_C}{dt} = -\frac{1}{RC}v_C$$

   The solution to the homogeneous equation will be of the form $Ae^{-t/(RC)}$, where A is a constant not known at present.

2. In (3.3), V_1 is the input forcing function. The input is of the form $V_1e^{0 \cdot t}$. The solution is also of the same form as the input; therefore, for a constant input, the solution is a constant. Let us assume the solution to be V_2. Then, plugging into (3.3) we simply obtain $V_1 = V_2$. In other words, the forced response is V_1.

3. The complete solution is now of the form $V_1 + Ae^{-t/(RC)}$. At $t = 0$, $v_C(0)$ is V_0. This means,

   $$V_1 + Ae^0 = V_0$$

   The value of A is obtained as $V_0 - V_1$. Our final solution is now ready.

   $$v_C(t) = \begin{cases} V_1 + (V_0 - V_1)e^{-\frac{t}{RC}}, \text{ for } t \geq 0 \\ V_0 \text{ for } t < 0 \end{cases} \tag{3.4}$$

$i(t)$ can now be evaluated. There are two ways to obtain $i(t)$. According to the circuit, $i(t)$ is the current through the capacitor. This makes $i(t) = C dv_C(t)/dt$. Alternately, $i(t)$ is the current through the resistor, where the potential across the resistor is $V_1 - v_C(t)$ after the switch closes, for $t \geq 0$. Not surprisingly, both methods to obtain $i(t)$ will be consistent.

$$i(t) = \begin{cases} 0 \text{ for } t < 0 \\ \frac{V_1 - V_0}{R} e^{-\frac{t}{RC}} \text{ for } t \geq 0 \end{cases}$$

Observation 3.1. A quick evaluation of (3.4) at $t = 0$ will show that the solution is continuous, that is, if we approach the solution from both the right and the left, towards $t = 0$, the limits are the same.

$$\lim_{t \to 0^+} v_C(t) = \lim_{t \to 0^-} v_C(t) = V_0$$

Although **this cannot be generalized**, most solutions for the voltage across a capacitor or for the current through an inductor, will be continuous. This is because the voltage across a capacitor, often, does not change instantaneously. An instantaneous change in the voltage across a capacitor will require Cdv/dt current, instantaneously infinite current. This is certainly possible, but occurs in special scenarios only.

In our solution of the homogeneous equation, we obtained proportionality to $e^{-t/(RC)}$. The product RC clearly has dimensions of time; its unit will be seconds. RC compares with the time; a small RC product will take less time, a large RC product will take more time to attain the same value. As such, the RC product is known as the **time constant** of the circuit.

$v_C(t)$ is evaluated and plotted in Octave in the script given in Prog. 3.1.

Program 3.1. seriesRC.m: Evaluation and plotting

```
V0 = 0; V1 = 1;
t = [0:1e-3:10]; % Time from 0 to 10s in steps of 1ms
RC = [0.01 0.04 0.16 0.64 2.56] % Different RC time constants
for k=1:length(RC) % For each time constant
  vC = V1 + (V0-V1)*e.^(-t/RC(k)); % vC(t)
  plot(t, vC); hold on; % Plot, but do not erase
end; % End of for loop
```

Fig. 3.2 shows $v_C(t)$, the voltage across the capacitor, for different RC time-constants in the circuit. Initially the capacitor is charged to V_0, and slowly it charges from V_0 to V_1 through the resistor.

How much time does the process take? The time it takes to reach the final value is infinite. However, an engineering approximation is to report the time taken to reach 90% of the target value.

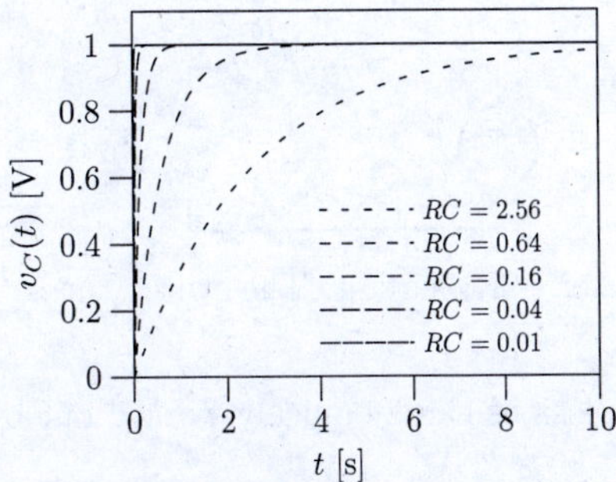

Figure 3.2: Voltage across the capacitor for a series RC circuit, for different RC time-constants. The plot is from Prog. 3.1.

Definition 3.1 (Rise or fall time): When the input is a step function, the rise and fall time are defined to be the time taken for the output to reach 90% of the target value. When the output is rising, it is referred to as the rise time. When the output is falling, it is referred to as the fall time.

The rise time, from an inspection of (3.4), is clearly the time for $e^{-t/(RC)}$ to reach 0.1. It takes approximately $2.3RC$ seconds to settle to 90% of the target value. As Fig. 3.2 clearly shows, the rise time decreases as the time constant of the circuit decreases.

3.2.2 Initial and final conditions

The solutions to the different circuit expressions can also be deduced with the help of a short-cut. The short-cut requires the knowledge of initial, final conditions and the time constant of the circuit.

In Fig. 3.1, the initial condition, $v_C(0^+)$, is V_0. As $t \to \infty$, by inspection it can be seen that $v_C \to V_1$. The time constant of the circuit is RC. The complete expression for $v_C(t)$ can now be written out as:

$$v_C(t) = V_1 + (V_0 - V_1)e^{-\frac{t}{RC}}$$

Now, let us try to work out $i(t)$ using this short-cut. At $t = 0^+$, the voltage across the capacitor continues to remain at V_0. The current through the resistor is therefore $(V_1 - V_0)/R$. As $t \to \infty$, the voltage across the capacitor reaches V_1 and the current through the resistor reduces to 0. From this information we can write out the expression for $i(t)$.

$$i(t) = \frac{V_1 - V_0}{R} e^{-\frac{t}{RC}}$$

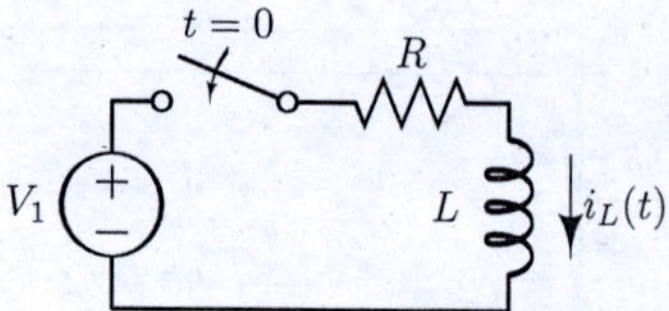

Figure 3.3: A series-RL circuit. The switch turns ON at $t = 0$.

All other circuit expressions can be quickly worked out by inspection.

3.2.3 Series RL circuit

A series-RL circuit is shown in Fig. 3.3. For $t < 0$, the loop is open, and the current in the loop is guaranteed to be zero. At $t = 0$, the switch is turned ON. A mesh equation written for $t \geq 0$ is as follows:

$$V_1 = i_L(t)R + L\frac{di_L(t)}{dt}$$

We can solve this differential equation using the three-step technique of section 3.1.6.

1. The homogeneous equation is:

   $$\frac{di_L(t)}{dt} = -\frac{R}{L}i_L(t)$$

   As such, its solution is of the form $Ae^{-tR/L}$.

2. The forcing function is a constant. The forced response will also be a constant. If $i_L(t)$ is a constant, I, its derivative is zero. The particular equation will therefore be:

   $$V_1 = IR + L \cdot 0$$

   The particular solution is, therefore, a current V_1/R.

3. The complete solution is $V_1/R + Ae^{-tR/L}$. At $t = 0^+$, we know the current through the inductor is 0. Therefore A is $-V_1/R$. The final solution is:

   $$i_L(t) = \begin{cases} 0 \text{ for } t < 0 \\ \frac{V_1}{R}(1 - e^{-\frac{t}{L/R}}) \text{ for } t \geq 0 \end{cases} \quad (3.5)$$

Notice the formatting of (3.5). We have placed t in the numerator and L/R in the denominator of the exponent. This is to make it appear similar to (3.4). **As such, the time constant of the circuit is L/R.**

Once we understand the time-constant of the R-L circuit, we can also write out the solution by evaluating initial and final conditions.

For example, the voltage across the inductor at $t = 0^+$ is V_1. As $t \to \infty$, the current through the inductor will stop changing; the voltage voltage across the inductor will be its derivative, that is, 0. Therefore, the expression for the voltage across the inductor, $v(t)$, is given as:

$$v(t) = V_1 e^{-\frac{t}{L/R}} \text{ for } t \geq 0$$

Remember, the time constant of an RC circuit is RC, while the time constant of an RL circuit is L/R. The reader is encouraged to check the dimensions of RC and L/R; both should have units of seconds.

3.3 Other first order RC and RL circuits

3.3.1 Order of a circuit

Circuits with only one capacitor or only one inductor will be first order. Such circuits will require the solving of a first-order linear differential equation.

In circuit theory, we like framing our differential equations with the voltage across a capacitor as a variable. The current through the capacitor depends on the voltage across. The voltage across the capacitor is referred to as a **state variable**. Similarly, differential equations are framed with the current through an inductor as a variable. The voltage across the inductor depends on this current. The current through an inductor is a **state variable**.

A complex circuit with several capacitors and inductors will have many state variables. Every capacitor voltage will contribute one state variable; every inductor current will contribute one state variable. However, there may be a situation where the voltage across a capacitor, or the current through an inductor depends on other state variables in the circuit. These **dependent** variables are **not** counted as state variables.

A circuit with N state variables is an Nth order circuit, and requires solving an Nth order linear differential equation. The following example will elucidate this discussion.

Example 3.1. What is the order of the following circuit?

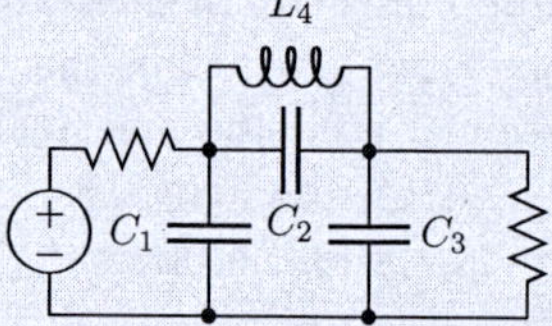

There are three capacitors, C_1, C_2 and C_3. There is one inductor, L_4. The voltages across the three capacitors are related to each other; the voltage across C_2 is the difference of the voltage across C_1 and the voltage across C_3. As such, between the capacitors, there are only 2 independent state variables. The inductor current provides a third independent state variable. The order of the circuit is three.

3.3.2 Step response of a first order circuit

When excited by a step-function (a constant for $t > 0$), all first-order circuits can be solved with the help of initial, final conditions, and by evaluating their time constants.

How do we evaluate the time constant of a complex circuit? To start with, let us isolate the single inductor or capacitor in the circuit. The driving point impedance of the remainder of the circuit, looking back from this inductor/capacitor, may be evaluated. The time constant of the circuit is now simply $R_N C$ or L/R_N.

How do we evaluate the final condition of a complex circuit? As the time $t \to \infty$, for a constant or step input, the current through an inductor, or the voltage across a capacitor will stop changing. As such, the derivative of this will become zero. The voltage across an inductor will become zero, or the inductor will behave as a short circuit. The current through a capacitor will become zero, or the capacitor will behave like an open circuit. The short-circuit current, or the open-circuit voltage will need to be solved for.

With these strategies in mind, let us attempt the evaluation of a few first order circuits. A large part of the evaluation is to be performed simply, by inspection.

Example 3.2. The circuit in the schematic below was set up a long time back, and at $t = 0$, the switch is flipped from one position to the other.

Find an expression for the voltage across the capacitor C_3, as a function of time.

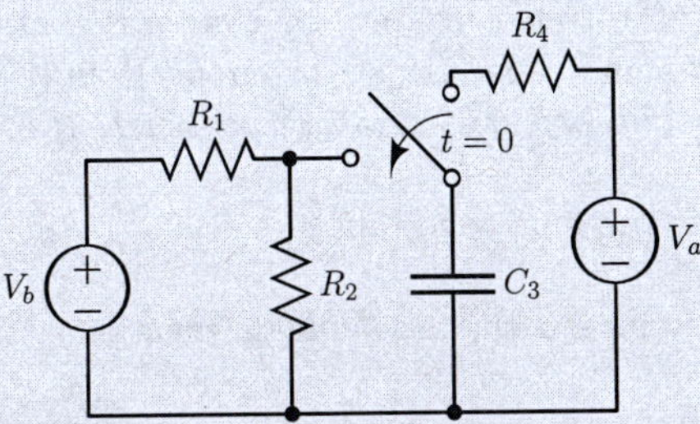

Let us assume the datum is the bottom node.

The only state variable in the circuit is the voltage across the capacitor, $v_C(t)$. For $t < 0$, the circuit would already have reached steady-state, and $\dot{v_C} = 0$. At steady-state, the drop across R_4 would be 0, and the voltage across the capacitor would be V_a. At $t = 0^-$, $v_C(0^-) = V_a$.

For $t \geq 0$, looking back from C_3, the driving point impedance of the circuit is $R_1 \parallel R_2$. As such, the time-constant of the circuit is $\tau = (R_1 \parallel R_2)C_3$.

As $t \to \infty$, the current through the capacitor reduces to zero, and the C_3 behaves as an open circuit. R_1 and R_2 simply form a potential divider across V_b. Therefore, the voltage across the capacitor as $t \to \infty$ is $V_b R_2/(R_1 + R_2)$.

Finally, we can write out the complete solution for the voltage across the capacitor.

$$v_C(t) = \begin{cases} V_a, \text{ for } t \leq 0, \\ V_b \frac{R_2}{R_1+R_2}(1 - e^{-t/\tau}) + V_a e^{-t/\tau} \text{ for } t > 0. \end{cases}$$

$v_C(t)$ $V_b \frac{R_2}{R_1+R_2}$ V_a 0 time, t

The graph above shows how v_C evolves with time. Till $t = 0$ v_C is V_a. For $t > 0$, the capacitor exponentially charges up to $V_b R_2/(R_1 + R_2)$.

Thumb rule (Shortcut). One can quickly evaluate the step response of a first order circuit (a circuit with only one state variable) using the following steps.

1. For the output variable, first, identify the initial condition.

2. Next, identify the time constant. One can arrive at this quickly if we null the input source and look at the circuit from the output quantity. The time constant will be of the form, either L/R or RC.

3. Third, identify the final condition.

4. Then simply write out the time-domain function.

Example 3.3. Analyze the first order RL circuit shown below. Find $i_L(t)$.

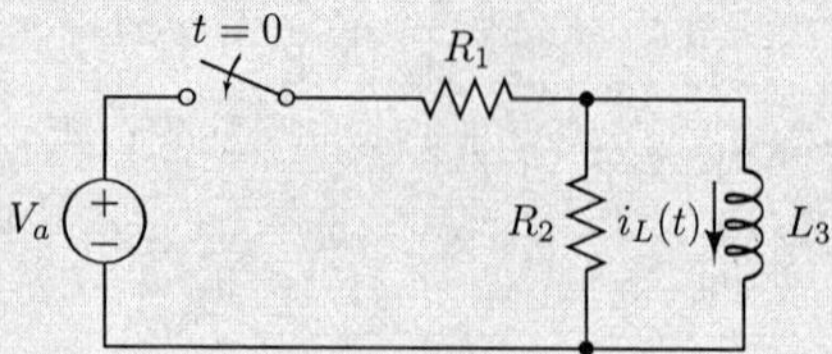

1. Before the switch is ON, at steady state, $\dot{i_L} = 0$. Therefore, the drop across the inductor is 0, the drop across R_2 is also 0, and i_L is 0, i.e., $i_L(0^-) = 0$.
2. While the switch is on, if we null the input source, we obtain L_3 in shunt with R_1 and R_2. This configuration gives us a time constant $\tau = L_3/(R_1 \parallel R_2) = L_3(R_1 + R_2)/(R_1 R_2)$.
3. As $t \to \infty$, the circuit goes into steady-state again. The final state will again have zero voltage drop across the inductor, and therefore, the inductor current $i_L(\infty) = V_a/R_1$.

Now we write out the time domain expression for $i_L(t)$.

$$i_L(t) = \begin{cases} 0, \text{ for } t \leq 0, \\ \frac{V_a}{R_1}(1 - e^{-t\frac{R_1 R_2}{L_3(R_1+R_2)}}), \text{ for } t > 0. \end{cases}$$

A common misconception among many is with instantaneously changing voltages and currents. All voltages and currents can instantaneously change; even when the voltage is across a capacitor, or the current is through an inductor. The following example is illustrative.

Example 3.4. In the following circuit, the input is a voltage source with the unit step function, $u(t)$. Is this circuit first order or second order? Solve for $v_{C1}(t)$ and $v_{C2}(t)$.

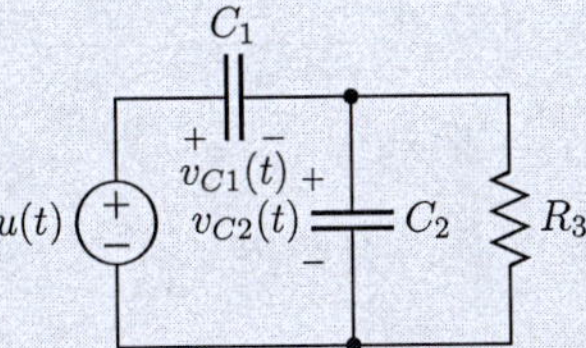

There is a loop with voltage sources and capacitors. Ordinarily, one would think there are two state variables, $v_{C1}(t)$ and $v_{C2}(t)$. However, these two are related to each other, as $v_{C2}(t) = u(t) - v_{C1}(t)$. The circuit is a first-order circuit.

Because there is a loop with voltage sources and capacitors, the initial conditions on the capacitors cannot all be pre-defined at the same time. Further, be prepared to encounter difficulties with the initial conditions.

For $t < 0$, there would be no charge on C_2 as it would discharge through the resistor R_3. Further, since $u(t)$ for $t < 0$ is 0, there would be no charge on C_1 either because the charge, if any, would be discharged through R_3. Therefore, $v_{C1}(0^-) = v_{C2}(0^-) = 0$. However, these voltages would not be held steady at $t = 0^+$ because, at this time, they need to satisfy the KVL with a different value of the voltage source. Therefore, both voltages will need to change to satisfy the new KVL instantaneously. We can obtain the instantaneous voltages $v_{C1}(0^+)$ and $v_{C2}(0^+)$ through charge sharing as $1 \cdot C_2/(C_1 + C_2)$ V and $1 \cdot C_1/(C_1 + C_2)$ V, respectively.

The time constant will be $R_3(C_1 + C_2)$, obtained by nulling the input source. The final voltage at steady state for v_{C2} is 0 volts. Therefore, the expression for $v_{C2}(t)$ as:

$$v_{C2}(t) = \begin{cases} 0, & \text{for } t < 0, \\ 1 \cdot \frac{C_1}{C_1+C_2} e^{-\frac{t}{R_3(C_1+C_2)}}, & \text{for } t \geq 0. \end{cases}$$

$v_{C1}(t)$
$\frac{C_1}{C_1+C_2}$
$v_{C2}(t)$
0
time, t

We can try to solve this mathematically as well. Assuming the bottom node to be the datum, we will have a single node equation for $t > 0$ and a single state-variable, v_{C2}.

$$C_1\frac{d}{dt}(v_{C2}-1)+C_2\frac{d}{dt}v_{C2}+v_{C2}/R_3=0$$

$$\text{or, } \frac{d}{dt}v_{C2}=-\frac{1}{R_3(C_1+C_2)}v_{C2}$$

The same result as before follows.

3.4 Second order networks: series, parallel RLC

In this section, we will analyze second-order circuits, i.e., circuits with two state variables. The same general method will be applied.

3.4.1 Series RLC circuits

Let us first analyze the circuit shown in Fig. 3.4. Let us assume the switch is turned ON at $t = 0$. Further, let us assume that the capacitor is initially discharged.

The two state variables in the circuit are $i_L(t)$ and $v_C(t)$, and as such, the circuit is clearly of second order. A single loop equation can be written for the circuit, for $t \geq 0$.

$$V_0 = i_L(t)R + L\frac{di_L(t)}{dt} + v_C(t) \tag{3.6}$$

A second equation gives us $i_L(t)$ in terms of $v_C(t)$.

$$i_L(t) = C\frac{dv_C(t)}{dt} \tag{3.7}$$

It is possible to plug (3.7) into (3.6) to obtain a single consolidated differential equation in one variable, $v_C(t)$.

$$V_0 = RC\dot{v_C} + LC\ddot{v_C} + v_C \tag{3.8}$$

In (3.8), the standard dot notation for derivatives has been used. Let us attempt solving this differential equation with the help of the same three-step technique of section 3.1.6.

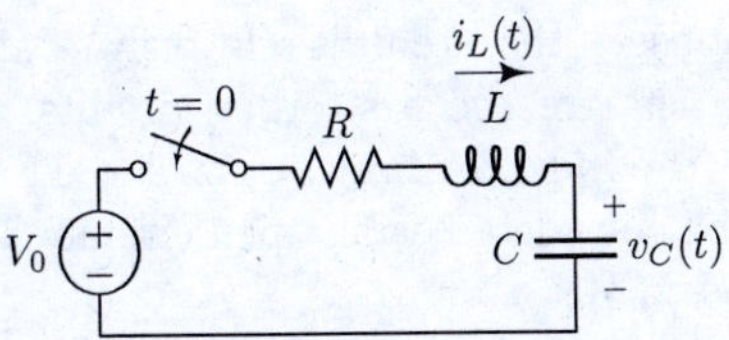

Figure 3.4: Series RLC circuit.

1. The homogeneous equation is:

$$LC\ddot{v_C} + RC\dot{v_C} + v_C = 0$$

The solution template for the second order (and all higher order) differential equation is also Ae^{st}. If we plug our template solution into the homogeneous equation, we obtain:

$$LCAs^2e^{st} + RCAse^{st} + Ae^{st} = 0$$

The factor Ae^{st} appears in all three terms and may be canceled.

$$LCs^2 + RCs + 1 = 0 \tag{3.9}$$

(3.9) is known as **the characteristic equation** of the system. For the second order characteristic polynomial equation in s, there are two possible values of s that can satisfy the equation.

$$s_1,\, s_2 = \frac{-RC \pm \sqrt{R^2C^2 - 4LC}}{2LC} = -\frac{R}{2L} \pm \sqrt{\left(\frac{R}{2L}\right)^2 - \frac{1}{LC}}$$

Depending on the value of the discriminant, the possible roots may be (1) distinct and real if the discriminant is positive, (2) real and equal if the discriminant is zero, (3) complex conjugate pairs if the discriminant is negative. Based on our discussion in section 3.1.1, it is understood that the solution will take a multitude of forms depending on the specific values of s calculated.

In general, there are two possible solutions to the homogeneous equation. The complete solution to the homogeneous equation is the sum of these two solutions.

Let us assume coefficients A_1 and A_2 for the two solutions. The complete solution to the homogeneous differential equation is:

$$v_C(t) = A_1e^{s_1t} + A_2e^{s_2t}$$

where s_1 and s_2 are the roots of the characteristic polynomial (3.9). This is the free response of the circuit.

2. The forced response of the circuit is relatively straightforward to obtain, and is just like that of the first-order circuits. For the applied DC forcing voltage, the inductor may be considered as a short circuit, the capacitor may be considered as an open circuit. The voltage across the capacitor settles to V_0.

3. The complete solution is $v_C(t) = V_0 + A_1 e^{s_1 t} + A_2 e^{s_2 t}$. Now we incorporate the initial conditions of the two state variables of the circuit. $v_C(0^+) = 0$ and $i_L(0^+) = 0$. From (3.7), the inductor current is $CA_1 s_1 e^{s_1 t} + CA_2 s_2 e^{s_2 t}$. Now we have two equations to work out values of A_1 and A_2. s_1 and s_2 are known from the roots of the characteristic equation.

$$\begin{aligned} V_0 + A_1 + A_2 &= 0 \\ A_1 s_1 + A_2 s_2 &= 0 \end{aligned}$$

A_1 and A_2 may be solved from the above set of simultaneous equations. All other voltages in the circuit can be computed from this solution.

Example 3.5. In the series RLC circuit of Fig. 3.4, R is 1 Ω, L is 1/2 H, and C is 1/3 F. For $V_0 = 1$ V, evaluate $i_L(t)$ and $v_C(t)$.

The characteristic equation is:

$$s^2/6 + s/3 + 1 = 0$$

The roots of the characteristic equation are $s_1,\ s_2 = -1 \pm j2.2361$. A_1 and A_2 form the pair of simultaneous equations (phrased as a matrix equation) below.

$$\begin{bmatrix} 1 & 1 \\ s_1 & s_2 \end{bmatrix} \begin{bmatrix} A_1 \\ A_2 \end{bmatrix} = \begin{bmatrix} -V_0 \\ 0 \end{bmatrix}$$

The Octave code-snippet below can be used to quickly solve.

```
octave:1> a = 1/6; b = 1/3; c = 1;
octave:2> s1 = (-b+sqrt(b^2-4*a*c))/2/a;
octave:3> s2 = (-b-sqrt(b^2-4*a*c))/2/a;
octave:4> A = inv([1 1; s1 s2])*[-1; 0]
A =

  -0.5000 + 0.2236i
  -0.5000 - 0.2236i

octave:5> t = 0:0.01:10;
octave:6> v = A(1)*e.^(s1*t)+A(2)*e.^(s2*t)+1;
octave:7> plot(t, v)
```

The Octave code snippet above clarifies the steps required to evaluate the complete waveform. The same steps can be performed algebraically. The graph below shows the waveform plotted.

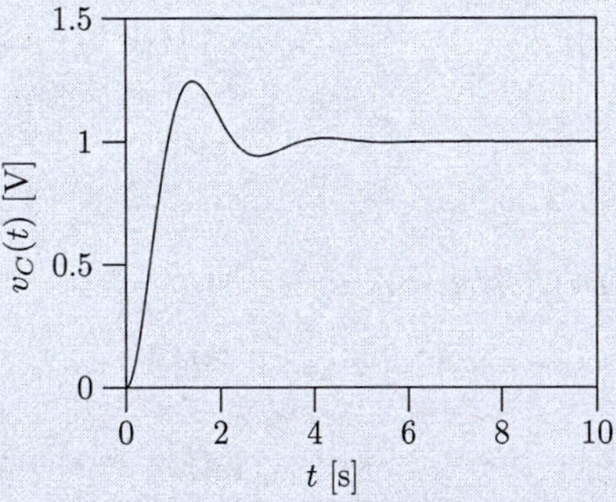

Let us step back for some time and understand a few details. s_1, s_2, A_1, A_2 computed in the previous example were complex. For other values of R, L, C, they could have been real. Are there any finer points that we need to be aware of? In the following sections, we will individually go through the cases when the discriminant in the roots of the characteristic polynomial is positive, negative and zero. The given Octave script works as long as the discriminant is non-zero.

3.4.2 Over-damped

In (3.9), the roots are real and distinct if $(RC)^2 > 4LC$, or if $R > 2\sqrt{L/C}$.

As such, two real values of s satisfy (3.9). It will be instructive to verify that both roots will be strictly negative. The free response, $A_1 e^{s_1 t} + A_2 e^{s_2 t}$, where both s_1 and s_2 are negative, is the sum of two damped exponentials. The response is known as an **over-damped** response.

Example 3.6. In Fig. 3.4, R is 6 Ω, L is 1 H, C is 1/3 F, V_0 is 1 V. Evaluate $v_C(t)$.

The procedure is the same as earlier. The characteristic equation is:

$$\frac{1}{3}s^2 + 2s + 1 = 0$$

The roots of the characteristic equation are $s_1 = -3+\sqrt{6}$ and $s_2 = -3-\sqrt{6}$. The values of A_1 and A_2 should be such that they satisfy:

$$A_1 + A_2 = -1, \text{ and, } (3-\sqrt{6})A_1 + (3+\sqrt{6})A_2 = 0$$

This gives $A_1 = -(\sqrt{3}+\sqrt{2})/2\sqrt{2}$ and $A_2 = (\sqrt{3}-\sqrt{2})/2\sqrt{2}$.

The final response is:

$$v_C(t) = 1 - \frac{\sqrt{3}+\sqrt{2}}{2\sqrt{2}}e^{-(3-\sqrt{6})t} + \frac{\sqrt{3}-\sqrt{2}}{2\sqrt{2}}e^{-(3+\sqrt{6})t}$$

3.4.3 Critically damped

In (3.9), the roots are real and repeated if $(RC)^2 = 4LC$, or if $R = 2\sqrt{L/C}$.

The solution of the homogeneous equation is slightly different in this case. The form of the solution is $(A_1 + tA_2)e^{st}$. With the coefficients in (3.9) positive, s is required to be negative. This response is **critically damped**.

Scan the QR-code, and learn more about how the form of the solution changes in the limit as s_1 and s_2 approach each other.

Example 3.7. In Fig. 3.4, R is $2\sqrt{3}\ \Omega$, L is 1 H, C is 1/3 F, V_0 is 1 V. Evaluate $v_C(t)$.

The characteristic equation is:

$$\frac{1}{3}s^2 + \frac{2}{\sqrt{3}}s + 1 = (\frac{s}{\sqrt{3}} + 1)^2 = 0$$

The single repeated root of the characteristic equation is at $s = -\sqrt{3}$.

$v_C(t)$ is therefore $1 + (A_1 + A_2 t)e^{-\sqrt{3}t}$. $v_C(0^+)$ evaluates to $1 + A_1$. Zero initial voltage across the capacitor implies the value of A_1 as -1. $i_L(t)$ is $Cdv_C(t)/dt$.

$$\therefore i_L(t) = 1/3 \cdot \big(-\sqrt{3}A_1 e^{-\sqrt{3}t} - \sqrt{3}A_2 t e^{-\sqrt{3}t} + A_2 e^{-\sqrt{3}t}\big)$$

At $t = 0^+$, $i_L(0^+)$ works out to $1/3(A_2 - \sqrt{3}A_1)$, which is zero. Therefore, A_2 is $-\sqrt{3}$.

Finally, we have the expression for $v_C(t)$ as:

$$v_C(t) = 1 - (1 + \sqrt{3}t)e^{-\sqrt{3}t}$$

3.4.4 Under-damped

The discriminant of the characteristic equation is negative in this case, and the two roots of the equation form a complex conjugate pair, given below.

$$s_1,\, s_2 = -\frac{R}{2L} \pm j\sqrt{\frac{1}{LC} - \left(\frac{R}{2L}\right)^2}$$

Let us simplify the expressions for the roots and denote the two roots as $-\sigma_0 \pm j\omega_0$. From the initial conditions we will have two equations in A_1 and A_2. As before,

$$A_1 + A_2 = -V_0, \text{ and, } A_1(\sigma + j\omega) + A_2(\sigma - j\omega) = 0$$

Since s_1 and s_2 are complex, the solutions to the simultaneous equations need to be complex. So A_1 and A_2 are also complex. The imaginary part of the first relationship is zero. Therefore, $\text{Im}[A_1] + \text{Im}[A_2] = 0$. Likewise, if we work out the imaginary part of the second relationship, we obtain:

$$\omega(\text{Re}[A_1] - \text{Re}[A_2]) + \sigma(\text{Im}[A_1] + \text{Im}[A_2]) = 0$$

Using the previous relation, that is, $\text{Im}[A_1] + \text{Im}[A_2] = 0$, we obtain $\text{Re}[A_1] = \text{Re}[A_2]$. Therefore A_1 and A_2 are conjugates of each other. Let us call A_1 and A_2 as $\alpha + j\beta$ and $\alpha - j\beta$ respectively.

Now the overall free response is:

$$\begin{aligned}
& (\alpha + j\beta)e^{-\sigma t + j\omega t} + (\alpha - j\beta)e^{-\sigma t - j\omega t} \\
= \;& e^{-\sigma t}\big(\alpha(e^{j\omega t} + e^{-j\omega t}) + j\beta(e^{j\omega t} - e^{-j\omega t})\big) \\
= \;& e^{-\sigma t}\big(2\alpha\cos(\omega t) - 2\beta\sin(\omega t)\big) \\
= \;& 2e^{-\sigma t}\big(\alpha\cos(\omega t) + \beta\sin(\omega t)\big)
\end{aligned}$$

The form of the solution can be recast as $e^{-\sigma t}(A\cos\omega t + B\sin\omega t)$. Additionally, the result can be rewritten by bringing $\sqrt{A^2 + B^2}$ out of the brackets.

$$\begin{aligned}
& e^{-\sigma t}(A\cos\omega t + B\sin\omega t) \\
= \;& (\sqrt{A^2 + B^2})e^{-\sigma t}\left(\frac{A}{\sqrt{A^2 + B^2}}\cos\omega t + \frac{B}{\sqrt{A^2 + B^2}}\sin\omega t\right) \\
= \;& A_0 e^{-\sigma t}(\cos\phi\cos\omega t + \sin\phi\sin\omega t) \\
= \;& A_0 e^{-\sigma t}\cos(\omega t - \phi)
\end{aligned}$$

where $A_0 = \sqrt{A^2 + B^2}$ and ϕ is $\tan^{-1} B/A$.

Note that in all the forms of the solution, there are two unknown constants. If the solution is chosen as $e^{-\sigma t}(A\cos\omega t + B\sin\omega t)$, A and B are unknown constants. In the second form, A_0 and ϕ are unknown constants. We choose the form of the solution depending on convenience.

Example 3.8. In Fig. 3.4, R is 1 Ω, L is 1 H, C is 1/3 F, V_0 is 1 V. Evaluate $v_C(t)$.

The characteristic equation is:

$$\frac{1}{3}s^2 + \frac{1}{3}s + 1 = 0$$

The roots of the characteristic equation are:

$$s_1,\, s_2 = -\frac{1}{2} \pm j\frac{\sqrt{11}}{2}$$

Now the final solution is of the form:

$$v_C(t) = 1 + e^{-t/2}\Big(A\cos(\sqrt{11}t/2) + B\sin(\sqrt{11}t/2)\Big)$$

At $t = 0$, the $v_C(t) = 0$. Therefore $1 + A = 0$, or $A = -1$. The derivative of $v_C(t)$ will give $i_L(t)$.

$$i_L(0) = 1/3 \cdot (-A/2 + B\sqrt{11}/2) = 0, \text{ or, } B = A/\sqrt{11}$$

Finally, we have the complete solution for $v_C(t)$ as:

$$v_C(t) = 1 - e^{-t/2}\Big(\cos(\sqrt{11}t/2) + \frac{1}{\sqrt{11}}\sin(\sqrt{11}t/2)\Big)$$

The over-damped, under-damped and critically damped responses of an RLC circuit appear different. A comparison of the three $v_C(t)$ in the three examples-3.6, 3.7, 3.8 are shown in the Fig. 3.5.

3.4.5 Response metrics

Definition 3.2 (Characteristic equation): The characteristic equation is an equation in s that represents the homogeneous equation or the free response. We can obtain the characteristic equation by writing out the differential equations in the circuit with zero input, combining them, plugging in a solution Ae^{st}, and simplifying.

Definition 3.3 (Poles): The roots of the characteristic equation are the poles.

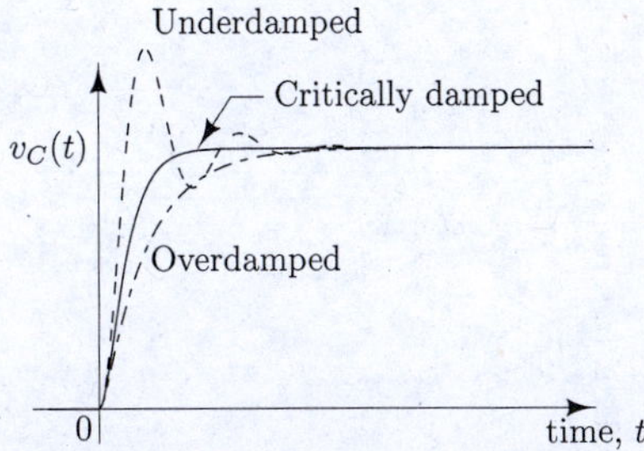

Figure 3.5: A comparison of over-damped, critically damped and under-damped responses. The critically damped response has no overshoot, but a very low rise time. The under-damped response is very sharp, but is accompanied with ringing. The over-damped response is slow but has no ringing.

The characteristic equation of a second-order system is of the following form:

$$\frac{s^2}{w_n^2} + 2\zeta\frac{s}{w_n} + 1 = 0$$

Solving this for s results in the following two poles:

$$s = -\zeta\omega_n \pm \omega_n\sqrt{\zeta^2 - 1}$$

ζ is known as the **damping factor**, and ω_n is known as the **natural frequency**. As long as the damping factor is more than 1, the two roots of the characteristic equation are real, and the system is **over-damped**. When the damping factor is exactly 1, the system is **critically damped**, and the two roots coincide. In this case, the solution to the differential equation becomes a little different, $(A_1t + A_2)e^{st}$. When the damping factor is less than 1, the system is **under-damped**, and the two roots become a complex conjugate pair. In that case, the response is still $A_1e^{s_1t} + A_2e^{s_2t}$, but this could be re-cast in the following form.

$$Ae^{-\zeta\omega_n t}\cos(\omega_n t\sqrt{1-\zeta^2} + \phi)$$

A and ϕ are two constants that we need to work out from the initial conditions and the forced response.

The step response of a second-order circuit has a few metrics. A sketch of the second-order circuit under-damped step-response is shown in Fig. 3.6.

All the metrics are defined for a unit-step input. The tolerance at the output of the step response is set from 90% to 110% of the target.

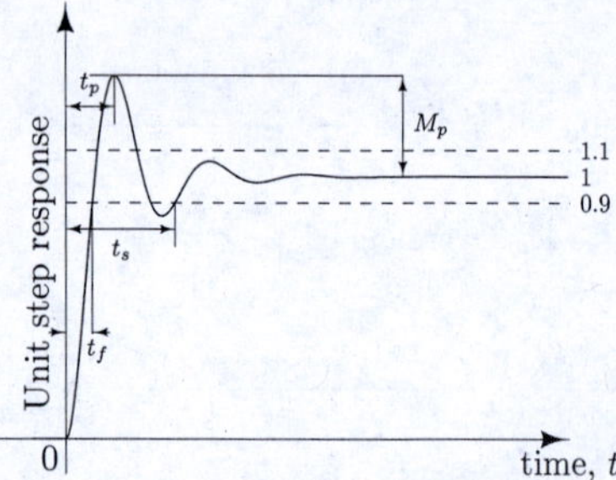

Figure 3.6: Metrics of the second-order under-damped response. t_p is the time at which the peak is attained; M_p is the peak overshoot. t_s is the settling time, t_f is the first time at which the response rises to 90% of the target.

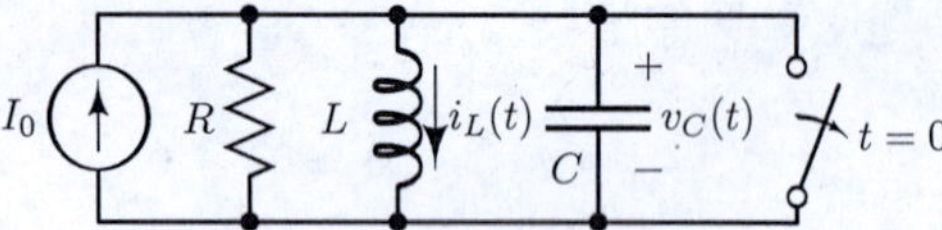

Figure 3.7: Parallel RLC circuit

- M_p is the **peak overshoot**, beyond the target of 100%. For over-damped and critically damped responses there is no overshoot and $M_p = 0$.

- t_p is the time at which the peak arrives. This parameter is not relevant for over-damped and critically damped responses.

- t_f is the first time at which the response reaches 90% of the target.

- t_s is the time beyond which the response strictly remains within the tolerance of 90% to 110%. For over-damped and critically damped responses, $t_s = t_f$. t_s is known as the **settling time**.

3.4.6 Parallel RLC

The parallel RLC circuit shown in Fig. 3.7 excited by a unit-step current source is identical to the dual of the series RLC circuit. For $t < 0$, the switch is ON, and the current in the current source flows through the switch. For $t \geq 0$, the switch is OFF, and there is a step current in the parallel combination of the resistor, capacitor and inductor. Let us assume $i_L(0^-) = 0$.

The circuit in Fig. 3.7 is the exact dual of the circuit in Fig. 3.4. The conductance in the parallel RLC circuit is to be numerically equal to the resistance in the series RLC circuit. The inductance in the parallel RLC circuit is to be numerically equal to the capacitance in the series RLC circuit. The capacitance in the parallel RLC circuit is to be numerically equal to the inductance in the series RLC circuit. If we consider the bottom node in Fig. 3.7 as the datum, there is a single node-voltage equation that we can formulate, as opposed to a single mesh current equation in Fig. 3.4.

$$I_0 = v_C(t) \cdot \frac{1}{R} + i_L(t) + C\frac{dv_C(t)}{dt}, \text{ and, } v_C(t) = L\frac{di_L(t)}{dt}$$

We combine the two equations by replacing the value of $v_C(t)$ obtained in the second equation into the first equation. This gives us a single second-order linear differential equation in one variable, namely, $i_L(t)$.

$$I_0 = LC\ddot{i}_L + \frac{L}{R}\dot{i}_L + i_L \tag{3.10}$$

(3.10) is the same as the differential equation of the series RLC circuit, (3.8), with R replaced by $1/R$, L replaced by C, C replaced by L, and V_0 replaced by I_0. As such, the solution will also be the same, as long as the correspondences are observed.

The characteristic equation, for the parallel RLC circuit, is as follows.

$$LCs^2 + \frac{L}{R}s + 1 = 0$$

s_1 and s_2 are the two roots of the characteristic equation. The free response of the circuit is $A_1e^{s_1t} + A_2e^{s_2t}$. As $t \to \infty$, the inductor behaves as a short-circuit, the capacitor as an open-circuit. As such, the forced response is $i_L(\infty) = I_0$. The complete solution is of the form:

$$i_L(t) = I_0 + A_1e^{s_1t} + A_2e^{s_2t}$$

At $t = 0^+$, $i_L(0^+) = 0$ (given assumption) and $v_C(0^+) = 0$ (switch has just opened). As such, we have two equations in A_1 and A_2 to satisfy the initial conditions.

$$A_1 + A_2 = -I_0, \text{ and, } s_1A_1 + s_2A_2 = 0$$

The values of A_1 and A_2 are to be worked out solving the two initial conditions.

3.5 Other second order circuits

Let us now analyze circuits that are not strictly series or parallel RLC. The general principle of analysis remains the same. We will still follow our three-step technique of section 3.1.6. First we will work out the free response without solving for the coefficients, then we will formulate the forced response, and finally we will combine the free and forced responses and evaluate the unknowns. A variant of the series RLC circuit follows.

Example 3.9. In the circuit below, after being closed for a long time, the switch is opened at $t = 0$. Obtain the complete expression for $v_C(t)$ for $t \geq 0$.

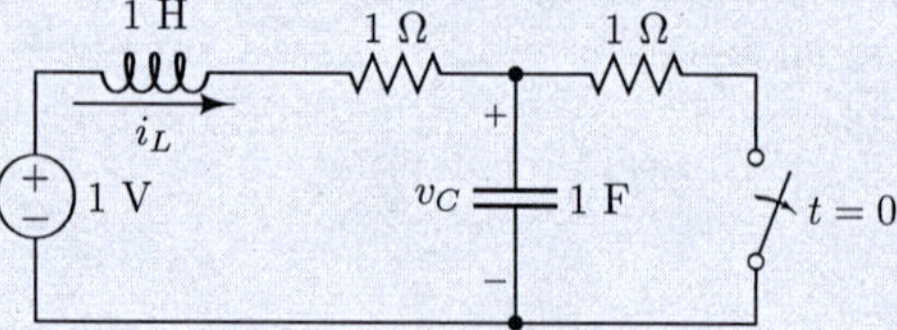

We can analyze the circuit with the help of a single mesh equation.

1. For $t \geq 0$, the switch is open, and the mesh equation for the circuit is:

$$1 \cdot \dot{i_L} + 1 \cdot i_L + v_C = 1$$

Further, $i_L = 1 \cdot \dot{v_C}$. The mesh equation simplifies to:

$$\ddot{v_C} + \dot{v_C} + v_C = 1$$

The characteristic equation of the homogeneous response works out to:

$$s^2 + s + 1 = 0$$

There are two possible values of s that satisfy the characteristic equation. $s = -1/2 \pm j\sqrt{3}/2$.

2. The input is a constant voltage. Therefore the forced response should also be a constant voltage. As $t \to \infty$, the circuit reaches a constant steady-state, and $v_C(\infty) = 1$, $i_L(\infty) = 0$.

3. Next, $v_C(t) = 1 + Ae^{-t/2}\cos(t\sqrt{3}/2 + \phi)$, where A and ϕ are constants from the solution of the free response. We need to evaluate these constants with the help of the initial conditions. For $t < 0$, the switch was closed, and the circuit was in steady-state. Therefore, $v_C(0^-)$ is 0.5 V, $i_L(0^-)$ is 0.5 A. At $t = 0^+$, $1 + A\cos\phi = 0.5$. Further, at

$t = 0^+$, $i_L = \dot{v_C} = -A\cos\phi/2 - \sqrt{3}A\sin\phi/2 = 0.5$. The conditions give $A = 1/\sqrt{3}$ and $\phi = 7\pi/6$.

$$v_C(t) = 1 + \frac{1}{\sqrt{3}}e^{-t/2}\cos(\frac{\sqrt{3}}{2}t + 7\pi/6)$$

The waveform corresponding to $v_C(t)$ is plotted below.

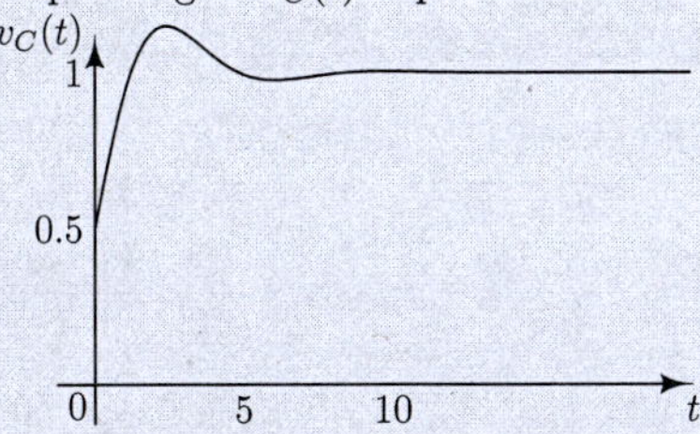

It is also possible to analyze second order circuits using the same short-cut technique that we studied for first-order circuits. In this case, though, we will need to work out two distinct "time constants", which may be in general, complex. Once we have s_1, s_2, with the help of the initial conditions, we can write out the expression for the waveform.

Example 3.10. In the following circuit, R is 1 Ω, L is 1/2 H, C is 1/3 F, V_0 is 1 V, V_1 is 2 V. The two switches open and close simultaneously at $t = 0$. Derive an expression for $i_R(t)$, as indicated in the schematic.

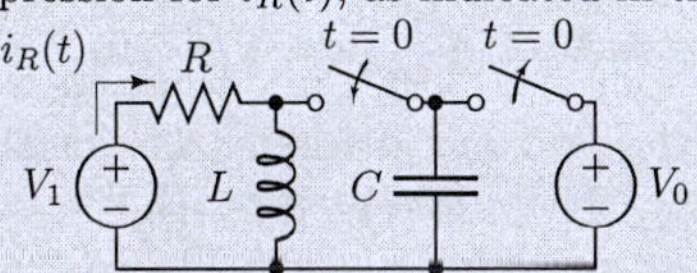

1. For $t \geq 0$, the circuit appears to be a parallel RLC circuit when the voltage source V_1 is nulled. As such, the characteristic equation is:

$$LCs^2 + \frac{L}{R}s + 1 = 0$$

This gives two poles, namely s_1 and s_2 that satisfy the characteristic equation. $s_1, s_2 = -1.5 \pm j1.94$. The system is under-damped.

2. As $t \to \infty$, the inductor appears as a short circuit and the capacitor appears as an open circuit. The current $i_R(\infty)$ is V_1/R. This gives us the form of the solution.

$$i_R(t) = V_1/R + e^{-1.5t}(A_1\cos 1.94t + A_2\sin 1.94t)$$

3. Now let us work out the initial conditions. Two initial conditions are required; let us work out $i_R(0^+)$ and $\dot{i_R}(0^+)$. At $t = 0^+$, the voltage across the capacitor is V_0. So $i_R(0^+) = (V_1 - V_0)/R$. After all, $i_R = (V_1 - v_C)/R$. In that case, $\dot{i_R} = -\dot{v_C}/R$. But $C\dot{v_C}$ is the current through the capacitor. The initial current through the inductor, $i_L(0^+)$, is V_1/R. Therefore, the initial current through the capacitor is $i_R(0^+) - i_L(0^+)$, which is $-V_0/R$. This gives us the initial $\dot{i_R}$ as $V_0/(R^2C)$.

$$
\begin{aligned}
\frac{di_R}{dt} &= \frac{d}{dt}\frac{V_1 - v_C}{R} &&= -\frac{1}{R}\frac{dv_C}{dt} \\
= -\frac{1}{RC}i_C &= -\frac{1}{RC}(i_R - i_L) &&= -\frac{1}{RC}\left(\frac{V_1 - V_0}{R} - \frac{V_1}{R}\right), \text{ at } t = 0 \\
& &&= \frac{V_0}{R^2C}, \text{ at } t = 0
\end{aligned}
$$

The two initial conditions are now ready. From the form of the solution, at $t = 0^+$, i_R is $V_1/R + A_1$, while $\dot{i_R}$ is $-1.5A_1 + 1.94A_2$. Now we have two equations in A_1 and A_2.

$$
\frac{V_1}{R} + A_1 = \frac{V_1 - V_0}{R}, \text{ and, } -1.5A_1 + 1.94A_2 = \frac{V_0}{R^2C}
$$

This gives $A_1 = -1$ and $A_2 = 0.77$. Finally,

$$
i_R(t) = 2 + e^{-1.5t}(-\cos 1.94t + 0.77 \sin 1.94t)
$$

One needs to work out the initial conditions for the quantity of interest and its derivative. In the previous example, the initial conditions of interest were $i_R(0^+)$ and $\dot{i_R}(0^+)$. These two along with the two time constants and the final value of the quantity of interest will allow us to construct the full time-domain response.

The following example does not fall into either series or parallel RLC categories. However, the analysis technique remains the same. First, we will work out the characteristic equation and its roots. Next, we will obtain the forced response. Finally, we will combine the forced and free responses and evaluate the unknowns with the help of initial conditions.

Example 3.11. In the following circuit, V_1 is 1 V, V_0 is 2 V. R_1 and R_2 are 1 Ω and 2 Ω respectively. L and C are 1/2 H and 1/3 F respectively. Find an expression for $i_C(t)$.

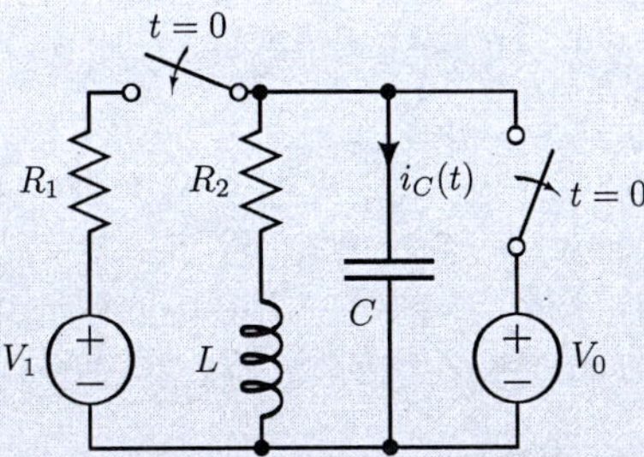

The circuit is neither of the familiar series and parallel RLC circuits. As such, we will first have to work out the characteristic equation. Let us consider the current through the inductor as i_L going downwards, and the voltage across the capacitor as v_C from top to bottom. We will use the node voltage method, with the bottom node as the datum. The node voltage method gives us the following:

$$\frac{v_C - V_1}{R_1} + i_L + C\frac{dv_C}{dt} = 0$$

$$i_L + \frac{L\frac{di_L}{dt} - v_C}{R_2} = 0$$

The second equation gives us v_C in terms of i_L and $\dot{i_L}$. Plugging this expression for v_C from the second equation into the first equation, we obtain a single differential equation in i_L, as follows.

$$LC\ddot{i_L} + (\frac{L}{R_1} + R_2C)\dot{i_L} + (1 + \frac{R_2}{R_1})i_L = \frac{V_1}{R_1}$$

For the characteristic equation, the forcing function on the right hand side is to be nulled. The poles of the system are at $s_1, s_2 = -3.5 \pm j2.40$.

Now, let us obtain the initial i_C and $\dot{i_C}$. At $t = 0^-$, the current through L is V_0/R_2, and the voltage across C is V_0. The current through L and the voltage across C will remain the same at $t = 0^+$. The current through V_1, going upwards, is $(V_1 - v_C)/R_1$. A part of this current becomes i_L, the remainder is i_C.

$$\therefore i_C = \frac{V_1 - v_C}{R_1} - i_L = \frac{V_1 - V_0}{R_1} - \frac{V_0}{R_2}, \text{ at } t = 0^+$$

$$\text{Also, } \dot{i_C} = -\frac{\dot{v_C}}{R_1} - \dot{i_L} = -\frac{i_C}{R_1C} - \frac{v_L}{L} = -\frac{i_C}{R_1C} - \frac{v_C - i_LR_2}{L}$$

At $t = 0^+$, v_C is 2 V, i_L is 1 A, i_C is -2 A, v_L is 0 V. $\dot{i_C}$ is 6 A/s at $t = 0^+$.

As $t \to \infty$, the current through the capacitor settles to 0. Therefore the complete expression for $i_C(t)$ is:

$$\begin{aligned} i_C(t) &= e^{-3.5t}(A_1 \cos 2.40t + A_2 \sin 2.40t) \\ \text{where, } -2 &= A_1, \text{ from } i_C(0^+) \\ \text{and, } 6 &= -3.5A_1 + 2.40A_2 \text{ from } \dot{i_C}(0^+) \\ \therefore A_2 &= -0.417 \end{aligned}$$

Finally, we have:

$$i_C(t) = -e^{-3.5t}(2 \cos 2.40t + 0.417 \sin 2.40t)$$

3.6 Unit summary

- Every **independent** inductor current and every **independent** capacitor voltage form a state variable. The number of state variables in a circuit give the order of the circuit.

- Typically, a first order circuit will have only one inductor or only one capacitor. It is possible to construct first order circuits that do not obey this.

- Typically, the capacitor voltage and the inductor current is continuous at all times. It is possible to construct circuits where this is not true, especially in light of the previous remark.

- A signal of the form $Be^{s_0 t}$ covers DC signals, sinusoids, rising and damped exponentials, rising and damped sinusoids, for different values of A and s_0. In circuit theory, the input waveform is always chosen as one of these.

- Every waveform in the circuit is the sum of a forced response and a free response. The forced response to an input $Be^{s_0 t}$ is $Ce^{s_0 t}$. A constant forcing input will result in a forced response that is a constant. A sine wave as the forcing input will result in a forced response that is a sine wave.

- The free response of a circuit is computed for an input of 0. The single differential equation in s, formed to evaluate the free response of a circuit is known as the characteristic equation. The roots of the characteristic equation are the poles.

- A system with distinct poles at s_1 and s_2 will have a free response of $A_1 e^{s_1 t} + A_2 e^{s_2 t}$. This can be extended to higher orders.

- If a system is excited by the input $Ae^{s_0 t}$, and if the system has poles at $\{s_k\}$, then the response of the system is $Ce^{s_0 t} + \sum_k A_k e^{s_k t}$. C and $\{A_k\}$ are coefficients that need to be obtained from the initial (and final) conditions.

- For a step input, at steady state, inductors behave as short circuits and capacitors behave as open circuits.

- The poles of a second order system determine if the system is over-damped, under-damped, or critically damped. If s_1 and s_2 are real and distinct, the system is over-damped. If the two poles are identical, the system is critically damped. If the poles are complex conjugate pairs, the system is under-damped.

- A critically damped system has a slightly different free response. In this limiting case, the free response is $(A_1 + A_2 t)e^{s_{(1,2)} t}$.

- An under-damped system will exhibit overshoot and ringing in the step response. The ringing will damp out with time. The form of the free response is expressed as $e^{\sigma t}(A_1 \cos \omega t + A_2 \sin \omega t)$, where $\sigma \pm j\omega$ are the two complex conjugate poles. The free response can also be expressed as $Ae^{\sigma t} \cos(\omega t + \phi)$. These two forms are both re-phrasings of the original form of the free response, $A_1 e^{s_1 t} + A_2 e^{s_2 t}$, where the coefficients and poles are complex.

- In a series RLC circuit, more resistance increases damping and a circuit changes from being under-damped to critically damped to over-damped as resistance increases. In a parallel RLC circuit, less resistance increases damping and a circuit changes from being under-damped to critically damped to over-damped as resistance decreases.

3.7 Exercises

Multiple choice type questions

3.1 A digital circuit drives a logic gate of capacitance 1 pF and has a source resistance of 50 Ω. What is the rise time of a digital pulse?

(a) 1.15 ps (b) 11.5 ps (c) 115 ps (d) 1.15 ns

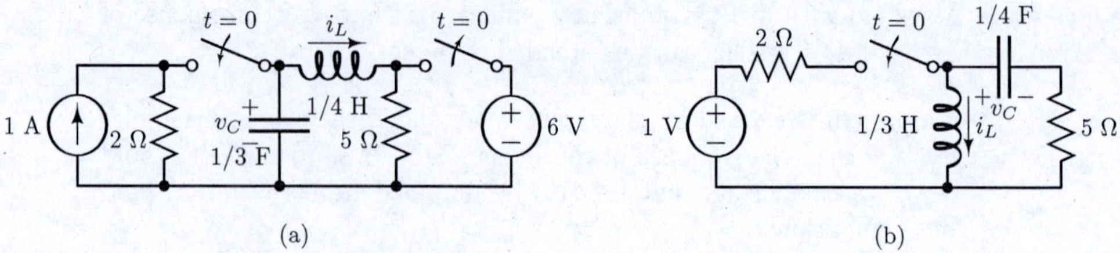

Figure 3.8: Schematics for questions 3.8, 3.9, 3.10, 3.11, 3.12, 3.13, 3.14, 3.15.

3.2 An RC circuit has a rise time of 1 ms when its output rises from 0 to 1 V. What is the rise time when the output rises from 3 V to 10 V?
(a) 1 ms (b) 3 ms (c) 7 ms (d) 10 ms

3.3 A switching power supply designed with an inductance of 1 mH drives a load of 100 Ω. What is the rise time of the power supply when it steps from 1 V to 5 V?
(a) 10 μs (b) 23 μs (c) 40 μs (d) 92 μs

3.4 A series RLC circuit has $L = 1$ mH and $C = 1$ μF. What is the value of R such that the circuit is critically damped?
(a) 31.5 Ω (b) 63 Ω (c) 100 Ω (d) 126 Ω

3.5 A parallel RLC circuit has $L = 1$ mH and $C = 1$ μF. What is the value of R such that the circuit is critically damped?
(a) 16 Ω (b) 31.5 Ω (c) 63 Ω (d) 126 Ω

3.6 A series RLC circuit has $L = 1$ mH and $C = 1$ μF. Which of the below values of R gives us an under-damped circuit?
(a) 31.5 Ω (b) 63 Ω (c) 100 Ω (d) 126 Ω

3.7 A parallel RLC circuit has $L = 1$ mH and $C = 1$ μF. Which of the below values of R gives us an under-damped circuit?
(a) 8 Ω (b) 12 Ω (c) 16 Ω (d) 31.5 Ω

3.8 In Fig. 3.8(a), what is $i_L(0^+)$?
(a) 0 (b) 0.5 A (c) 0.7 A (d) 1.2 A

3.9 In Fig. 3.8(a), what is $v_C(0^+)$?
(a) 2 V (b) 4 V (c) 6 V (d) 8 V

3.10 In Fig. 3.8(a), what is $\dot{i_L}(0^+)$?
(a) -24 A/s (b) 0 (c) 12 A/s (d) 24 A/s

3.11 In Fig. 3.8(a), what is $\dot{v_C}(0^+)$?
(a) -6 V/s (b) 0 (c) 3 V/s (d) 6 V/s

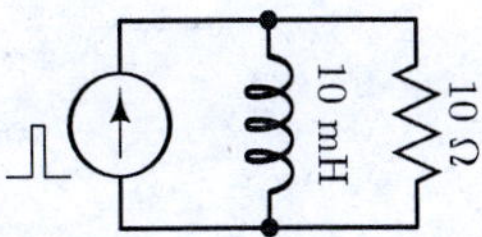

Figure 3.9: Circuit schematic for exercise 3.20.

3.12 In Fig. 3.8(b), what is $i_L(0^+)$?
(a) 0 (b) 0.14 A (c) 0.2 A (d) 0.5 A

3.13 In Fig. 3.8(b), what is $v_C(0^+)$?
(a) 0 (b) 0.28 V (c) 0.71 V (d) 1 V

3.14 In Fig. 3.8(b), what is $\dot{i_L}(0^+)$?
(a) 0 (b) 0.28 A/s (c) 1.14 A/s (d) 2.14 A/s

3.15 In Fig. 3.8(b), what is $\dot{v_C}(0^+)$?
(a) 0 (b) 0.57 V/s (c) 1 V/s (d) 1.14 V/s

Short answer type questions

3.16 In an nth order circuit, to arrive at the final step response solution, we will need to evaluate ______ initial and final conditions to solve for ______ coefficients.

3.17 A critically damped system has a damping factor, ζ, of ______.

3.18 In a second-order under-damped system, if the free response is $A_1e^{s_1t} + A_2e^{s_2t}$, then A_1 and A_2 are related as ______.

3.19 In a second-order under-damped system, if we consider the free response to be $A_1e^{s_1t}+A_2e^{s_2t}$, A_1 and A_2 will always work out to be complex conjugates. A second-order system should have two unknown coefficients. If A_1 and A_2 are conjugates of each other, then which are the two unknown coefficients?

Numericals and long answer type questions

3.20 A current source pushes current into an R-L load, as given in Fig. 3.9. From 0 A, the current rises to 1 A at 1 ms, remains 1 A till 1.5 ms, and at 1.5 ms drops to 0 A. The initial current in the inductor at $t = 0$ is i_0. Work out the current in the inductor as a function of time. Find the current at 1.5 ms. What is i_0 if the current at 1.5 ms is i_0?

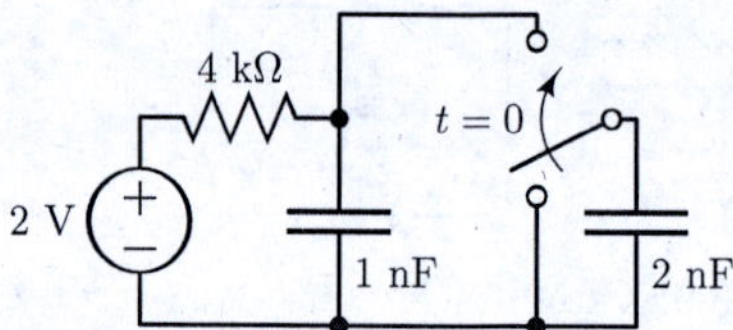

Figure 3.10: Circuit schematic for exercise 3.21.

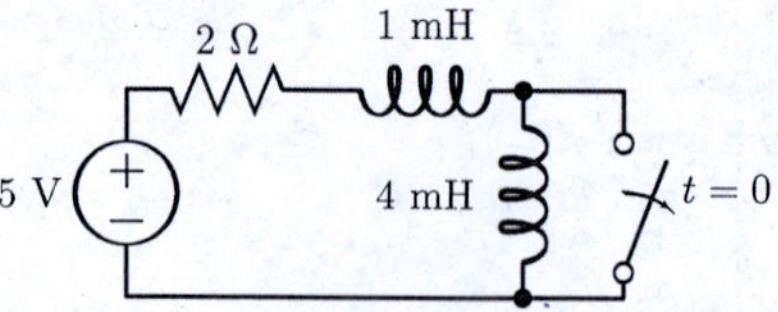

Figure 3.11: Schematic for exercise 3.22.

3.21 In the circuit schematic of Fig. 3.10, the switch (this is a single-pole-double-throw or SPDT switch) is flipped at $t = 0$. The 2 nF capacitor was initially discharged. At $t = 0^+$, the 2 nF capacitor is placed in shunt with the 1 nF capacitor. Find the voltage across the capacitors as a function of time.

3.22 In the circuit shown in Fig. 3.11, the switch is opened at $t = 0$. Evaluate the current through the inductors as a function of time.

3.23 In the circuit schematic of Fig. 3.12, the switch is turned ON at $t = 0$. Evaluate $i(t)$ as a function of time. *Hint: One can easily split the current into the sum of two first-order currents.*

3.24 In the circuit configurations of Fig. 3.13, $v_S(t)$ is $u(t) - u(t-1)$. Assume all capacitors are discharged at $t = 0^-$. Sketch the profiles of $v_L(t)$ for the four configurations with reasoning.

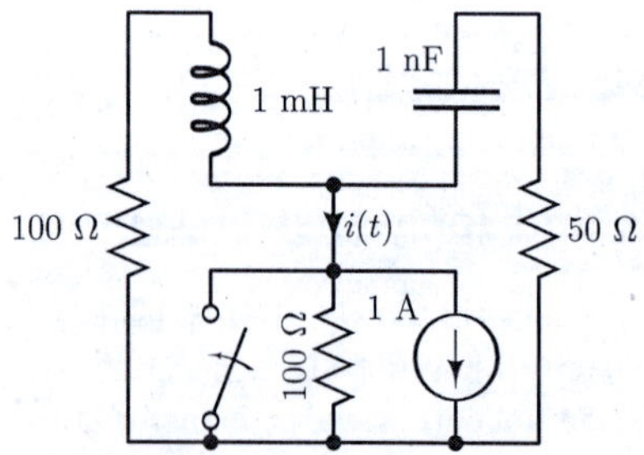

Figure 3.12: Schematic for exercise 3.23.

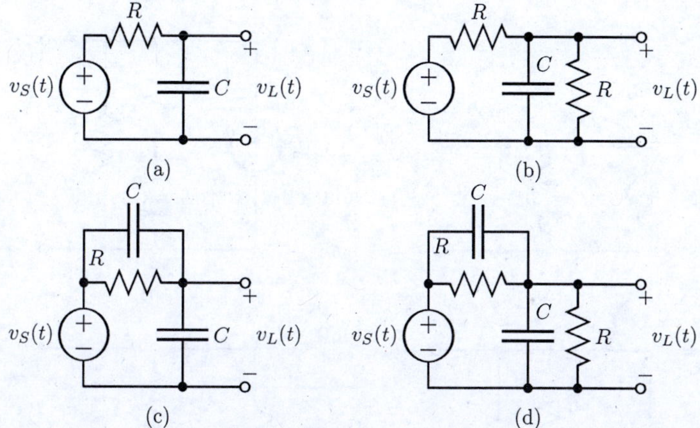

Figure 3.13: Circuit configurations for exercise 3.24.

3.25 In the circuit configurations of Fig. 3.14, $i_S(t) = u(t) - u(t-1)$. Assume all inductors have zero current at $t = 0^-$. Sketch the profiles of $v_L(t)$ for the four configurations with reasoning.

3.26 Consider the second-order circuits of Fig. 3.15. $v_S(t)$ in each case is $u(t) - u(t-1)$. Do all four circuits have the same poles? Find out why a few of the circuits do have the same poles. Let us assume $R = 10\ \Omega$, $C = 10\ \mu\text{F}$, and $L = 10$ mH. Assume the capacitor is initially discharged, and the initial current through the inductor is zero. Evaluate the voltage across the capacitor, in each case, as a function of time. Re-evaluate with $L = 1$ mH, C, and R as before. Again, re-evaluate with $L = 10$ mH, $C = 25\ \mu\text{F}$, and R as before.

3.27 In the circuit of Fig. 3.16, the switch is closed at $t = 0$. Find the current through the 10 mH inductor, left to right, as a function of time. Assume the circuit had reached a steady-state well before the switch was closed.

3.28 Consider the circuit of Fig. 3.17(a). $v_1(t)$ is $50u(t)$ volts, while $v_2(t)$ is $100u(-t)$ volts. Find an expression for the voltage across the capacitor as a function of time. Now consider the circuit of Fig. 3.17(b). Is the behavior of this circuit the same as that of Fig. 3.17(a)? Give reasons. If the behavior is different, work out the voltage across the capacitor as a function of time.

3.29 In the circuit shown in Fig. 3.18, $i_S(t) = u(t) \times 5$ mA. Find the expression for $i_C(t)$.

3.30 In the circuit schematic of Fig. 3.19, evaluate the voltage across the 0.5 H inductor as a function of time.

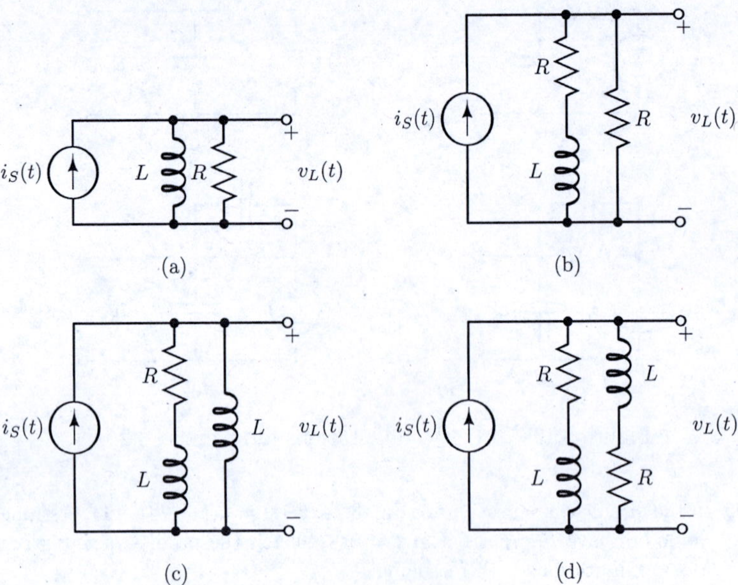

Figure 3.14: Circuit configurations for exercise 3.25.

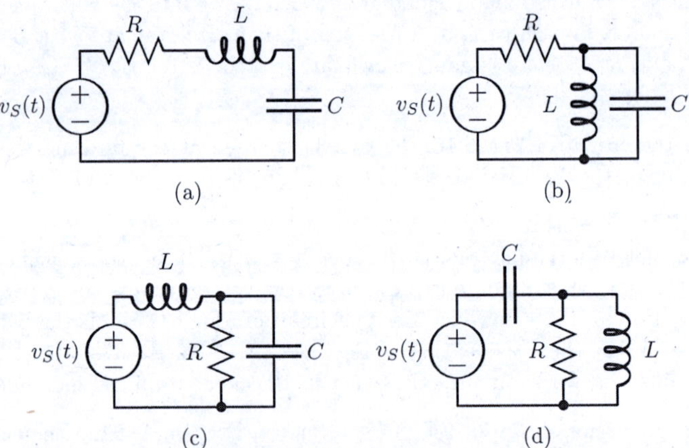

Figure 3.15: Circuit schematics for exercise 3.26.

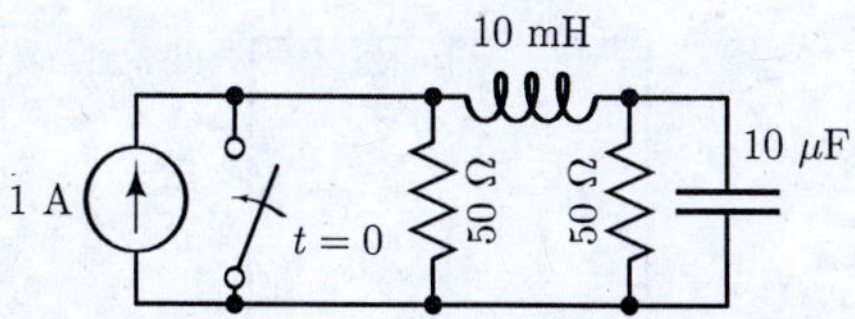

Figure 3.16: Circuit schematic for exercise 3.27.

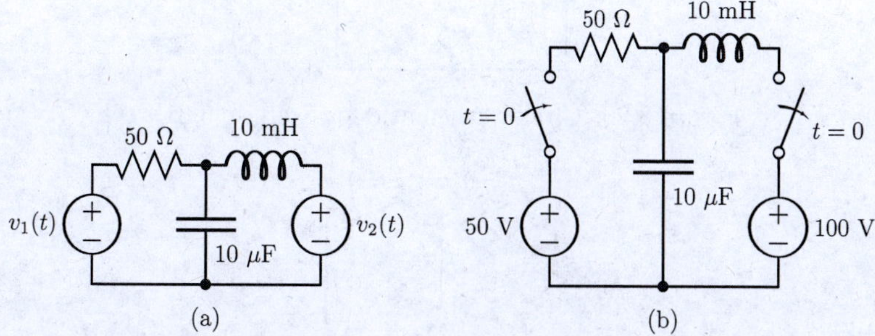

Figure 3.17: Circuit schematics for exercise 3.28.

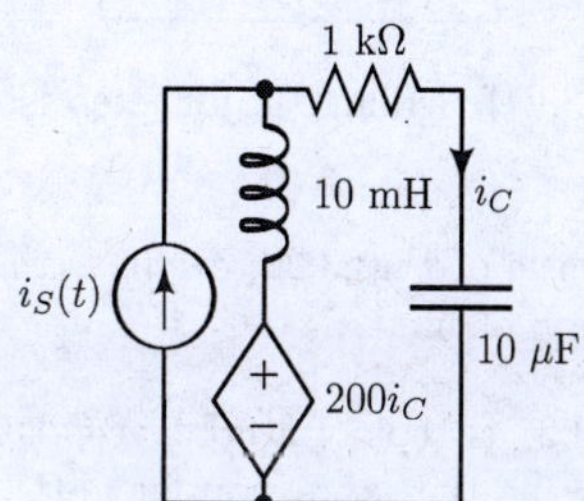

Figure 3.18: Circuit diagram for exercise 3.29.

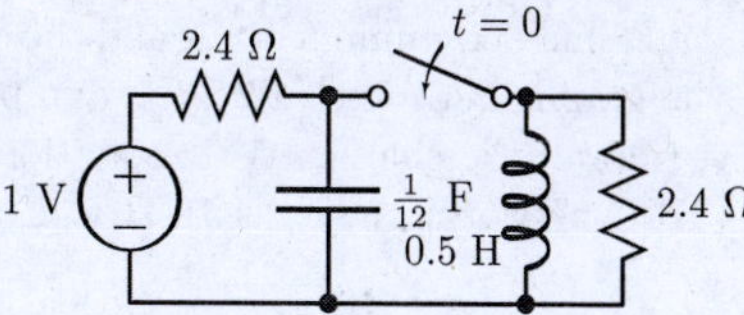

Figure 3.19: Circuit schematic for exercise 3.30.

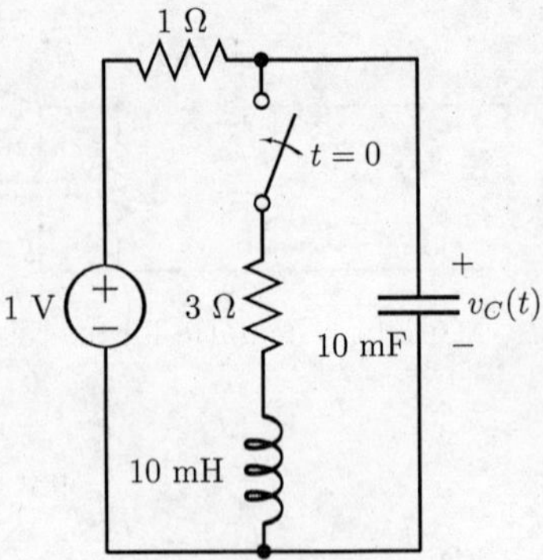

Figure 3.20: Circuit schematic for exercise 3.31.

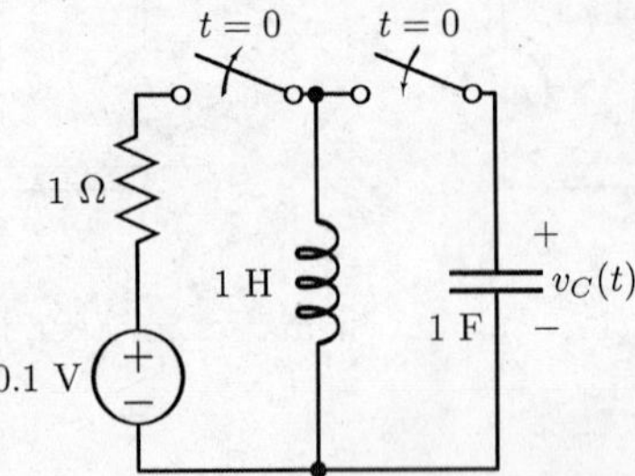

Figure 3.21: Schematic for exercise 3.32.

3.31 In the circuit schematic of Fig. 3.20, evaluate the voltage $v_C(t)$ across the capacitor as a function of time.

3.32 The circuit of Fig. 3.21 is that of a crude DC-DC converter. Assume that $v_C(0^-)$ is -5 volts. Find the expression for $v_C(t)$. What is the largest $|v_C|$? If the switches are reversed when $|v_C|$ is the largest, and a cycle is repeated, what will be the new largest $|v_C|$?

3.33 Fig. 3.22 shows the unit-step response of a second order RLC circuit. The times at which the maxima and minima occur and the corresponding voltages are indicated on the graph. Estimate (a) ζ, the damping factor, (b) ω_n, the natural frequency, (c) and the characteristic equation of the system. If the circuit is a series RLC network, and R is 10 Ω, can you predict the values of L and C?

3.34 What is the optimal value of ζ, the damping factor, that leads to the fastest settling time for a step input?

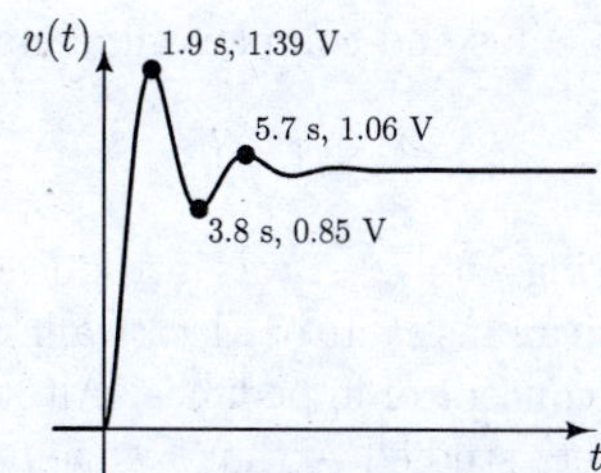

Figure 3.22: Second order RLC response for exercise 3.33.

Know more

Historical profile

Leonard Euler was a Swiss mathematician and engineer of the 18th century. His contributions were groundbreaking in mathematical analysis, complex number theory, graph theory, and topology. With several contributions in fluid dynamics, optics, astronomy, mechanics, his career was primarily between the Imperial Russian Academy of Sciences and the Berlin Academy. The real number e denoting the base of the natural logarithm, is named after Euler. Pierre-Simone Laplace had once said, "Read Euler, read Euler, he is the master of us all." Euler is regarded as one of the greatest mathematicians ever.

Leonard Euler[1]

Euler standardized much of the mathematics that we use today, including the notations of $f(x)$, $\sum$, use of x, y, z as unknowns, use of a, b, c as constants etc. Trigonometry was standardized by Euler, the use of π, e and i as $\sqrt{-1}$ were also standardized by Euler. Euler combined Leibniz's differential calculus with Newton's method of fluxions into modern calculus. One of the most famous identities in mathematics that combines geometry,

[1]Photograph taken from the public domain.

algebra, trigonometry, calculus and complex numbers is the Euler's identity:

$$e^{j\pi} + 1 = 0$$

The EE notation of using j for $\sqrt{-1}$ was introduced by Charles Steinmetz. At a meeting of the American Institute of Electrical Engineers in 1893, Arthur Kennelly introduced the concept of impedance. Although not present at this meeting, Charles Steinmetz studied Arthur Kennelly's paper and suggested the use of j as a standard for all electrical engineers.

Understand in depth

Study the theory of differential equations from a standard Mathematics text book. In particular, the sections on existence and uniqueness theorems for initial value problems, will help one understand the theory of differential equations.

- Advanced Engineering Mathematics, by Erwin Kreyszig, John Wiley & Sons Inc., chapters 1, 2 and 3, with a focus on sections 1.7, 2.6, and 3.2.

Modified nodal analysis (MNA) for dynamical circuits is a general technique to analyze any RLC circuit of any order. MNA is an algorithm used by circuit simulators for transient analysis. In combination with a state-space representation and the matrix exponential, MNA can be used to numerically analyze any circuit.

Scan the QR-code and read more about MNA, the state-space representation, the matrix exponential, and numerical analysis of the transient behavior of circuits.

Every pole of a circuit corresponds to a mode at which the circuit responds. In addition, each mode of the input results in a mode of response of the circuit. For an input with one mode, an nth order circuit responds in $n + 1$ modes. In each of these modes, KVL, KCL, and device properties are individually satisfied. The technique results in the Laplace-transform method of solving dynamical circuits; this will be discussed in Unit-5.

Unit 4

Sinusoidal steady-state analysis

Unit specifics

In this unit the student will study the following:

- The sine as a rotating phasor and phasor diagrams
- Impedances and admittances
- Mutual inductance, the dot convention, transformers
- AC circuit analysis
- Power in AC circuits, complex power, rms values
- Three phase circuits

Rationale

In this unit, we will study circuit analysis under sinusoidal excitation. All linear circuit elements, namely, resistors, capacitors, inductors, inductors with mutual coupling, dependent and independent voltage and current sources will be considered in this unit.

There are a few premises of this study:

1. All independent sources in the circuit are sinusoidally oscillating **coherently** at **precisely the same frequency**.
2. The circuit was set up a long time ago, and we do not need to worry about **any initial conditions**. In other words, the circuit is in a steady state. The free response of the circuit is to be ignored.

3. The circuit does not oscillate by itself, without any input source.

The conditions are not met by default. Some of these conditions are very difficult to meet. Even against these difficulties, this study will make the given assumptions.

Pre-requisites

- Physics: The physics behind inductors is needed for the study of mutually coupled inductors and transformers.
- Mathematics: Basic calculus, trigonometry and complex algebra.
- Software: We will continue using Octave for numerical analysis.

Unit outcomes

The list of outcomes of this unit are as follows.

U4-O1: Be able to analyze an AC circuit comfortably.

U4-O2: Be able to visualize the sinusoid as a rotating phasor. Be able to draw phasor diagrams and simplify complex problems intuitively.

U4-O3: Be able to re-work all network theorems for AC circuits. Gain the intuitions granted by the network theorems for all circuits.

U4-O4: Understand power in AC circuits. Gain the intuition behind complex power.

U4-O5: Appreciate the design of three-phase circuits. Be able to analyze three-phase circuits.

Unit-4 outcomes	**Expected mapping with course outcomes** *(1: Weak, 2: medium, and 3: strong correlation)*					
	CO-1	**CO-2**	**CO-3**	**CO-4**	**CO-5**	**CO-6**
U4-O1	3	-	3	-	-	1
U4-O2	-	3	2	-	-	-
U4-O3	-	3	1	1	1	-
U4-O4	-	2	1	3	1	-
U4-O5	2	1	1	3	-	-

4.1 Preliminaries

Theorem 4.1 (Linearity): A circuit with arbitrary connections of linear circuit elements (resistors, capacitors, inductors, mutual inductance, linear dependent sources) is a linear system.

Proof. Tableau analysis of a circuit has three parts:

$$\begin{aligned} \mathbf{A}\cdot\mathbf{I} &= \mathbf{0}, \text{ which is valid for a network with L, C, M} \\ \mathbf{V}-\mathbf{A}^T\mathbf{E} &= \mathbf{0}, \text{ which is also valid for a network with L, C, M} \\ \mathbf{MV}+\mathbf{NI} &= \mathbf{U}, \text{ which needs to be modified for L, C, M.} \end{aligned}$$

For an inductor L_k in branch k, $v_k - L_k\dot{i_k} = 0$. For a capacitor C_k in branch k, $i_k - C_k\dot{v_k} = 0$. For a mutual inductance M between branches p and q, with L_p in branch p and L_q in branch q, the equations are $v_p - L_p\dot{i_p} \pm M\dot{i_q} = 0$ and $v_q - L_q\dot{i_q} \pm M\dot{i_p} = 0$. If we create an operator, D, for the derivative, these device equations could be represented in the form $\mathbf{MV}+\mathbf{NI} = \mathbf{U}$, with D operators embedded within the $\mathbf{M}$ and $\mathbf{N}$ matrices.

Let $\mathbf{e_1}$, $\mathbf{v_1}$, and $\mathbf{i_1}$ be the circuit's response for a specific input $\mathbf{u_1}$. Further, let $\mathbf{e_2}$, $\mathbf{v_2}$, and $\mathbf{i_2}$ be the circuit's response to another input $\mathbf{u_2}$. Then,

$$\begin{bmatrix} \mathbf{0} & \mathbf{0} & \mathbf{A} \\ -\mathbf{A}^T & \mathbf{1} & \mathbf{0} \\ \mathbf{0} & \mathbf{M} & \mathbf{N} \end{bmatrix} \cdot \begin{bmatrix} \mathbf{e_1} \\ \mathbf{v_1} \\ \mathbf{i_1} \end{bmatrix} = \begin{bmatrix} \mathbf{0} \\ \mathbf{0} \\ \mathbf{u_1} \end{bmatrix}$$

$$\begin{bmatrix} \mathbf{0} & \mathbf{0} & \mathbf{A} \\ -\mathbf{A}^T & \mathbf{1} & \mathbf{0} \\ \mathbf{0} & \mathbf{M} & \mathbf{N} \end{bmatrix} \cdot \begin{bmatrix} \mathbf{e_2} \\ \mathbf{v_2} \\ \mathbf{i_2} \end{bmatrix} = \begin{bmatrix} \mathbf{0} \\ \mathbf{0} \\ \mathbf{u_2} \end{bmatrix}$$

We can add the RHS and LHS of the two equations.

$$\begin{bmatrix} \mathbf{0} & \mathbf{0} & \mathbf{A} \\ -\mathbf{A}^T & \mathbf{1} & \mathbf{0} \\ \mathbf{0} & \mathbf{M} & \mathbf{N} \end{bmatrix} \cdot \begin{bmatrix} \mathbf{e_1}+\mathbf{e_2} \\ \mathbf{v_1}+\mathbf{v_2} \\ \mathbf{i_1}+\mathbf{i_2} \end{bmatrix} = \begin{bmatrix} \mathbf{0} \\ \mathbf{0} \\ \mathbf{u_1}+\mathbf{u_2} \end{bmatrix}$$

The addition could be performed because the derivative operator is distributive, i.e., $Di_1 + Di_2$ is nothing but $D(i_1 + i_2)$. Homogeneity can also be proved the same way.

The above proves the linearity requirement. ∎

Theorem 4.2 (Time invariance): A circuit with arbitrary linear circuit elements (resistors, capacitors, inductors, mutual inductance, linear dependent sources) is time-invariant.

Proof. Like in the proof for linearity, let us embed a differentiation operator, D, within the $\mathbf{M}$ and $\mathbf{N}$ matrices. Let $\mathbf{e}(t)$, $\mathbf{v}(t)$, $\mathbf{i}(t)$ be the solution for the circuit when $u(t)$ is the input.

$$\begin{bmatrix} \mathbf{0} & \mathbf{0} & \mathbf{A} \\ -\mathbf{A}^T & \mathbf{1} & \mathbf{0} \\ \mathbf{0} & \mathbf{M} & \mathbf{N} \end{bmatrix} \cdot \begin{bmatrix} \mathbf{e}(t) \\ \mathbf{v}(t) \\ \mathbf{i}(t) \end{bmatrix} = \begin{bmatrix} \mathbf{0} \\ \mathbf{0} \\ \mathbf{u}(t) \end{bmatrix}$$

Then, if we delay the solution by T seconds, we obtain $\mathbf{e}(t-T)$, $\mathbf{v}(t-T)$, $\mathbf{i}(t-T)$. If we apply the tableau equation to this set of results, we obtain:

$$\begin{bmatrix} \mathbf{0} & \mathbf{0} & \mathbf{A} \\ -\mathbf{A}^T & \mathbf{1} & \mathbf{0} \\ \mathbf{0} & \mathbf{M} & \mathbf{N} \end{bmatrix} \cdot \begin{bmatrix} \mathbf{e}(t-T) \\ \mathbf{v}(t-T) \\ \mathbf{i}(t-T) \end{bmatrix} = \begin{bmatrix} \mathbf{0} \\ \mathbf{0} \\ \mathbf{u}(t-T) \end{bmatrix}$$

The above proves that the system is time-invariant. ∎

Definition 4.1 (Dirac delta function, impulse function): The delta function is defined as $\delta(t)$, where:

$$\delta(t) = \begin{cases} 0, & \text{for all } t \neq 0 \\ \infty, & \text{for } t = 0 \end{cases}$$

$$\text{and, } \int_{-\infty}^{\infty} \delta(t)dt = 1.$$

Any function of time, $x(t)$, can be described as a sum of scaled and shifted impulse functions. Since the impulse function is non-zero only at a single value of time, this sum of impulse functions works out as an integral.

$$x(t) = \int_{-\infty}^{\infty} x(\tau)\delta(t-\tau)d\tau \tag{4.1}$$

For example, if we want to compute $x(t)$ at $t = t_0$, we plug $t = t_0$ into (4.1). The expression works out as the integral of $x(\tau)\delta(t_0-\tau)$. However, $\delta(t_0-\tau)$ is zero for all τ, except at t_0. So this works out to $x(t_0)\int \delta(t_0-\tau)d\tau$, which works out to $x(t_0)$.

Definition 4.2 (Impulse response): If the input to an (linear and time invariant) LTI system is $\delta(t)$, then the corresponding output of the system, $h(t)$, is defined as the impulse response of the LTI system.

The impulse response, $h(t)$, is defined for LTI systems. If the input to the system is $5\delta(t-t_1) - 3\delta(t-t_2) + 7\delta(t+t_3)$, because the system is LTI, the corresponding output is $5h(t-t_1) - 3h(t-t_2) + 7h(t+t_3)$.

If the input to an LTI system with impulse response $h(t)$ is $x(t)$, and the output is $y(t)$, then:

$$\begin{aligned} x(t) &= \int_{-\infty}^{\infty} x(\tau)\delta(t-\tau)d\tau \\ \text{So, } y(t) &= \int_{-\infty}^{\infty} x(\tau)h(t-\tau)d\tau \end{aligned} \tag{4.2}$$

For convergence of (4.2), $x(t)$ and $h(t)$ have to be absolutely integrable. It is fair to assume that the input $x(t)$ is absolutely integrable. However, there could be a system where $h(t)$ is not absolutely integrable. In the initial set of premises, as given in the rationale at the beginning of the unit, we are looking at systems in which initial conditions a long time ago do not affect the system at present. In other words, we are looking at a system where the response to initial conditions a long time back decays with time. Such a system will have an absolutely integrable $h(t)$.

The integration variable τ can be changed to $\tau' = t - \tau$. Then (4.2) changes to:

$$y(t) = \int_{-\infty}^{\infty} h(\tau')x(t-\tau')d\tau' \tag{4.3}$$

The two equations, (4.2) and (4.3), are equivalent because the second came from the first by a simple variable change. (4.2) and (4.3) are known as the **convolution integral**.

Theorem 4.3 (Fundamental theorem of sinusoidal steady-state analysis): In a linear time-invariant (LTI) system, if the input is a sinusoidal function, the output is also a sinusoidal function at the same angular frequency.

Proof. Let us assume the LTI system has an input $x(t) = Ae^{j\omega t}$. If the system has an impulse response $h(t)$, then the output is

$$\begin{aligned} y(t) &= \int_{-\infty}^{\infty} h(\tau)Ae^{j\omega(t-\tau)}d\tau \\ \text{or, } y(t) &= Ae^{j\omega t}\int_{-\infty}^{\infty} h(\tau)e^{-j\omega\tau}d\tau \\ \text{or, } y(t) &= Ae^{j\omega t}\cdot K, \text{ where } K = \int_{-\infty}^{\infty} h(\tau)e^{-j\omega\tau}d\tau, \text{ a complex constant.} \end{aligned}$$

The above result says that the output $y(t)$ is proportional to $e^{j\omega t}$.

A real input should result in a real output. An imaginary input should result in an imaginary output. The superposition of a real and imaginary input will result in a complex output. Therefore, if the input to the system

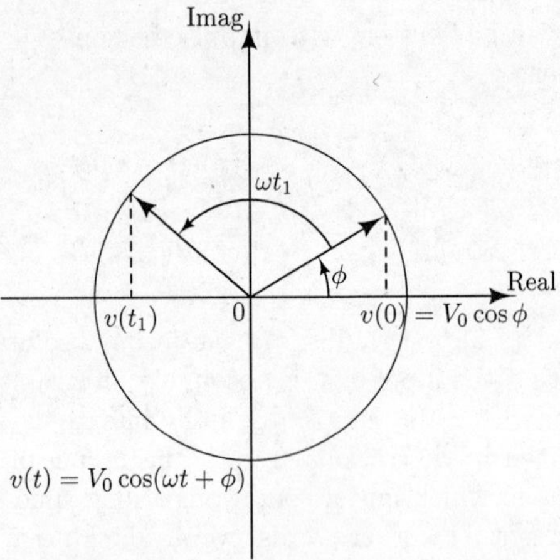

Figure 4.1: The phasor is a rotating complex vector. The actual voltage is the projection of the phasor on the real axis.

is only the real part of $x(t)$, i.e., $\cos\omega t$, then the system's output will be the real part of $y(t)$. The real part of $y(t)$ is a sinusoid with angular frequency ω.

■

Corollary 4.1. Suppose the input to an LTI circuit (with resistors, inductors, capacitors, mutual inductance, and linear dependent sources) is a sinusoidal waveform with angular frequency ω. In that case, node potentials, branch voltages, and branch currents in the circuit are sinusoidal with the same angular frequency ω.

A linear circuit excited with a sinusoidal input at ω will have sinusoidal voltages and currents at ω. The amplitudes and phases of these different voltages and currents in the circuit are different, the frequencies are the same.

4.2 The phasor

Definition 4.3 (Phasor): A sinusoidal waveform at an angular frequency ω will be of the form $v(t) = V_0 \cos(\omega t + \phi)$, where V_0 and ϕ are real constants. To this waveform, we can associate a complex number of the form $\bar{V} = V_0\angle\phi = V_0 \cos\phi + jV_0 \sin\phi$. We define $\bar{V}$ as a phasor.

Fig. 4.1 describes the idea behind a phasor. The phasor itself rotates with time. It has a certain magnitude (radius) and an angle that it subtends at $t = 0$. The actual voltage or current is the projection of this phasor on the real axis. We define a phasor by its radius (or magnitude, or amplitude of the sinusoid) V_0, and its starting angle ϕ. The vector is rotating at an angular velocity of ω, its projection on the real axis is nothing but $V_0 \cos(\omega t + \phi)$. The position of the phasor at $t = 0$ is the definition of the phasor. Since all voltage phasors and current phasors in the circuit rotate at an angular velocity of ω, we omit the ωt portion of the complex number and only report the vector at $t = 0$.

Example 4.1. What is the phasor corresponding to a current waveform, $i(t) = 2\sin(100\pi t + 30°)$ A?

When $100\pi t = 60°$, $i(t)$ is 2 A. This is the largest positive value of $i(t)$. At $t = 0$, $i(t)$ is $2\sin 30°$ or 1 A. The phasor can be visualized as a rotating complex vector, as shown below.

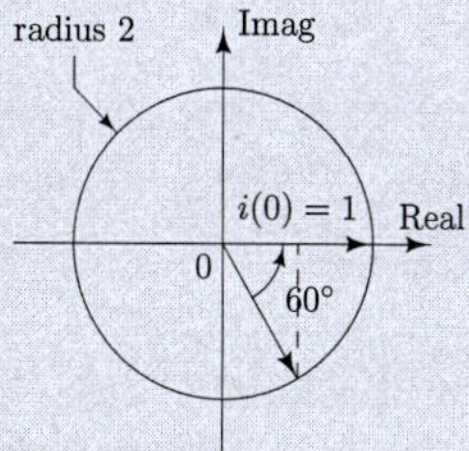

The phasor can be represented as a complex number, $2\angle - 60°$, or as $2e^{-j\pi/3}$, or as $1 - j\sqrt{3}/2$.

Observation 4.1. The function $\sin \omega t$ has a value of 1 when ωt is 90°, $\cos \omega t$ has a value of 1 when ωt is 0. As such, sin **lags** behind cos by 90°. cos **leads** sin by 90°.

Example 4.2. What is the angle between two phasors, $v(t)$ and $i(t)$, where, $v(t) = -10\cos(\omega t - 60°)$ V, and $i(t) = 5\sin(\omega t + 60°)$ A?

$v(t)$ is -10 V when ωt is 60°. So, $v(t)$ is 10 V when ωt is 240°. Therefore, $\bar{V} = 10\angle - 240°$, or $\bar{V} = 10\angle 120°$.

$i(t)$ is 5 A when ωt is 30°. Therefore, $\bar{I} = 5\angle - 30°$, or $\bar{I} = 5\angle 330°$.

The angle between the two phasors is 150°.

Theorem 4.4 (Phasor addition): The phasor of the sum of two voltages is the sum of the two phasors corresponding to each voltage. If $v_1(t) = V_1\cos(\omega t + \phi_1)$ and $v_2(t) = V_2\cos(\omega t + \phi_2)$, then the phasor corresponding to $v_1(t) + v_2(t)$ is nothing but $V_1\angle\phi_1 + V_2\angle\phi_2$.

Proof. Let us define $V_3 = V_1\cos\phi_1 + V_2\cos\phi_2$ and $V_4 = V_1\sin\phi_1 + V_2\sin\phi_2$.

$$\begin{aligned}
v_1(t) + v_2(t) &= V_1\cos(\omega t + \phi_1) + V_2\cos(\omega t + \phi_2) \\
&= \mathrm{Re}[V_1\exp(j\omega t + \phi_1) + V_2\exp(j\omega t + \phi_2)] \\
&= \mathrm{Re}[\exp(j\omega t)(V_1\cos\phi_1 + V_2\cos\phi_2 + j(V_1\sin\phi_1 + V_2\sin\phi_2))] \\
&= \mathrm{Re}[\exp(j\omega t)(V_3 + jV_4)]
\end{aligned}$$

So the phasor for $v_1(t) + v_2(t)$ is the complex number $V_3 + jV_4$. Now, if we compare this to $V_1\angle\phi_1 + V_2\angle\phi_2$, we get:

$$\begin{aligned}
V_1\angle\phi_1 + V_2\angle\phi_2 &= (V_1\cos\phi_1 + V_2\cos\phi_2) + j(V_1\sin\phi_1 + V_2\sin\phi_2) \\
&= V_3 + jV_4
\end{aligned}$$

■

Example 4.3. Three branches form a loop. The voltage drop across the first branch is $v_1(t) = 1\cos(\omega t + 45°)$. The voltage drop across the second branch is $v_2(t) = -2\sin(\omega t + 30°)$. What is the voltage drop across the third branch?

Let us convert the voltages into their corresponding phasors. v_1 is 1 when ωt is $-45°$. $\therefore \bar{V_1} = 1\angle 45°$. v_2 is -2 when ωt is $60°$. v_2 is 2 when ωt is $-120°$. $\therefore \bar{V_2} = 2\angle 120°$.

$$\begin{aligned}
\bar{V_3} &= -(\bar{V_1} + \bar{V_2}) = -(1\angle 45° + 2\angle 120°) \\
&= -(\frac{1}{\sqrt{2}} + j\frac{1}{\sqrt{2}} - 1 + j\sqrt{3}) = 0.293 - j2.439 \\
&= 2.457e^{-j1.45} \\
\therefore v_3(t) &= 2.457\cos(\omega t - 1.45)
\end{aligned}$$

(Note: The angle given in the answer is in radians, not degrees.)

Corollary 4.2. We can write KVL and KCL in terms of phasors.

Theorem 4.5 (Phasors obey Tellegen's theorem): If $\{\bar{V}_k\}$ are the phasors corresponding to all the branch voltages, and $\{\bar{I}_k\}$ are the phasors corresponding to all the branch currents, then:

$$\sum_1^k \bar{V}_k \bar{I}_k = 0, \qquad \sum_1^k \bar{V}_k \bar{I}_k^{\,*} = 0,$$

$$\sum_1^k \bar{V}_k^{\,*} \bar{I}_k = 0 \quad \text{and} \quad \sum_1^k \bar{V}_k^{\,*} \bar{I}_k^{\,*} = 0.$$

That is, the phasors and their conjugates obey Tellegen's theorem.

Proof. At $\omega t = 0$, the voltages are $\{\mathrm{Re}[\bar{V}_k]\}$ and the currents are $\{\mathrm{Re}[\bar{I}_k]\}$. At $\omega t = \pi/2$, the voltages are $\{\mathrm{Im}[\bar{V}_k]\}$ and currents are $\{\mathrm{Im}[\bar{I}_k]\}$. We apply Tellegen's theorem between these voltages and currents.

$$\begin{aligned}\therefore \ & \sum \mathrm{Re}[\bar{V}_k]\,\mathrm{Re}[\bar{I}_k] = \sum \mathrm{Re}[\bar{V}_k]\,\mathrm{Im}[\bar{I}_k] \\ = \ & \sum \mathrm{Im}[\bar{V}_k]\,\mathrm{Re}[\bar{I}_k] = \sum \mathrm{Im}[\bar{V}_k]\,\mathrm{Im}[\bar{I}_k] = 0\end{aligned}$$

So, for example,

$$\begin{aligned}\sum \bar{V}_k \bar{I}_k^{\,*} &= \sum (\mathrm{Re}[\bar{V}_k] + j\,\mathrm{Im}[\bar{V}_k])(\mathrm{Re}[\bar{I}_k] - j\,\mathrm{Im}[\bar{I}_k]) \\ &= \sum \mathrm{Re}[\bar{V}_k]\,\mathrm{Re}[\bar{I}_k] - j\sum \mathrm{Re}[\bar{V}_k]\,\mathrm{Im}[\bar{I}_k] \\ &\quad + j\sum \mathrm{Im}[\bar{V}_k]\,\mathrm{Re}[\bar{I}_k] + \sum \mathrm{Im}[\bar{V}_k]\,\mathrm{Im}[\bar{I}_k] \\ &= 0, \text{ because the four terms are individually 0.}\end{aligned}$$

Similarly, we can prove that the other voltage and current phasor combinations also follow Tellegen's theorem. ■

4.3 Impedances and admittances

4.3.1 Inductor

If the current through an inductor L is $i_L(t) = I_L \cos(\omega t)$, then the voltage across the inductor is

$$v_L(t) = L\dot{i}_L(t) = -I_L \omega L \sin(\omega t) = I_L \omega L \cos(\omega t + \pi/2)$$

Representing in terms of phasors, for a current $\bar{I}_L = I_L \angle 0$, the voltage is $\bar{V}_L = I_L \omega L \angle \pi/2 = j\omega L \bar{I}_L$. The current-voltage relationship implies that the inductor behaves like a resistor of value $j\omega L$.

Definition 4.4 (Impedance): We define the ratio of the branch-voltage phasor to the branch-current phasor in the sinusoidal steady-state as impedance. Impedance is a generalized form of resistance and is generally complex.

4.3.2 Capacitor

If the voltage across a capacitor C is $v_C(t) = V_C \cos(\omega t)$, then the current through the capacitor is:

$$i_C(t) = C\dot{v}_C(t) = -V_C\omega C \sin(\omega t) = V_C\omega C \cos(\omega t + \pi/2)$$

Representing in terms of phasors, for a voltage $\bar{V_C} = V_C\angle 0$, the current is $\bar{I_C} = V_C\omega C\angle\pi/2 = j\omega C\bar{V_C}$. The voltage-current relationship implies that the capacitor behaves like a conductor of value $j\omega C$.

Definition 4.5 (Admittance): We define the ratio of the branch-current phasor to the branch-voltage phasor in the sinusoidal steady-state as admittance. Admittance is a generalized form of conductance, and is generally complex.

Corollary 4.3. Two impedances, Z_1 and Z_2, when in series, can be combined as $Z_1 + Z_2$. Two admittances, Y_1 and Y_2, when in shunt, can be combined as $Y_1 + Y_2$.

Proof. If a current phasor $\bar{I}$ flows through the Z_1 in series with Z_2, the total voltage drop will be $Z_1\bar{I} + Z_2\bar{I}$, which is nothing but $\bar{I}(Z_1 + Z_2)$. Similarly, if a voltage phasor $\bar{V}$ is applied across two parallel admittances Y_1 and Y_2, then the total current will be $Y_1\bar{V} + Y_2\bar{V}$ which is nothing but $\bar{V}(Y_1 + Y_2)$. ■

Definition 4.6 (Resistance, reactance): The real part of an impedance is its resistance. The imaginary part of an impedance is its reactance.

Definition 4.7 (Conductance, susceptance): The real part of admittance is conductance. The imaginary part of admittance is susceptance.

Example 4.4. Evaluate the driving point impedance of the following network. Assume the circuit is to be operated at a frequency of 1 rad/s. Further, assume R_1 is 1 Ω, C_2 is 1/2 F, L_3 is 1/3 H, C_4 is 1/4 F, L_5 is 1/5 H.

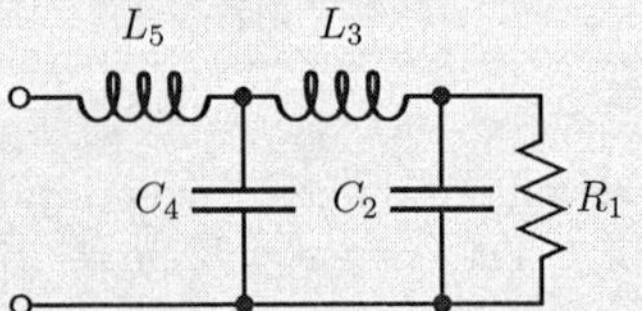

Let us solve this with the help of Octave.

```
octave:1> [R1, C2, L3, C4, L5] = deal(1, 1/2, 1/3, 1/4, 1/5);
octave:2> w=1;
octave:3> Y1 = 1/R1;
octave:4> Y2 = Y1 + j*w*C2;
octave:5> Z2 = 1/Y2;
octave:6> Z3 = Z2 + j*w*L3;
octave:7> Y3 = 1/Z3;
octave:8> Y4 = Y3 + j*w*C4;
octave:9> Z4 = 1/Y4;
octave:10> Z5 = Z4 + j*w*L5
Z5 =  0.745149 - 0.012160i
octave:11> abs(Z5)
ans = 0.7452
octave:12> angle(Z5)
ans = -0.016318
octave:13> angle(Z5)*180/pi
ans = -0.93495
```

We first worked out the parallel admittance (`Y2`) of R_1 and C_2. Then we converted the result into an impedance (`Z2`) and placed it in series with L_3. The result (`Z3`) was converted to an admittance (`Y3`), and placed in shunt with C_4, to give `Y4`. The admittance was converted to an impedance (`Z4`) and placed in series with L_5 to give `Z5`.

$$Z_5 = 0.745 - j0.012 = 0.745e^{-j0.016} = 0.745\angle - 0.93°$$

Observation 4.2. The complex equivalent of resistance is **impedance** and is given in ohms (Ω). The complex equivalent of conductance is **admittance** and is given in siemens (S).

An inductor of value L at an angular frequency of ω has an impedance of $j\omega L$. A capacitor of value C at an angular frequency of ω has an admittance of $j\omega C$.

All series-parallel rules are followed after replacing inductors and capacitors with their corresponding impedances.

Example 4.5. In the series-RLC circuit shown below, evaluate the voltage $\bar{V}$. The current source is a sinusoidal source of 1 A, the impedances of the resistor, the inductor, and the capacitor, are 2 Ω, $j3$ Ω, and $-j4$ Ω respectively.

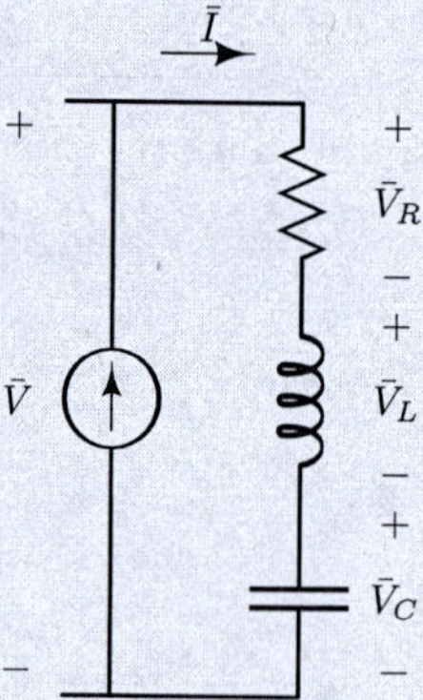

Let us consider $\bar{I}$ as a phasor with a reference direction of 0°. Then $\bar{V}_R$ is aligned with $\bar{I}$ and is just 2× the magnitude of $\bar{I}$. $\bar{V}_L$ **leads** $\bar{I}$ by 90° and has a magnitude 3× that of $\bar{I}$. $\bar{V}_C$ **lags** behind $\bar{I}$ and has a magnitude 4× that of $\bar{I}$. The complete voltage $\bar{V}$ is the sum of $\bar{V}_R$, $\bar{V}_L$ and $\bar{V}_C$ and can be graphically computed with the help of vector addition, as shown below.

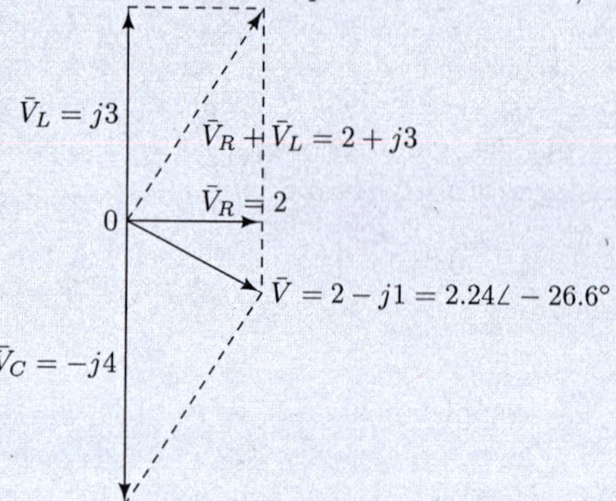

The diagram above is known as a **phasor diagram**. Each phasor is rotating with time. The diagram only depicts the phasors at $t = 0$.

Finally, the voltage across the combination of the resistor, capacitor and inductor is $\bar{V} = 2 - j1$ V, or $2.24\angle - 26.6°$.

Phasor diagrams like the one in the previous example give intuition about the circuit. They are used as aids in analysis, although they are not as powerful as direct numerical computation. In the following example, let us twist the series-RLC circuit a little bit and apply a voltage source instead of a current source.

Example 4.6. In the series-RLC circuit shown below, evaluate the current $\bar{I}$. The voltage source is a sinusoidal source of 1 V, the impedances of the resistor, the inductor, and the capacitor, are 2 Ω, $j3$ Ω, and $-j4$ Ω respectively.

Let us still consider the current phasor, $\bar{I}$, as the reference phasor. Although its angle is not 0°, let us draw a time snapshot at a time when its angle is 0°. The phasors $\bar{V}_R$, $\bar{V}_L$ and $\bar{V}_C$ are in the direction of, 90° ahead of, and 90° behind $\bar{I}$, respectively. The phasor diagram is shown below, and is no different from the phasor diagram in the previous example.

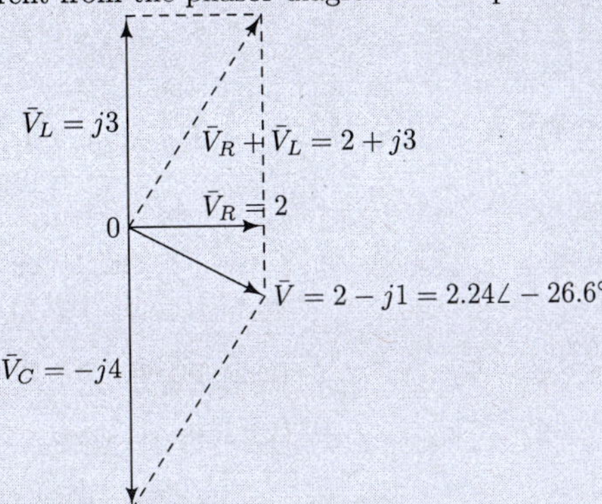

If we take the reference phasor as that of $\bar{I}$, then $\bar{V}$ is lagging by 26.6° and has a magnitude 2.24× that of the current. Clearly, if the reference phasor is the voltage, the current will **lead** by 26.6° and will be 2.24× smaller.

$$\therefore \bar{I} = \frac{1}{2.24}\angle 26.6° = 0.4 + j0.2 \text{ A}$$

The computation is straightforward with the help of complex numbers.

$$\bar{I} = \bar{V}/(Z_R + Z_L + Z_C) = 1/(2 + j3 - j4) = 1/(2 - j1) = 0.4 + j0.2 \text{ A}$$

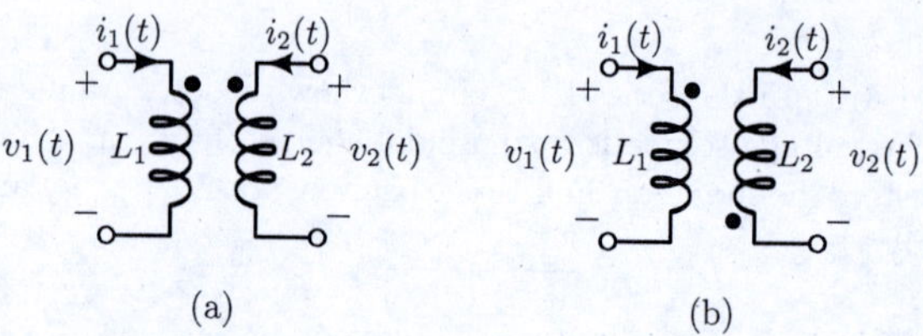

Figure 4.2: Mutual inductance. (a) The coupled terminals are both placed at the positive terminals, and (b) the coupling is between the positive and negative terminals.

4.4 Mutual inductance

4.4.1 The dot convention

A pair of mutually coupled inductors form a two-port circuit component with three parameters, L_1, L_2, and M. The mutually coupled inductors of Fig. 4.2(a) are related by the following equations in the time domain. The dots mark the terminal-pair that has been coupled. If the coupled terminals are also the positive terminals of the two-port, the relationships are straightforward.

$$\begin{aligned} v_1(t) &= L_1\dot{i_1}(t) + M\dot{i_2}(t) \\ v_2(t) &= M\dot{i_1}(t) + L_2\dot{i_2}(t) \end{aligned}$$

In the phasor domain, i_1 and i_2 are $I_1\cos(\omega t)$ and $I_2\cos(\omega t)$, respectively. The mutual inductance equations resolve to the following:

$$\begin{aligned} \bar{V_1} &= j\omega L_1\bar{I_1} + j\omega M\bar{I_2} \\ \bar{V_2} &= j\omega M\bar{I_1} + j\omega L_2\bar{I_2} \end{aligned}$$

In Fig. 4.2(b), the coupling is between the positive and negative terminals of the two ports. In this case, the signs of $\bar{V_2}$ and $\bar{I_2}$ need to be reversed.

$$\begin{aligned} \bar{V_1} &= j\omega L_1\bar{I_1} - j\omega M\bar{I_2} \\ -\bar{V_2} &= j\omega M\bar{I_1} - j\omega L_2\bar{I_2} \\ \text{or, } \bar{V_2} &= -j\omega M\bar{I_1} + j\omega L_2\bar{I_2} \end{aligned}$$

In this case, the mutual coupling simply becomes negative.

The rule can also be stated as follows:

1. If currents in both the coils are entering at the dot (or both leaving at the dot), the sign for M will be the same as that of L.
2. If current in one coil is entering at the dot, and current in the other coil is leaving at the dot, the sign for M will be opposite that of L.

Example 4.7. Evaluate the driving point impedance looking into the two circuits shown below. L_1 and L_2 are 1/3 H and 1/4 H respectively. The mutual inductance between the two inductors, M, is 1/5 H. The operating frequency of the circuit is 100 rad/s.

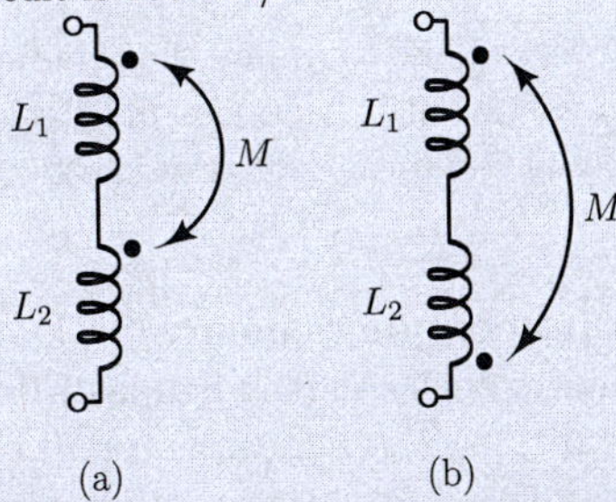

To evaluate the driving point impedance, let us push a current $\bar{I}$ through the network, and evaluate the voltage. Let us denote $\bar{V_1}$ as the voltage across L_1 from top to bottom, and $\bar{V_2}$ as the voltage across L_2 from top to bottom.

Case (a):

$$\begin{aligned} \bar{V_1} &= j\omega L_1\bar{I} + j\omega M\bar{I} \\ \bar{V_2} &= j\omega M\bar{I} + j\omega L_2\bar{I} \\ \therefore\ \bar{V_1} + \bar{V_2} &= j\omega(L_1 + L_2 + 2M)\bar{I} \end{aligned}$$

The driving point impedance of the circuit in (a) is $j\omega(L_1 + L_2 + 2M)$, or $j98.33\ \Omega$.

Case (b):

$$\begin{aligned} \bar{V_1} &= j\omega L_1\bar{I} - j\omega M\bar{I} \\ \bar{V_2} &= -j\omega M\bar{I} + j\omega L_2\bar{I} \\ \therefore\ \bar{V_1} + \bar{V_2} &= j\omega(L_1 + L_2 - 2M)\bar{I} \end{aligned}$$

The driving point impedance of the circuit in (b) is $j\omega(L_1 + L_2 - 2M)$, or $j18.33\ \Omega$.

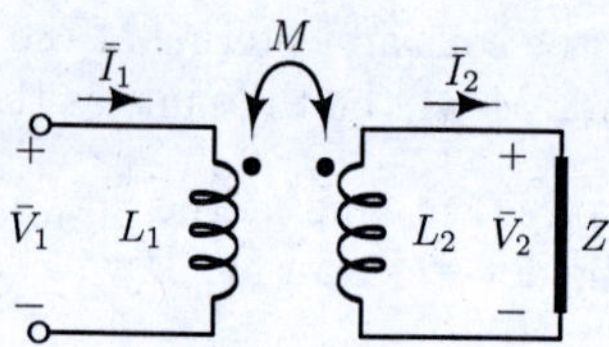

Figure 4.3: Two mutually coupled inductors are driven by a voltage $\bar{V}_1$. The second inductor drives a load Z.

4.4.2 The transformer

Mutual inductance is created by coupling the windings of two coils. The two coils individually possess self inductance. The relationship between the mutual inductance and the self inductances of the two coils is as follows:

$$M = k\sqrt{L_1 L_2} \tag{4.4}$$

In (4.4), k is known as the **coupling coefficient**, and $k \in [0, 1]$. When the two coils are not coupled to each other, k is 0; as the coupling between the coils becomes tighter and tighter, k increases from 0 to 1, in the limit.

In the limiting case, when $k = 1$, the mutual inductances form an ideal transformer. Let us analyze.

In the circuit of Fig. 4.3, a voltage $\bar{V}_1$ is applied across L_1. L_1 and L_2 are mutually coupled with a mutual inductance of M. L_2 is connected across an arbitrary impedance Z. The voltage across Z is $\bar{V}_2$. $\bar{I}_1$ is the current in L_1 and $\bar{I}_2$ is the current in Z.

$\bar{V}_1$ and $\bar{V}_2$ are written in terms of $\bar{I}_1$ and $\bar{I}_2$ as follows.

$$\begin{aligned}
\bar{V}_1 &= j\omega L_1\bar{I}_1 - j\omega M\bar{I}_2 \\
\bar{V}_2 &= j\omega M\bar{I}_1 - j\omega L_2\bar{I}_2 = \bar{I}_2 Z \\
\text{or, } j\omega M\bar{I}_1 &= \bar{I}_2(j\omega L_2 + Z) \\
\therefore\ \bar{I}_2 &= \frac{j\omega M}{j\omega L_2 + Z}\bar{I}_1 \\
\therefore\ \bar{V}_1 &= (j\omega L_1 + \frac{\omega^2 M^2}{j\omega L_2 + Z})\bar{I}_1 \\
\text{or, } \bar{I}_1 &= \frac{j\omega L_2 + Z}{j\omega L_1(j\omega L_2 + Z) + \omega^2 M^2}\bar{V}_1 \\
\therefore\ \bar{I}_2 &= \frac{j\omega M}{j\omega L_1(j\omega L_2 + Z) + \omega^2 M^2}\bar{V}_1 \\
\therefore\ \bar{V}_2 &= \frac{j\omega M Z}{j\omega L_1(j\omega L_2 + Z) + \omega^2 M^2}\bar{V}_1
\end{aligned}$$

In the situation where $k = 1$, $M = \sqrt{L_1 L_2}$, or $M^2 = L_1 L_2$. Plugging in this value, we obtain:

$$\bar{V}_2 = \frac{j\omega M Z}{-\omega^2 L_1 L_2 + j\omega L_1 Z + \omega^2 L_1 L_2}\bar{V}_1 = \frac{M}{L_1}\bar{V}_1 = \sqrt{\frac{L_2}{L_1}}\bar{V}_1$$

In short, if we apply $\bar{V}_1$ across L_1, we obtain $\sqrt{L_2/L_1}\bar{V}_1$ across L_2, regardless of the value of Z.

A quick recall of the physics of an inductor is now required. The inductance of a coil changes with the number of turns in the coil. The flux generated in the coil is proportional to the n, the number of turns. The total flux linked into the coil is n times the flux generated. As such, the inductance is proportional to n^2, where n is the number of turns in the coil.

Scan the QR-code, and learn more about the physics of an inductor. Also learn more about mutual inductances and how coupling is introduced between coils.

If the turns in L_1 is given by n_1, and the turns in L_2 is given by n_2, then L_1 is proportional to n_1^2, L_2 is proportional to n_2^2. Overall, we now have the following for $\bar{V}_2$:

$$\bar{V}_2 = \sqrt{\frac{L_2}{L_1}}\bar{V}_1 = \frac{n_2}{n_1}\bar{V}_1 \tag{4.5}$$

(4.5) is the equation for an ideal transformer. If the number of turns in the primary is n_1 and the turns in the secondary is n_2, $\bar{V}_2/\bar{V}_1 = n_2/n_1$.

When Z is a short circuit, $\bar{I}_2/\bar{I}_1 = M/L_2 = \sqrt{L_1/L_2} = n_1/n_2$. Notice, this is the inverse ratio. The current in an arm of the transformer is inversely proportional to the number of turns in the arm.

4.5 AC circuit analysis

We will now use phasors to analyze circuits. We will replace every inductor L with its equivalent impedance, $j\omega L$. We will replace every capacitor C with its equivalent admittance, $j\omega C$, or impedance, $1/(j\omega C)$. After this, we will analyze the circuit as if all inductors and capacitors are like resistors.

Observation 4.3. In AC circuits, voltages are oscillating sinusoidally. Is it important to denote the + and − signs around the voltage sources and for potential differences? Yes, it is important! At $t = 0$, the voltage between two nodes has both magnitude and a sign. An opposite sign denotes a phase that is 180° off.

Thumb rule. Replace every inductor, L, with $j\omega L$. Replace every capacitor, C, with $-j/(\omega C)$. Now perform sinusoidal steady-state circuit analysis. Use complex numbers in polar or Cartesian forms, depending on convenience.

Example 4.8. Analyze the circuit shown below using the mesh method.

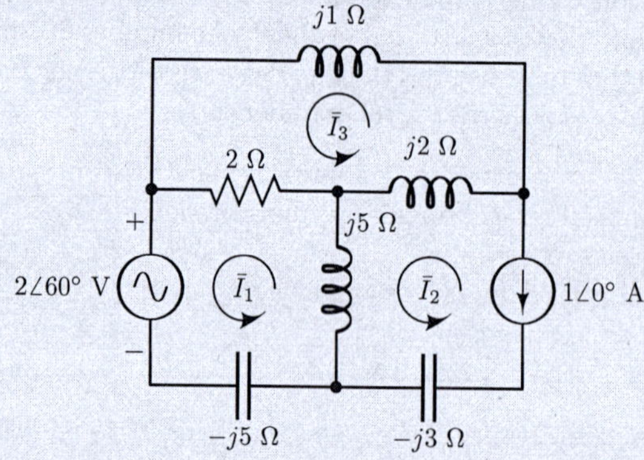

$\bar{I}_2$ is nothing but $-1\angle 0 = -1$ A.

$$\begin{bmatrix} 2 & -2 \\ -2 & 2+j3 \end{bmatrix} \cdot \begin{bmatrix} \bar{I}_1 \\ \bar{I}_3 \end{bmatrix} = \begin{bmatrix} -j5 - 2(\cos 60^\circ + j\sin 60^\circ) \\ -j2 \end{bmatrix}$$

$$\text{or, } \begin{bmatrix} \bar{I}_1 \\ \bar{I}_3 \end{bmatrix} = \begin{bmatrix} 1/2 - j/3 & -j/3 \\ -j/3 & -j/3 \end{bmatrix} \begin{bmatrix} -1 - j6.732 \\ -j2 \end{bmatrix} = \begin{bmatrix} -3.411 - j3.033 \\ -2.911 + j0.333 \end{bmatrix}$$

The result gives $i_1(t) = 4.564\cos(\omega t - 138^\circ)$ A, and $i_3(t) = 2.93\cos(\omega t + 173^\circ)$ A. Of course, $i_2(t) = -\cos(\omega t)$. We may compute all other branch currents and branch voltages as required.

Complex numbers need to be freely converted between Cartesian and polar forms while handling phasors. It is understood that an effective numerical analysis program like Octave, or an efficient calculator that can handle complex numbers, is essential for solving numericals.

Common mistakes while solving numericals involve (1) conversion between degrees and radians, and (2) conversion between radians per second and hertz. While solving a numerical, double-check that all trigonometric computations are correctly entered in radians (in case you are using Octave).

If you are using a calculator, the mode of the calculator can be changed from radians to degrees. In case you are operating with degrees, all angles (sometimes including ωt) need to be changed to degrees.

Let us solve an instance of a circuit that has mutual inductance as part of the network. We will slightly modify the circuit in example 4.8 with mutual coupling between the three inductors. In example 4.9, there are three inductors that have been physically wound together. As a result, there is a mutual inductance between the first and second inductor, a mutual inductance between the second and third inductor, and another mutual inductance between the third and first inductor.

Example 4.9. In the circuit shown below the three inductors are coupled with each other. The values of the mutual inductances are indicated in parentheses. Analyze the circuit.

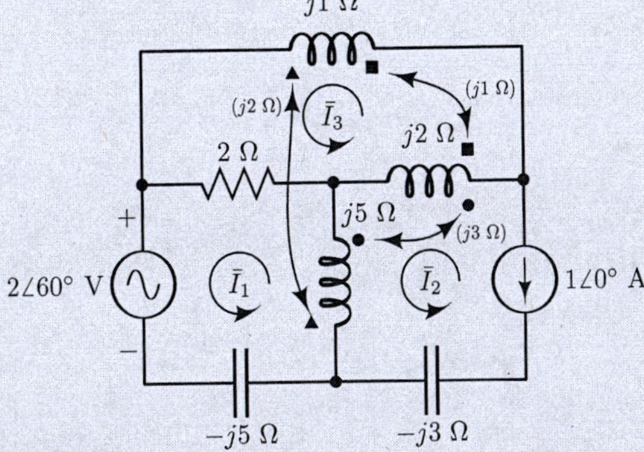

We will use the mesh method to solve the circuit. The mesh currents, $\bar{I}_1$, $\bar{I}_2$, $\bar{I}_3$, are as indicated in the schematic. Note that while the terminal-pairs with dots are coupled to each other, the terminal-pairs without dots are also coupled to each other. Also, verify that the mutual inductances given in the circuit result in coupling coefficients that are ≤ 1. The mesh equations are written below.

$$2\angle 60^\circ - j5\bar{I}_1 + j5(\bar{I}_1 - \bar{I}_2) + \underbrace{j3(\bar{I}_3 - \bar{I}_2) - j2\bar{I}_3}_{\text{Mutual inductance terms}} + 2(\bar{I}_1 - \bar{I}_3) = 0$$

$$j2(\bar{I}_3 - \bar{I}_2) \underbrace{- j\bar{I}_3 + j3(\bar{I}_1 - \bar{I}_2)}_{\text{Mutual inductance terms}} + j\bar{I}_3 + \underbrace{j(\bar{I}_2 - \bar{I}_3) + j2(\bar{I}_2 - \bar{I}_1)}_{\text{Mutual inductance terms}} + 2(\bar{I}_3 - \bar{I}_1) = 0$$

$$-\bar{I}_2 = 1\angle 0^\circ$$

The mesh equations can be compressed into the following matrix equation.

$$\begin{bmatrix} 2 & -2+j1 \\ -2+j1 & 2+j1 \end{bmatrix} \begin{bmatrix} \bar{I}_1 \\ \bar{I}_3 \end{bmatrix} = \begin{bmatrix} -2e^{j\pi/3} - j8 \\ -j2 \end{bmatrix}$$

Notice that with the value of $\bar{I}_2$ plugged into the equation system, the matrix is symmetric. Symmetry in the matrix will be maintained as long as there are no dependent sources, even when there are mutual inductances in the network. This can often be used as a sanity check. Finally $\bar{I}_1$ and $\bar{I}_3$ are as follows.

```
octave:1> I = inv([2, -2+j; -2+j, 2+j])*[-2*e^(j*pi/3)-8*j; -2*j]
I =
  -3.8122 - 1.5907i
  -3.9599 + 1.2954i
```

In the next example, we will analyze a circuit using the node voltage method.

Example 4.10. Analyze the circuit shown below using the node voltage method.

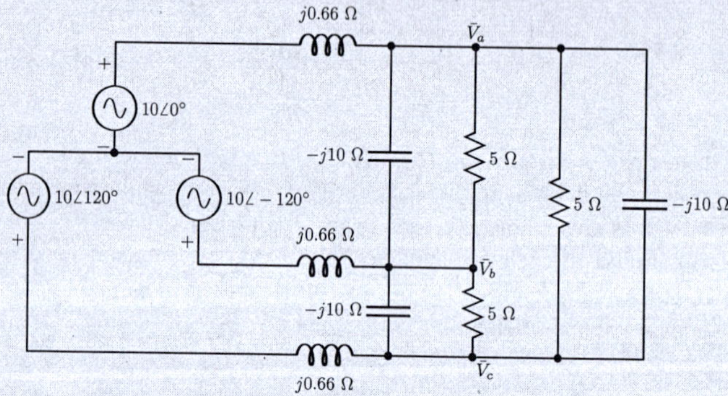

We have considered the central node (negative terminal) of the three voltage sources as the datum. The three unknown node voltages are labeled as $\bar{V}_a$, $\bar{V}_b$, and $\bar{V}_c$. The following are the node voltage equations.

$$\begin{bmatrix} 0.4-j1.3 & -0.2-j0.1 & -0.2-j0.1 \\ -0.2-j0.1 & 0.4-j1.3 & -0.2-j0.1 \\ -0.2-j0.1 & -0.2-j0.1 & 0.4-j1.3 \end{bmatrix} \cdot \begin{bmatrix} \bar{V}_a \\ \bar{V}_b \\ \bar{V}_c \end{bmatrix} = \begin{bmatrix} -j15 \\ -13+j7.5 \\ 13+j7.5 \end{bmatrix}$$

$$\text{or,}\quad \begin{bmatrix} \bar{V}_a \\ \bar{V}_b \\ \bar{V}_c \end{bmatrix} = \begin{bmatrix} 0.22 + j0.66 & -0.11 & -0.11 \\ -0.11 & 0.22 + j0.66 & -0.11 \\ -0.11 & -0.11 & 0.22 + j0.66 \end{bmatrix} \cdot \begin{bmatrix} -j15 \\ -13 + j7.5 \\ 13 + j7.5 \end{bmatrix}$$

$$\text{or,}\quad \begin{bmatrix} \bar{V}_a \\ \bar{V}_b \\ \bar{V}_c \end{bmatrix} = \begin{bmatrix} 10 - j5 \\ -9.33 - j6.16 \\ -0.67 + j11.16 \end{bmatrix} = \begin{bmatrix} 11.18\angle -26.6^\circ \\ 11.18\angle -146.6^\circ \\ 11.18\angle 93.4^\circ \end{bmatrix}$$

We can now compute all the branch voltages and branch currents.

4.5.1 Network theorems

Superposition theorem: In Theorem 4.1, we showed that superposition is valid for all linear time-invariant circuits under all conditions. The superposition theorem is valid in the sinusoidal steady state.

Tellegen's theorem: We used Tellegen's theorem in Theorem 4.5. Tellegen's theorem is valid for phasors because the real parts of the phasors are the voltages and currents at $t = 0$, and the imaginary parts of the phasors are the voltages and currents at $\omega t = \pi/2$.

Thévenin's, Norton's theorems: Thévenin's and Norton's theorems are valid with phasors. We can convert all inductors, capacitors, and mutual inductances to their impedance forms. We can now calculate the open-circuit voltage, short-circuit current, and driving point impedance as complex phasors.

Reciprocity theorem: We can apply the reciprocity theorem when a two-port network has passive components only, i.e., resistors, capacitors, inductors, and mutual inductances.

Compensation theorem: The compensation theorem will apply in the sinusoidal steady state. Impedances will be used in place of resistances.

Maximum power transfer theorem: The maximum power transfer theorem will be discussed in section 4.6.3, after we have worked out the power absorbed in terms of phasors, in section 4.6. In the sinusoidal steady state, maximum power will be transferred from a source with series impedance Z_S to a load of impedance Z_L, when $Z_L = Z_S^*$, that is, when the load impedance is the conjugate of the source impedance. This will be proved in section 4.6.3.

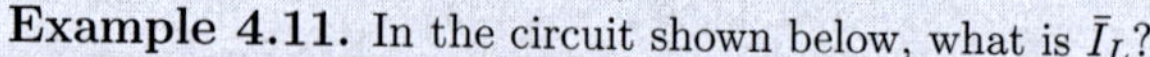

Example 4.11. In the circuit shown below, what is $\bar{I}_L$?

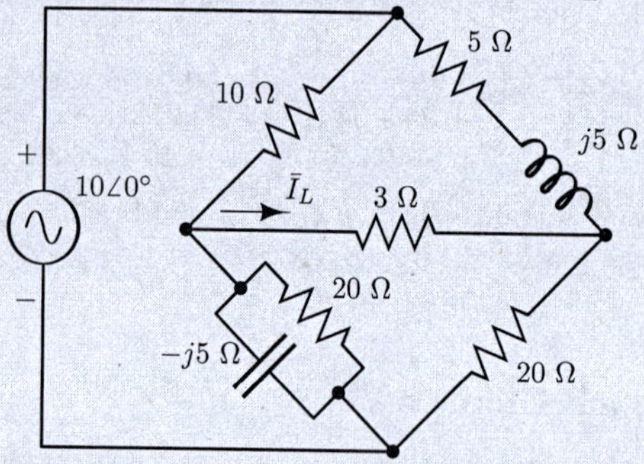

The 3 Ω resistor is considered as a load and pulled out. The remaining one-port network can now be modeled using Thévenin's theorem. First, we need to evaluate the open-circuit voltage. Using potential division, $\bar{V}_T$ works out to:

$$\bar{V}_T = 10\left(\frac{-j5 \parallel 20}{-j5 \parallel 20 + 10} - \frac{20}{20 + 5 + j5}\right) = -5.29 - j1.66$$

Next, we null the input voltage source (it becomes a short circuit) and we find the driving point resistance. The overall Z_T works out to:

$$Z_T = (10 \parallel (-j5 \parallel 20)) + (20 \parallel (5 + j5)) = 7.02 - j0.12$$

The model of the circuit without the 3 Ω load is $\bar{V}_T$ in series with Z_T. Therefore the current $\bar{I}_L$ is:

$$\bar{I}_L = \bar{V}_T/(Z_T + 3) = -0.53 - j0.17 = 0.55\angle -162°$$

Example 4.12. The three-port network shown below, $\mathcal{N}$, is linear and passive. A few experimental observations were made.

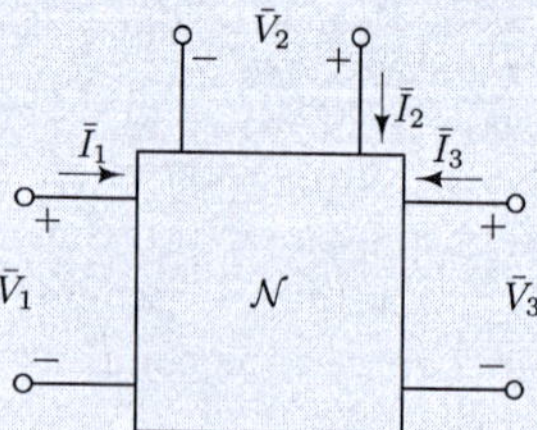

When $\bar{V}_1 = 1\angle 0°$, and ports 2 and 3 are short circuits,

1. $\bar{I}_1 = 0.1\angle 0°$,
2. $\bar{I}_2 = 0.2\angle 60°$,
3. and $\bar{I}_3 = 0.3\angle 30°$.

What is $\bar{I}_1$, when $\bar{V}_1$ is 5 V, $\bar{V}_2$ is $(3 + j4)$ V and $\bar{V}_3$ is $(3 - j4)$ V?

Since the network is linear and passive, we can apply superposition, homogeneity, and reciprocity.

$\bar{V}_1$	$\bar{V}_2$	$\bar{V}_3$	$\bar{I}_1$	Comment
$1\angle 0°$ V	0	0	$0.1\angle 0°$ A	
0	$1\angle 0°$ V	0	$0.2\angle 60°$ A	Reciprocity
0	0	$1\angle 0°$ V	$0.3\angle 30°$ A	Reciprocity
5 V	0	0	$(0.5 + j0)$ A	Homogeneity
0	$(3 + j4)$ V	0	$(-0.39 + j0.92)$ A	Homogeneity
0	0	$(3 - j4)$ V	$(1.38 - j0.59)$ A	Homogeneity
5 V	$(3 + j4)$ V	$(3 - j4)$ V	$(1.49 + j0.33)$ A	Superposition

4.6 Power computations with phasors

4.6.1 Phasor convention

Before we start our discussion on power in the sinusoidal steady state, we are going to re-define phasors. This is the standard convention for phasors. The convention for phasors used earlier is to be discarded.

Convention. [Phasor convention] Note the standard convention for phasors. For a current $i(t) = \sqrt{2}I_0 \cos \omega t$, the phasor is I_0. The magnitude of the phasor is the root-mean-squared (rms) value of the time-domain signal, or $\sqrt{2}$ times smaller than the amplitude. The scaling has been done so that the power in (4.7) relates easily to the power at DC. This convention for phasors is the standard, and is used henceforth.

4.6.2 Power in an element

Let us compute the power consumed by a device, where the current through the device is $i(t) = \sqrt{2} \cdot I\cos(\omega t + \phi_1)$, and the voltage across the device is $v(t) = \sqrt{2} \cdot V\cos(\omega t + \phi_2)$. The phasor corresponding to $i(t)$ is $I\angle\phi_1$, or, $Ie^{j\phi_1}$, or, $I\cos\phi_1 + jI\sin\phi_1$. The phasor corresponding to $v(t)$ is $V\angle\phi_2$, or, $Ve^{j\phi_2}$, or, $V\cos\phi_2 + jV\sin\phi_2$.

The average power consumed over one period is given by:

$$\begin{aligned} P &= \frac{\omega}{2\pi}\int_0^{2\pi/\omega} v(t)i(t)dt = \frac{\omega}{\pi}\cdot VI\cdot\int_0^{2\pi/\omega}\cos(\omega t+\phi_1)\cos(\omega t+\phi_2)dt \\ &= VI\frac{\omega}{2\pi}\int_0^{2\pi/\omega}(\cos(2\omega t+\phi_1+\phi_2)+\cos(\phi_2-\phi_1))dt \\ &= VI\frac{\omega}{2\pi}\int_0^{2\pi/\omega}\cos(\phi_2-\phi_1)dt \\ &= VI\cos(\phi_2-\phi_1) \end{aligned} \tag{4.6}$$

The average power over one period also happens to be the real part of the product of the phasor corresponding to $v(t)$ and the conjugate of the phasor corresponding to $i(t)$.

$$P = VI\cos(\phi_2 - \phi_1) = \text{Re}(Ve^{j\phi_2}\cdot Ie^{-j\phi_1}) = \text{Re}(\bar{V}\cdot\bar{I}^*) \tag{4.7}$$

$P = \text{Re}(\bar{V}\cdot\bar{I}^*)$ is also known as the "active power" consumed by the device. The imaginary part, $P_r = \text{Im}(\bar{V}\cdot\bar{I}^*)$, is called the "reactive power" consumed by the device. If we split the device into a real resistance and a reactance, the resistance will consume the active power. The reactance will store energy.

In Theorem 4.5, we had proved that $\sum \bar{V}_k\bar{I}_k^* = 0$ with the help of Tellegen's theorem. The sum of all complex powers in a circuit is 0. This implies that the sum of active powers is 0, and **the sum of reactive powers is 0**.

$\cos(\phi_2 - \phi_1)$ is known as the **power factor**. If the current leads the voltage, i.e., if the phase of the current is more than the phase of the voltage, the power factor is said to be leading. If the phase of the current lags behind the phase of the voltage, the power factor is lagging. The power factor leads in case of a capacitive load and lags in case of an inductive load.

Definition 4.8 (Power factor): If the voltage across a load and the current through the load differ in phase by ϕ, the power factor is $\cos\phi$. If the current leads the voltage, the power factor is **leading**. If the current lags behind the voltage, the power factor is **lagging**. Power factor is dimensionless. (If the real power lags behind the apparent power, the power factor is lagging.)

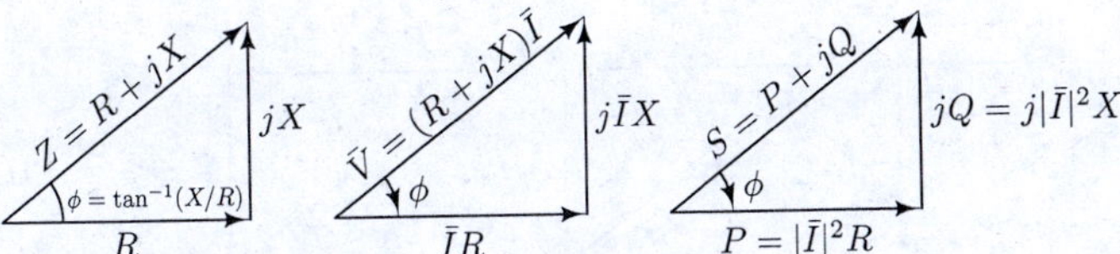

Figure 4.4: The power triangle for an inductive impedance. P is proportional to R, the real part of the impedance. Q is proportional to X and S is proportional to Z. P lags behind S.

Capacitive loads have leading power factor, while inductive loads have a lagging power factor.

In a practical circuit we can only have voltage sources. This is why we always use the phase of a voltage as the reference. The phase of the current is referred to the phase of the voltage, either as leading in case of capacitive circuits, or lagging in case of inductive circuits.

Definition 4.9 (Active power): The active power is the power consumed and is given by $\mathrm{Re}[\bar{V}\bar{I}^*]$. The unit for active power is **watt [W]**.

Definition 4.10 (Reactive power): The reactive power is the energy cycling through the storage elements of the circuit per cycle and is $\mathrm{Im}[\bar{V}\bar{I}^*]$. The unit for reactive power is **var [var]**.

Definition 4.11 (Apparent power): The current magnitude and voltage magnitude product is the apparent power and is $|\bar{V}\bar{I}^*| = |\bar{V}| \cdot |\bar{I}|$. The unit for apparent power is **volt ampere [V·A]**.

Definition 4.12 (Complex power): The complex power is the complex product, $\bar{V}\bar{I}^*$. The unit for complex power is **volt ampere [V·A]**.

In short, the complex power, S, is $\bar{V}\bar{I}^*$. The real part of S is P, the active power, or the real power. The imaginary part of S is Q, the reactive power. The magnitude of S is the apparent power.

Let us force a current of $(1\angle 0)$ A through an inductive impedance $Z = R + jX$. Fig. 4.4 shows a phasor diagram. The current is in the direction of R. The voltage across the impedance, $(Z \cdot 1)$ V, will be in the direction of Z. The complex power, S, is $\bar{V}\bar{I}^*$. In this case, $\bar{I}^*$ is $(1\angle 0)$ A and $\bar{V}$ is (Z) V. The real power, P, is the power in R, the reactive power, Q, is the reactive power in jX. The current vector, $(1\angle 0)$, lags behind the voltage vector, in line with Z. The power factor is $\cos\phi$, lagging. The active power P lags behind the complex power S. An inductive load results in positive var and a lagging power factor. **An inductor always consumes positive reactive power in var.**

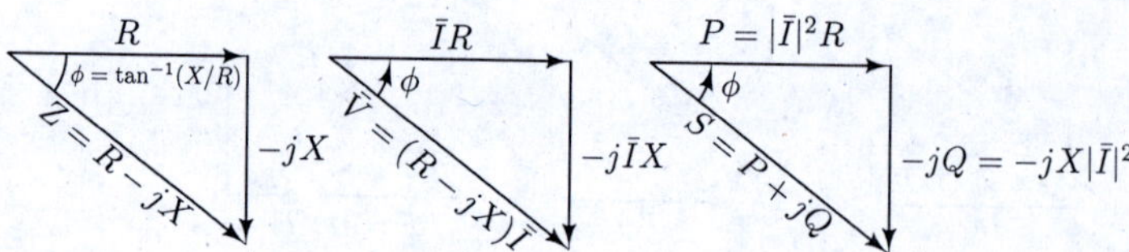

Figure 4.5: The power triangle for a capacitive impedance. P leads S.

If we force a current of $\bar{I}$ through a capacitive impedance, $R - jX$, the voltage across the impedance is $(R + jX)\bar{I}$. Fig. 4.5 shows phasor diagrams for this case. The complex power, S, is $(R - jX)|\bar{I}|^2$. The real power, P is $R|\bar{I}|^2$, while the reactive power, Q, is $-X|\bar{I}|^2$. The reactive power in a capacitive load is negative. The current vector (in the direction of R) is ahead of the voltage vector, and the power factor is $\cos\phi$, leading. Note that the active power P leads the complex power S. **A capacitor always consumes negative power in var. Alternately, remember that a capacitor generates or delivers var; an inductor consumes var.**

Convention. [Capitalization in units] By now, you would have noticed that the abbreviations for the units are sometimes written in capital letters, and sometimes in small. Refer to [6] for the correct standard. The unit is in a capital letter if it is the name of a person. That is why 'V', 'A', 'H', 'F', 'Ω', 'W' are all capital letters. On the other hand, the unit of distance, meter, is written as 'm'. The unit for apparent power is written as 'V·A' because the 'V' and the 'A' are names of important people.

All units, when written in expanded form, are to be written in lower case. The unit for conductance, siemens, is written as 'S', while the unit for time, seconds, is written as 's'. Can you explain why decibels are abbreviated as dB?

Example 4.13. In each of the following cases, find the complex power. (1) Apparent power is 100 V·A, power factor is 0.8 lagging. (2) Reactive power is 100 var, apparent power is 125 V·A. (3) Active power is 100 W, power factor is 0.8 leading. (4) Reactive power is −100 var, power factor is 0.8.

1. The active power is $100 \times 0.8 = 80$ W. By Pythagoras' theorem,

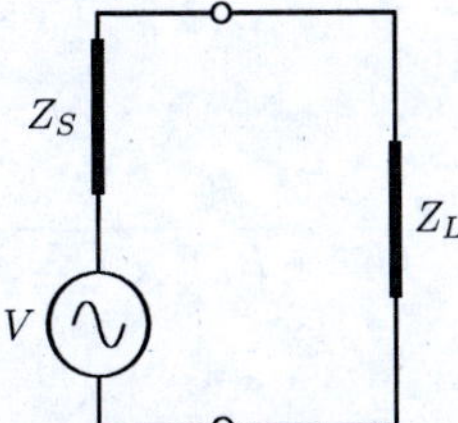

Figure 4.6: Source trying to deliver power to a load

the reactive power must be of magnitude 60 V·A. The power factor is lagging; the active power must be lagging behind the apparent power; the reactive power is positive; +60 var. Finally, $S = (80 + j60)$ V·A.

2. The reactive power is 100 var and the apparent power is 125 V·A. By Pythagoras' theorem, the active power must be 75 W. Finally, $S = (75 + j100)$ V·A.

3. The active power is 100 W and the power factor is 0.8 leading. Therefore the apparent power must be 125 V·A; the reactive power must be −75 var. Finally, $S = (100 - j75)$ V·A.

4. The reactive power is −100 var and the power factor is 0.8. If $\cos\phi$ is 0.8, then $\sin\phi$ is 0.6 and $\tan\phi$ is 0.75. Therefore the active power is 133.33 W. Finally, $S = (133.33 - j100)$ V·A.

4.6.3 Maximum power transfer theorem

Fig. 4.6 shows a voltage source of source impedance Z_S, trying to deliver power to a load, Z_L. **The basic premise of the rest of the discussion is that the source impedance is fixed.** No matter what, one is not allowed to change the source impedance.

Theorem 4.6 (Maximum power transfer with phasors): A voltage source with source impedance Z_S delivers maximum power to a load when the load impedance, Z_L, is Z_S^*.

Proof. The current through the circuit is $V/(Z_S + Z_L)$. The voltage across Z_L is $VZ_L/(Z_S + Z_L)$. Therefore the power delivered to Z_L is:

$$S_L = V\frac{Z_L}{Z_S + Z_L} \cdot \frac{V^*}{Z_S^* + Z_L^*} = |V|^2 \frac{Z_L}{|Z_S + Z_L|^2}$$

If Z_S is $R_S + jX_S$, and if Z_L is $R_L + jX_L$, the power delivered to the load Z_L will become:

$$S_L = |V|^2 \frac{R_L + jX_L}{(R_S + R_L)^2 + (X_S + X_L)^2}$$

The real (active) part of this power is:

$$P_L = \mathrm{Re}(S_L) = |V|^2 \frac{R_L}{(R_S + R_L)^2 + (X_S + X_L)^2} \tag{4.8}$$

If we need maximum power delivered to the Z_L, the derivatives of (4.8) with respect to R_L and X_L will have to be 0. The derivative with respect to R_L being 0 implies:

$$\begin{aligned} \frac{d}{dR_L}\left(\frac{R_L}{(R_S + R_L)^2 + (X_S + X_L)^2}\right) &= 0 \\ \Rightarrow\ 2R_L(R_S + R_L) - (R_S + R_L)^2 &= 0 \\ \Rightarrow\ R_L &= R_S \end{aligned} \tag{4.9}$$

Further, the derivative with respect to X_L being 0 implies:

$$\begin{aligned} \frac{d}{dX_L}\left(\frac{R_L}{(R_S + R_L)^2 + (X_S + X_L)^2}\right) &= 0 \\ \Rightarrow\ 2R_L(X_L + X_S) &= 0 \\ \Rightarrow\ X_L &= -X_S \end{aligned} \tag{4.10}$$

(4.9) and (4.10) indicate that $Z_L = R_S - jX_S$, or in other words, Z_L has to be Z_S^*. ■

Further, the maximum power delivered to the load is:

$$P_{max} = \frac{|V|^2}{4R_S} \tag{4.11}$$

where R_S is the real part of both Z_S and $Z_L = Z_S^*$ for maximum power transfer.

4.6.4 Reactive power

The instantaneous power in a branch is $v(t)i(t)$, where $v(t)$, $i(t)$ are the branch voltage and current, respectively. If $i(t) = \sqrt{2}I_0 \cos \omega t$, and $v(t) =$

$\sqrt{2}V_0\cos(\omega t+\phi)$, then the instantaneous power is $p(t)$.

$$\begin{aligned} p(t) &= 2V_0I_0\cos\omega t\cos(\omega t+\phi) && (4.12)\\ &= V_0I_0\cdot(\cos\phi+\cos(2\omega t+\phi)) \\ &= V_0I_0\cdot(\cos\phi+\cos\phi\cos(2\omega t)-\sin\phi\sin(2\omega t)) \\ &= V_0I_0\cos\phi\cdot(1+\cos(2\omega t))-V_0I_0\sin\phi\cdot\sin(2\omega t) && (4.13) \end{aligned}$$

(4.13) shows that the instantaneous power is effectively the sum of two quantities. The first term is the active power, always greater than 0, and of average value $V_0I_0\cos\phi$. The second term, the reactive power, has a zero average and an amplitude of $V_0I_0\sin\phi$.

The impedance relates the current and voltage phasors. The phasor for the current is I_0. The phasor for the voltage is $V_0\angle\phi = V_0\cos\phi + jV_0\sin\phi$. If the impedance is $R+jX$, I_0R is $V_0\cos\phi$, and I_0X is $V_0\sin\phi$. (4.13) can now be rewritten as:

$$p(t) = \underbrace{I_0^2R\cdot(1+\cos(2\omega t))}_{\text{Resistive power}} - \underbrace{I_0^2X\cdot\sin(2\omega t)}_{\text{Reactive power}} \qquad (4.14)$$

The first term in (4.14) is proportional to R and is the instantaneous power delivered to the resistor. The second term in (4.14) is proportional to X and is the instantaneous reactive power. The circuit stores energy in the reactive elements and dissipates energy in the resistive elements. The rate of change of the stored energy is the reactive power.

Example 4.14. In the circuit shown below, find the impedance of the capacitor such that maximum power is transferred from the source to the load. Evaluate the power absorbed by the load and its power factor.

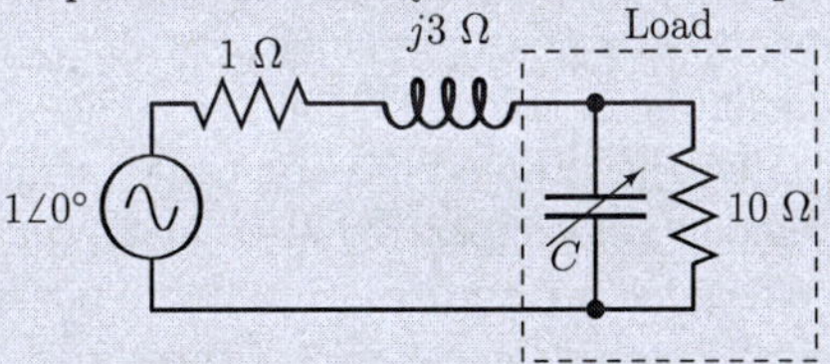

Looking back from the load, the Thévenin equivalent impedance is $(1+j3)$ Ω. Therefore, maximum power will be absorbed by the load when its impedance is $(1-j3)$ Ω. If the impedance of the capacitor is $-jX$, then $-jX \parallel 10 = (1-j3)$. If we work with admittance, we obtain $j/X+0.1 = 1/(1-j3) = 0.1+j0.3$. This gives us X as 0.3 Ω. The impedance of the capacitor should be $-j0.3$ Ω.

When the impedance of the capacitor is $-j0.3\ \Omega$, the impedance of the load is $(1 - j3)\ \Omega$. The total impedance in the circuit is $2\ \Omega$; the current provided by the source is 0.5 A. The voltage across the load is therefore $(0.5 - j1.5)$ V. The power absorbed by the load is $(0.25 - j0.75)$ VA. The active power absorbed by the load is 0.25 W, the reactive power absorbed by the load is -0.75 var.

If the angle of the complex power is ϕ, the power factor is $\cos\phi$. In our case, the angle of the complex power is $\tan^{-1}(-3)$. Therefore, the power factor is $1/\sqrt{10}$, leading (since the load is capacitive).

The apparent power is essential to power engineers because it relates to the current magnitude that the transmission cable needs to support. The transmission cable can support a larger amplitude of current only if it has a larger diameter. The apparent power directly relates to the diameter of the transmission cable, and therefore, to the cost of the (copper in the) transmission cable.

The reactive power is also significant to power engineers. Reactive power requirement implies extra current that the transmission cable needs to support, even when the current is not used as active power. Reactive power can easily be compensated for in the load by adding opposite storage elements, e.g., we can compensate an inductive load with extra capacitance and a capacitive load with extra inductance. Compensation can reduce the apparent power and lead to a saving in copper.

4.7 Three phase circuits

Three-phase (in general, poly-phase) circuits were independently invented by several engineers, including Nikola Tesla, in the 1880s. Three phase power generation is simple. Further, three-phase power transmission is economical as it reduces the amount of copper required in power transmission. All modern power systems generate and transmit three-phase power.

We can model a three-phase generator as three voltage sources, namely $\bar{V}_r = V_0\angle 0°$, $\bar{V}_y = V_0\angle 120°$, and $\bar{V}_b = V_0\angle 240°$. All voltages are with respect to a common neutral terminal. An example of a three-phase supply (and three-phase load) is in Example 4.10. The three phases of a three-phase system are color-coded as red, yellow and blue, for convenience. The load to a three-phase supply may be connected in two configurations, typically known as star and delta.

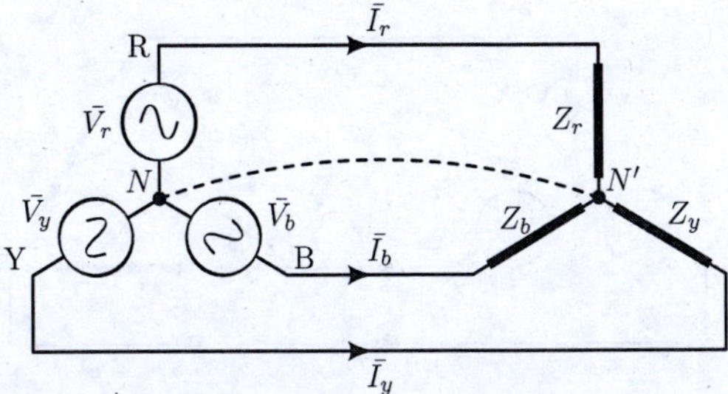

Figure 4.7: General three-phase configuration with a star-connected load.

4.7.1 Star-connected load

In a star-connected load, three loads are connected between a common terminal (known as neutral) and the three phases, $\bar{V}_r$, $\bar{V}_y$, and $\bar{V}_b$. Further, there may be a wire between the common terminal (neutral) of the three loads and the common terminal (neutral) of the three sources. Fig. 4.7 shows the general configuration. The wire NN' is the neutral connection.

If the three loads are Z_r, Z_y, and Z_b, respectively, then the currents in these three loads are $\bar{I}_r = \bar{V}_r/Z_r$, $\bar{I}_y = \bar{V}_y/Z_y$, and $\bar{I}_b = \bar{V}_b/Z_b$, respectively. The current through the neutral wire ($N'N$ of Fig. 4.7) is $\bar{I}_r + \bar{I}_y + \bar{I}_b$. If the three loads are equal (balanced, $Z_r = Z_y = Z_b = Z$), the current through the neutral is:

$$\bar{I}_n = \bar{V}_r/Z + \bar{V}_y/Z + \bar{V}_b/Z = V_0/Z(1\angle 0^\circ + 1\angle 120^\circ + 1\angle 240^\circ) = 0$$

The zero current through the neutral wire is the main attraction of the three-phase power delivery system. As long as the loads in the three phases are balanced, the current through the neutral wire will be zero. With slight imbalances in the three loads the neutral current will no longer be zero but much smaller than the currents in the three loads. As such, the copper requirements in the neutral wire are substantially reduced. A thin wire with low current carrying capacity can be used for the neutral wire. For the same load current requirements, a single-phase system will require roughly 2× the amount of copper. In a typical power delivery system, we connect the neutral terminals to the ground or Earth at both the load and the source. Current through the Earth will indicate an imbalance in the loads.

4.7.2 Delta-connected load

In a delta-connected load, three load impedances, Z_{ry}, Z_{yb}, and Z_{br}, are connected across $\bar{V}_{ry}$, $\bar{V}_{yb}$, and $\bar{V}_{br}$, respectively. In this case, there is no neutral

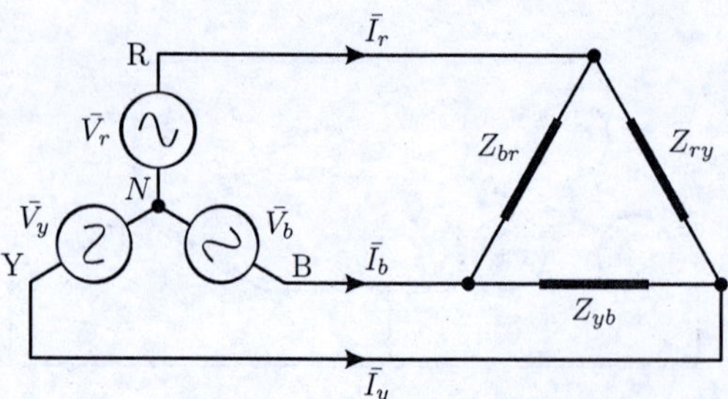

Figure 4.8: General three-phase configuration with a delta-connected load.

wire. However, the three loads are required to be floating, not referenced, not related to ground / Earth. Let us assume $\bar{V}_r$ is the reference, $\bar{V}_y$ is $120°$ ahead of $\bar{V}_r$, and $\bar{V}_b$ is $120°$ behind $\bar{V}_r$. The voltages across the three loads are $\bar{V}_{ry} = \bar{V}_r - \bar{V}_y = \sqrt{3}V_0\angle -30°$, $\bar{V}_{yb} = \bar{V}_y - \bar{V}_b = \sqrt{3}V_0\angle 90°$, and $\bar{V}_{br} = \bar{V}_b - \bar{V}_r = \sqrt{3}V_0\angle 210°$.

The rms values of the voltages across the three loads are $\sqrt{3}$ times the rms values of the voltages with respect to ground.

Fig. 4.8 shows a general configuration for a delta-connected load. The currents $\bar{I}_r$, $\bar{I}_y$, and $\bar{I}_b$, can easily be computed in the following manner.

$$\begin{aligned}
\bar{I}_r &= \bar{V}_{ry}/Z_{ry} - \bar{V}_{br}/Z_{br} \\
\bar{I}_y &= \bar{V}_{yb}/Z_{yb} - \bar{V}_{ry}/Z_{ry} \\
\bar{I}_b &= \bar{V}_{br}/Z_{br} - \bar{V}_{yb}/Z_{yb}
\end{aligned}$$

If the three impedances are equal, i.e., if the load is balanced, the three line currents, $\bar{I}_r$, $\bar{I}_y$, and $\bar{I}_b$ are $3V_0/Z$, $3V_0/Z\angle 120°$, and $3V_0/Z\angle -120°$, respectively.

The rms values of the three line currents are $\sqrt{3}$ times the rms values of the currents in the three loads.

4.7.3 Line and phase

A three-phase power system delivers power over three wires (lines) and possibly a neutral. The load has three phases and may be star-connected or delta-connected.

Definition 4.13 (Line current): The rms current in each line of the three-phase system is the line-current.

Definition 4.14 (Phase current): The rms current in each phase of the load is the phase current.

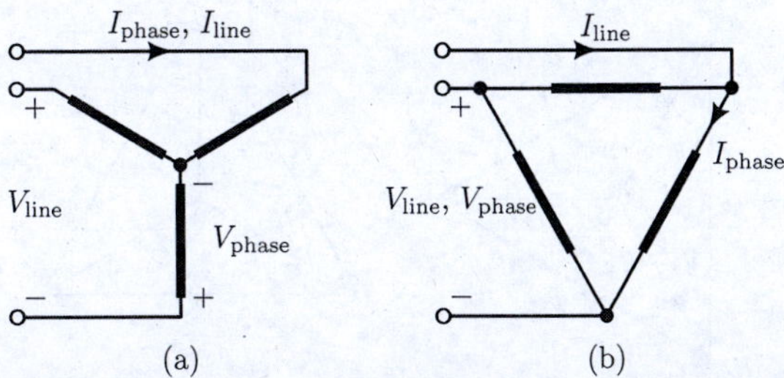

Figure 4.9: Line and phase voltages and currents. (a) The line current and phase current are the same in a star-connected load. (b) The line voltage and phase voltage are the same in a delta-connected load.

Definition 4.15 (Line voltage): The rms voltage between a pair of lines is the line voltage.

Definition 4.16 (Phase voltage): The rms voltage across each phase of the load is the phase voltage.

The definitions are depicted in the schematics of Fig. 4.9. In case of a star-connected load, the line and phase currents are the same, while the line voltage is $\sqrt{3}$ times the phase voltage. In case of a balanced delta-connected load, the line and phase voltages are the same, while the line current is $\sqrt{3}$ times the phase current.

4.7.4 Balanced loads

The load is balanced when the impedances in the three arms of the three-phase system are equal. Balanced loads are straightforward to work with, regardless of whether they are star or delta-connected.

Corollary 4.4. In a balanced three-phase star-connected load, the value of each impedance is Z. An equivalent delta-connected load will have an impedance of $3Z$ in each arm.

Corollary 4.5. In a balanced three-phase delta-connected load, the value of each impedance is Z. An equivalent star-connected load will have an impedance of $Z/3$ in each arm.

Proof. The two corollaries (4.4 and 4.5) are converses of each other; both are proved together.

Let us consider the three phase-voltages of a star-connected load, and call them $\bar{V}_r = V_0\angle 0°$, $\bar{V}_y = V_0\angle 120°$ and $\bar{V}_b = V_0\angle 240°$. Then, the three

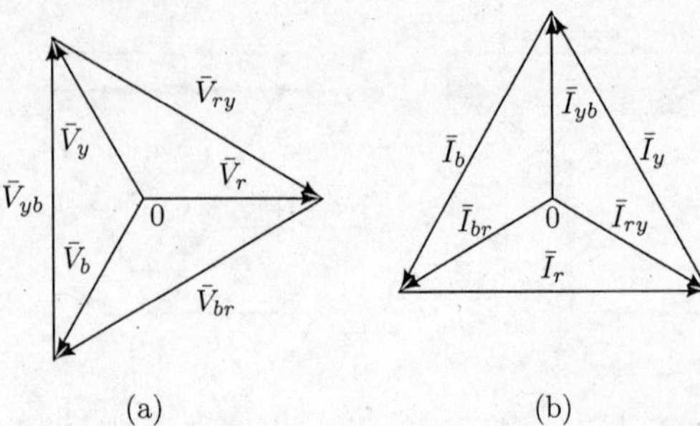

Figure 4.10: (a) Phasor diagram for the line and phase voltages in a star-connected three-phase system. (b) Phasor diagram for the line and phase currents in a delta-connected three-phase system.

line voltages are $\bar{V}_{ry} = \bar{V}_r - \bar{V}_y = \sqrt{3}V_0\angle-30°$, $\bar{V}_{yb} = \sqrt{3}V_0\angle 90°$, $\bar{V}_{br} = \sqrt{3}V_0\angle 210°$. The phase and line voltages can be quickly deduced from the phasor diagram in Fig. 4.10(a).

The line currents are $\bar{I}_r = V_0/Z$, $\bar{I}_y = (V_0/Z)\angle 120°$, $\bar{I}_b = (V_0/Z)\angle 240°$.

If we had a delta-connected load with Z' in each arm, the three line voltages would remain the same as before. The phase currents would now be $\bar{I}_{ry} = \bar{V}_{ry}/Z' = (\sqrt{3}V_0/Z')\angle-30°$, $\bar{I}_{yb} = (\sqrt{3}V_0/Z')\angle 90°$, and $\bar{I}_{br} = (\sqrt{3}V_0/Z')\angle 210°$.

The overall line currents for the delta-connected load would be $\bar{I}_r = \bar{I}_{ry} - \bar{I}_{br} = (3V_0/Z')\angle 0°$, $\bar{I}_y = (3V_0/Z')\angle 120°$, $\bar{I}_b = (3V_0/Z')\angle 240°$. This can be quickly deduced from the phasor diagram in Fig. 4.10(b).

Overall, it can be seen that if $Z' = 3Z$, the line currents in case of the star-connected and delta-connected loads are equivalent. ■

In summary, we can replace a balanced star-connected load of Z in each arm with a balanced delta-connected load with $3Z$ in each arm. How do we remember? The star connection is almost like a series connection; so the net impedance in a star connection increases. A delta connection is almost like a parallel connection; so the net impedance in a delta connection decreases. We use this hint to remember that if a star load is to be replaced with a delta, we need to replace with a larger impedance in each arm.

Example 4.15. In the following schematic, the three delta-connected impedances are equal, the three star-connected impedances are equal. The line voltage is $220\sqrt{3}$ V rms. What are the line currents? What is the overall power factor of the circuit?

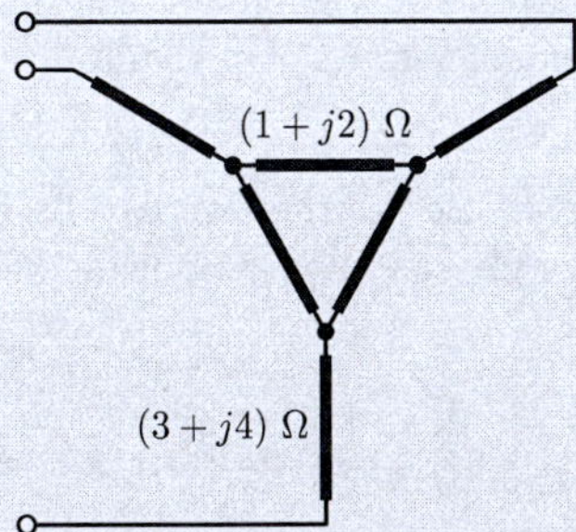

The delta-connected impedances can be modeled as impedances of value $(1/3 + j2/3)\ \Omega$, connected in star. Now in each phase there are two impedances in series forming a star. The two impedances combine as: $(3+j4)+(1/3+j2/3) = (10/3+j14/3)\ \Omega$. For each phase, the phase voltage is 220 V. Therefore the line current for the R-phase is $(22.3 - j31.2)$ A, or $(38.4\angle -54°)$ A.

The three line currents are $(38.4\angle -54°)$ A, $(38.4\angle 66°)$ A, $(38.4\angle 186°)$ A.

The power factor is $\cos(-54°)$ or 0.58 lagging.

4.7.5 Unbalanced three-phase networks

Three-phase networks which are not balanced can be analyzed using the regular node-voltage and mesh methods. Star-delta transformations to simplify the analyses can be used; such transformations are introduced in section 7.4. Fortescue's theorem is often used to transform an unbalanced three-phase system into a set of three balanced three-phase systems.

Scan the QR-code, and learn more about Fortescue's theorem, how to convert an unbalanced three-phase network into a set of three balanced three-phase networks with a positive, negative and zero sequence.

4.8 Unit summary

- A circuit may settle to its sinusoidal steady state if all the independent sources exciting the circuit operate at one frequency, ω.

- All voltages and currents in the circuit will be sinusoidal at the same frequency as the excitation, ω. Only amplitudes and phases will be different.

- The amplitude and phase information can be expressed as a complex number, known as a phasor. The magnitude of the complex number is the amplitude $/\sqrt{2}$.

- A circuit with R, L, C, M, and other linear controlled sources is a linear time-invariant (LTI) system.

- An inductor of value L can be replaced by an impedance of value $j\omega L$. A capacitor of value C can be replaced by an admittance of value $j\omega C$. Impedance is the complex form of resistance; admittance is the complex form of conductance.

- Impedance = resistance $+j\cdot$reactance. ($Z = R + jX$). Admittance = conductance $+j\cdot$susceptance. ($Y = G + jB$).

- First, we convert all circuit components to their impedance forms. Then, we can analyze the circuit with complex numbers.

- Power in any component is $\bar{V}\bar{I}^*$, where $\bar{V}$ and $\bar{I}$ are the voltage and current phasors for the component.

- $\bar{V}\bar{I}^* = S = P+jQ$. The real part of the complex power, S, is the active power, P; the imaginary part, Q, is the reactive power; the magnitude, $|S|$, is the apparent power. The total S consumed in a network is zero; individually, the total P and the total Q consumed in a network are zero (by Tellegen's theorem).

- Capacitive loads have a leading power factor, inductive loads have a lagging power factor. Capacitive loads consume negative reactive power. Inductive loads consume positive reactive power. A capacitor or an inductor does not consume any active power.

- For maximum power transfer, $Z_L = Z_S^*$.

- Three-phase power delivery saves the number of wires used (the total amount of copper used) and is, therefore, more efficient. The neutral conductor does not need to carry current in a balanced three-phase system because $I_0\angle 0 + I_0\angle 120° + I_0\angle -120° = 0$.

- A balanced star-connected load can be transformed into a balanced delta-connected load by multiplying each impedance by 3.

4.9 Exercises

Multiple choice type questions

4.1 What is the phase between $v_1(t)$ and $v_2(t)$, where $v_1(t) = 2\cos(\omega t - 40^\circ)$ and $v_2(t) = 3\sin(\omega t + 20^\circ)$?
(a) 20° (b) 30° (c) 60° (d) 110°

4.2 What is the phase between $v_1(t)$ and $v_2(t)$, where $v_1(t) = -2\cos(\omega t - 40^\circ)$ and $v_2(t) = 3\sin(\omega t + 20^\circ)$?
(a) 70° (b) 120° (c) 150° (d) 160°

4.3 What is the phase between $v_1(t)$ and $v_2(t)$, where $v_1(t) = 2\cos(\omega t - 40^\circ)$ and $v_2(t) = -3\sin(\omega t + 20^\circ)$?
(a) 70° (b) 120° (c) 150° (d) 160°

4.4 The phasor corresponding to $v(t) = 14.14\sin(100\pi t - 50^\circ)$ is:
(a) $10\angle -140^\circ$ (b) $10\angle 140^\circ$ (c) $10\angle -50^\circ$ (d) $10\angle 50^\circ$

4.5 What is the least possible effective inductance of two coupled inductors, L_1 and L_2, that have been placed in series?
(a) $L_1 + L_2$ (b) $L_1 - L_2$ (c) $(\sqrt{L_1} + \sqrt{L_2})^2$ (d) $(\sqrt{L_1} - \sqrt{L_2})^2$

4.6 What is the maximum possible effective inductance of two coupled inductors, L_1 and L_2, that have been placed in series?
(a) $L_1 + L_2$ (b) $L_1 - L_2$ (c) $(\sqrt{L_1} + \sqrt{L_2})^2$ (d) $(\sqrt{L_1} - \sqrt{L_2})^2$

4.7 What is the least possible effective inductance of two coupled inductors, L_1 and L_2, that have been placed in parallel?
(a) 0 (b) $\frac{L_1 L_2}{L_1 + L_2}$ (c) Smaller of L_1, L_2 (d) Larger of L_1, L_2

4.8 What is the maximum possible effective inductance of two coupled inductors, L_1 and L_2, that have been placed in parallel?
(a) $\frac{L_1 L_2}{L_1 - L_2}$ (b) $\frac{L_1 L_2}{L_1 + L_2}$ (c) Smaller of L_1, L_2 (d) Larger of L_1, L_2

4.9 A load has a power factor of 0.8 and draws 100 W. What is the apparent power?
(a) 60 V·A (b) 80 V·A (c) 112.5 V·A (d) 125 V·A

4.10 The current through and voltage across a load are $2\cos(100\pi t + 30^\circ)$ A and $-110\sin(100\pi t - 40^\circ)$ V respectively. What is the power dissipated by the load?
(a) 37 W (b) 103 W (c) 110 W (d) 206 W

4.11 The reactive power in a load is 80 var and the power factor is 0.6. What is the active power?
(a) 60 W (b) 80 W (c) 100 W (d) 120 W

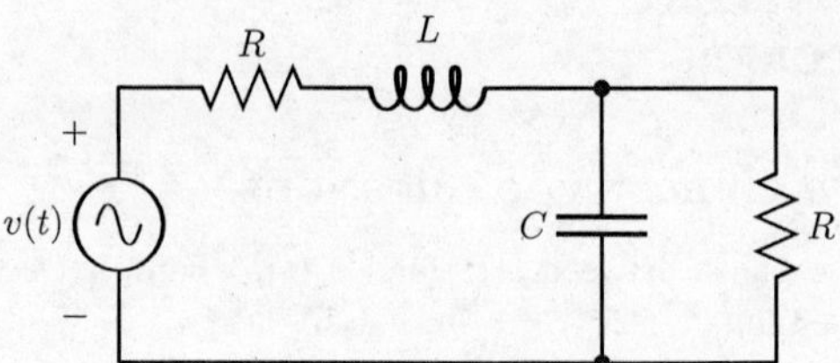

Figure 4.11: Circuit schematic for exercise 4.18.

4.12 The reactive power in a load is −50 var and the apparent power is 130 V·A. What is the active power?
(a) 80 W (b) 100 W (c) 120 W (d) 130 W

4.13 The power factor of a load is 0.8 lagging. If the resistive part of the load impedance is 80 Ω, what is the reactive part of the load?
(a) $-j60\ \Omega$ (b) $j60\ \Omega$ (c) $j64\ \Omega$ (d) $j100\ \Omega$

Short answer type questions

4.14 Model the mutual coupling of two inductors with the help of linear current-controlled voltage sources.

4.15 The unit for apparent power is ______, the unit for real power is ______, and the unit for reactive power is ______.

4.16 An inductive load will have a ______ power factor, while a capacitive load will have a ______ power factor. A purely resistive load will have a power factor of ______.

4.17 A series RLC network is excited by a sinusoidal source. The circuit is in the steady state. What is the frequency of the source such that the reactive power delivered by the source is zero?

Numericals

4.18 In the circuit of Fig. 4.11, $v(t)$ is $21\cos(100\pi t)$. R is 7 Ω, C is 690 μF, L is 14 mH. What is the phasor for the voltage across C? What is the active power generated by the input voltage source? What is the reactive power generated by the input voltage source?

4.19 In the circuit of Fig. 4.12, L_1 and L_2 are 2 mH and 3 mH, respectively. R_1 and R_2 are 4 Ω and 2 Ω, respectively. The input source, V_{in}, is a voltage

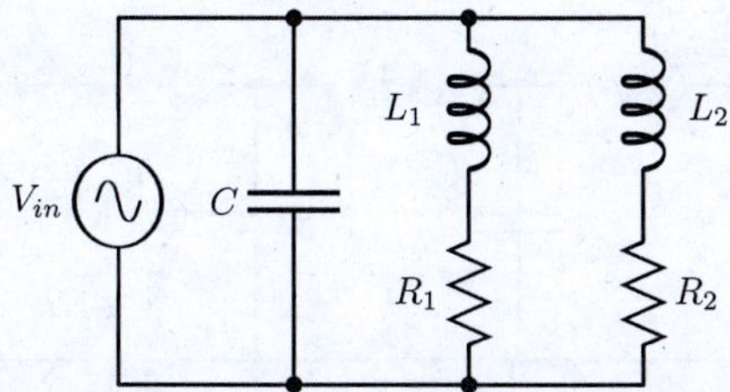

Figure 4.12: Circuit schematic for exercise 4.19.

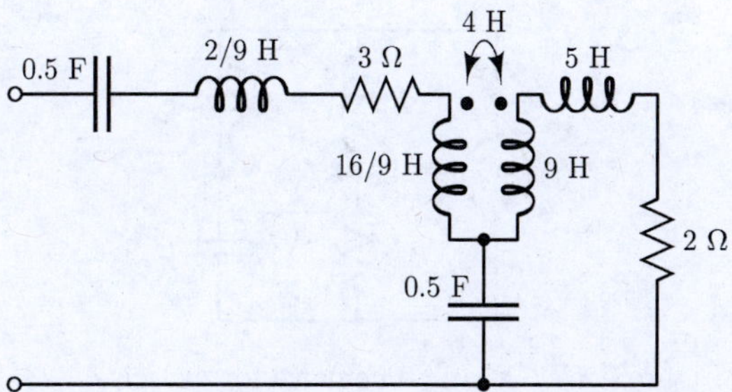

Figure 4.13: Circuit schematic for exercise 4.21.

source operating as a sinusoid at 100π rad/s. What should be C, such that the power delivered by V_{in} is purely active?

4.20 In the same circuit as in Fig. 4.12, L_1 and L_2 are mutually coupled, with a coupling coefficient of 0.82, the dots (not marked) are at the top of both inductors. As before, L_1 and L_2 are 2 mH and 3 mH; R_1, R_2 are 4 Ω and 2 Ω. The input operates at 100π rad/s. What is C such that the power delivered by V_{in} is purely active?

4.21 Find the impedance looking into the circuit of Fig. 4.13 at an angular frequency of 1 rad/s.

4.22 In the circuit of Fig. 4.14, $v(t) = \cos(\omega t)$ V. The values of L, C, and R are 0.75 H, 41.67 mF, and 3 Ω, respectively. What is the angular frequency, ω, such that $v_{out}(t)$ has an amplitude of $1/(2\sqrt{2})$ volts?

4.23 In the circuit of Fig. 4.15, R is 1 Ω, L is 3 H, C is 2 F. The input voltage $v(t) = 100\cos t$. If R_L is 1 Ω, what is the expression for $i(t)$?

4.24 In the circuit of Fig. 4.16, R is 6 Ω, X is 6 Ω. The voltage source is a phasor of $10\angle 0°$ V, while the current source is a phasor of $1 + j$ A. Find the phasor corresponding to I.

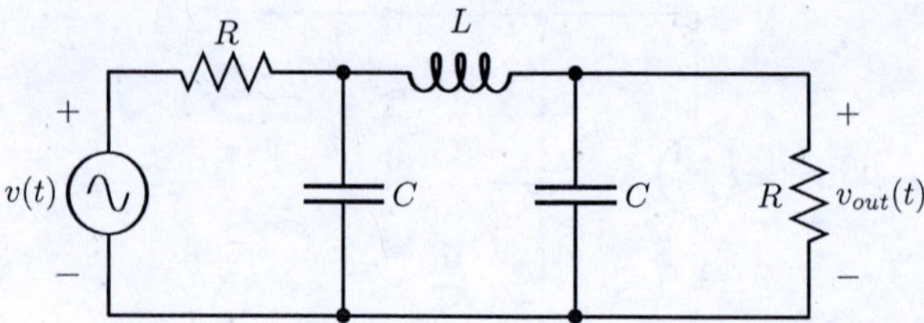

Figure 4.14: Circuit schematic for exercise 4.22.

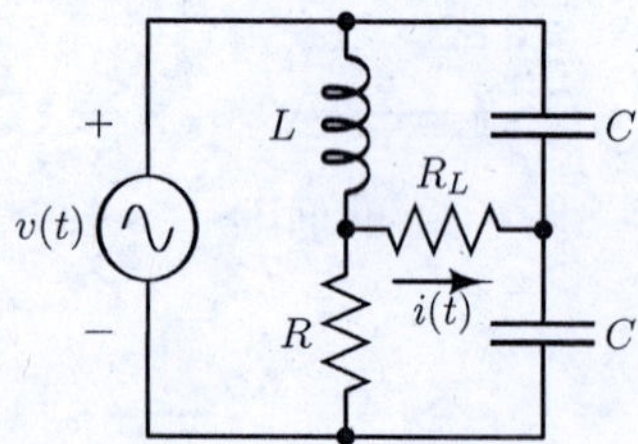

Figure 4.15: Circuit schematic for exercise 4.23.

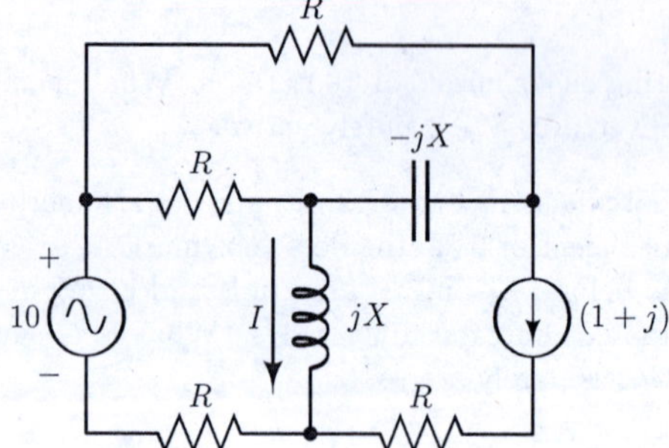

Figure 4.16: Circuit schematic for exercise 4.24.

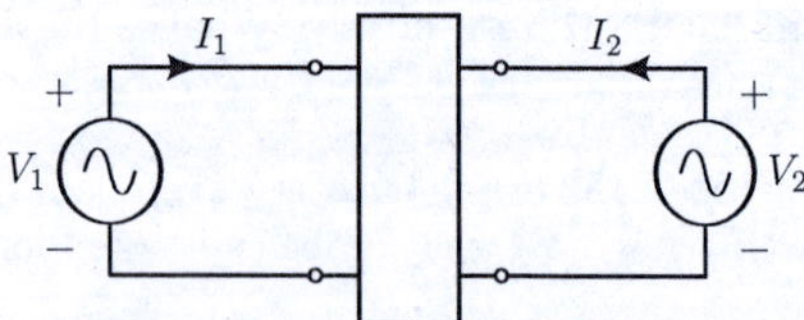

Figure 4.17: Circuit schematic for exercise 4.25.

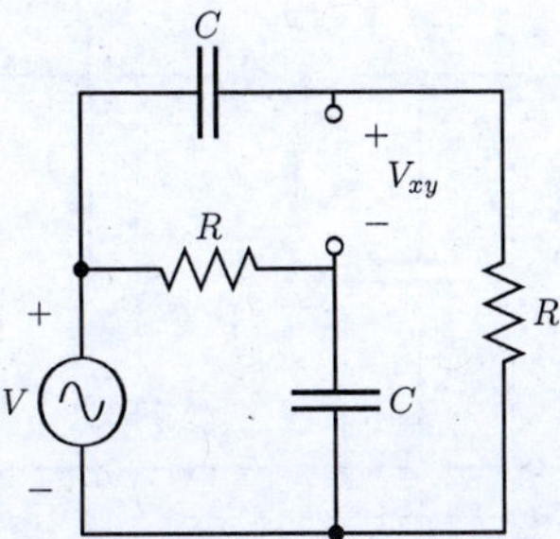

Figure 4.18: Circuit schematic for exercise 4.26.

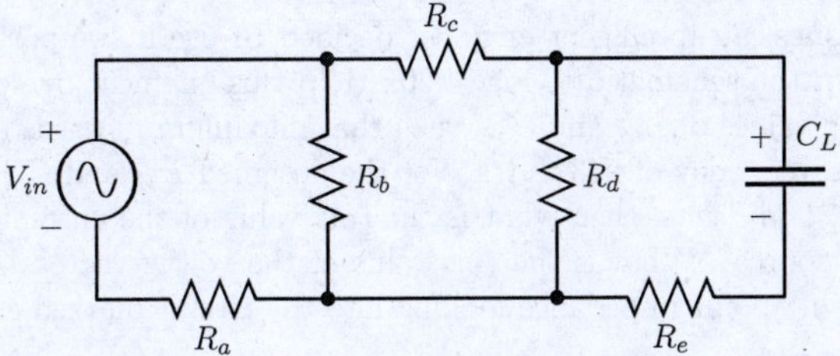

Figure 4.19: Circuit schematic for exercise 4.27.

4.25 In the circuit of Fig. 4.17, the two-port network shown is a linear circuit. When V_1 of $10\angle 0$ is applied, and V_2 is 0, the measured I_1 and I_2 are $2j+1$ A and $3j+4$ A, respectively. When V_2 of $10\angle 0$ is applied, and V_1 is 0, the measured I_1 and I_2 are $3j+1$ A and $3j+1$ A, respectively. Find the power (active, reactive) absorbed by the circuit when V_1 is $10\angle 0$ and V_2 is $20\angle 90°$.

4.26 In the schematic of Fig. 4.18, find the phasor V_{xy}. Solve the problem as a function of ω, i.e., the angular frequency of the input source.

4.27 In the schematic of Fig. 4.19, R_a, R_b, R_c, R_d, R_e, and C_L are 9 Ω, 7 Ω, 1 Ω, 8 Ω, 6 Ω and 5 mF respectively. Let us assume V_{in} is a phasor of $100\angle 0$ V, operating at an angular frequency of 100π rad/s. Evaluate the phasor corresponding to the voltage across C_L. Compute the reactive power generated by the voltage source. Compute the imaginary part of the current coming out of the plus terminal of V_{in}.

4.28 A pair of headphones has an input impedance that behaves like a resistor R_L in series with an inductor L_L. R_L and L_L are 13 Ω and 5 mH respectively. A reputed company designed a power amplifier with a source resistance, R_S, of 50 Ω. To transfer maximum power from the power amplifier to the

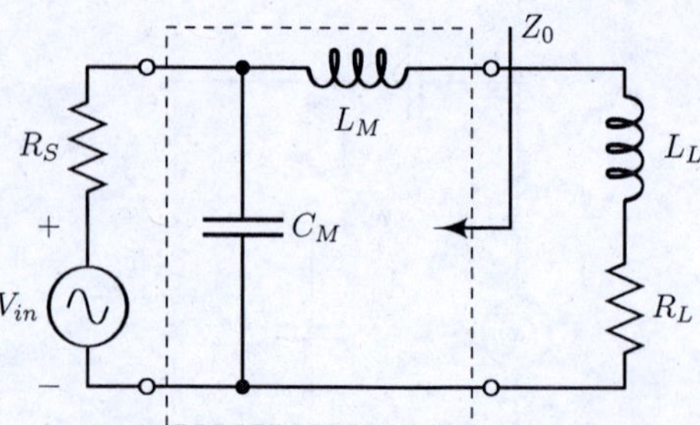

Figure 4.20: Circuit schematic for exercise 4.28.

headphones, an intelligent engineer decided to use a two-port network, as shown in the schematic of Fig. 4.20. Help this engineer by computing the requisite values of L_M and C_M, such that maximum power is transferred to R_L at a frequency of 953 rad/s. For the designed L_M and C_M, if the input V_{in} is 821 mV rms, then what is the rms value of the current through the voltage source? What is the rms value of the voltage across R_L? What is the driving point impedance looking into the circuit marked as Z_0?

4.29 A radio receiver antenna is modeled as a voltage source in series with 50 Ω. A tuning circuit is used to tune to the right frequency. The tuning circuit is an LC-shunt network with a variable capacitance. The circuit is further connected to an amplifier for further reception and digitization. The model of the amplifier is a load resistor of 50 Ω. The inductor in the LC-shunt network is 1 μH. What is the optimal capacitance in the LC-shunt network, such that the receiver is able to receive all FM signals from 80 MHz to 110 MHz? For this value of the capacitor, what is the voltage across the load at 80 MHz, at 95 MHz, and at 110 MHz? Is there an improvement if we optimize the value of the capacitor at 93.8 MHz?

4.30 A linear and passive three-port network has ports $\mathcal{P}_1$, $\mathcal{P}_2$, and $\mathcal{P}_3$. When $1\angle 0$ is applied at $\mathcal{P}_1$, the short-circuit currents measured at $\mathcal{P}_2$ and $\mathcal{P}_3$ are 3 A and 2.4 A, respectively. What will be the short-circuit current at $\mathcal{P}_1$, when $2.9\angle 60°$ and $4.9\angle 140°$ are applied at $\mathcal{P}_2$ and $\mathcal{P}_3$ at the same time?

4.31 A three-phase 50-Hz voltage source has $V_{an} = 230\angle 0°$ volts, $V_{bn} = 230\angle -120°$ volts, and $V_{cn} = 230\angle 120°$ volts. The voltage source feeds an inductive load with $Z = 5\angle 75°$ ohm, as shown in the schematic in Fig. 4.21. All voltages are rms quantities. A capacitor bank is installed to improve the power factor to 1.00, as seen by the source. Assume that the potential at n is the same as at n'. What is the three-phase apparent power (active plus reactive) absorbed by the inductive load Z? What is the value of C? What is the three-phase apparent power absorbed by the capacitor bank? What

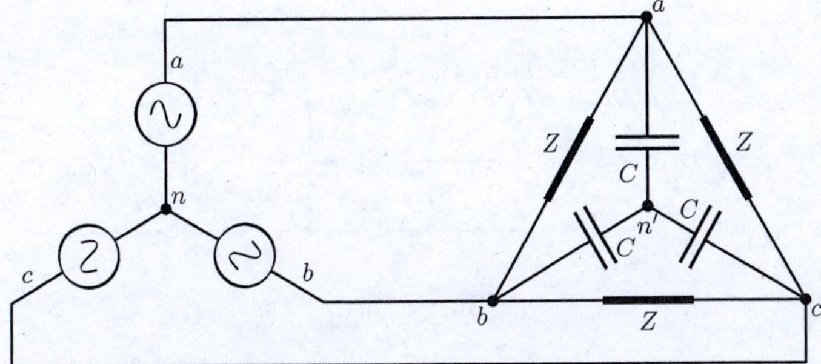

Figure 4.21: Three-phase circuit schematic for exercise 4.31.

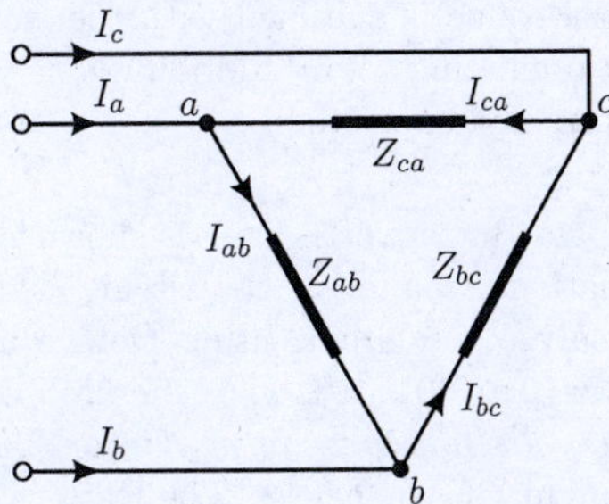

Figure 4.22: Three-phase schematic for the load, exercise 4.32.

is the three-phase apparent power supplied by the source?

4.32 A three-phase voltage source supplies current to a delta-connected load as shown in the circuit in Fig. 4.22. The line voltages are $V_a = 127\angle-30°$ V, $V_b = 127\angle-150°$ V, $V_c = 127\angle 90°$ V. The load impedances are $Z_{ab} = 10\Omega$, $Z_{bc} = R + jX$, $Z_{ca} = R - jX$, where $X \geq 0$. The rms values of the phase currents I_{ab}, I_{bc}, and I_{ca} are equal, and the total power dissipated is 4.84 kW. Calculate the phase voltages V_{ab}, V_{bc}, and V_{ca}. Calculate R and X. Calculate the phase currents I_{bc} and I_{ca}. Calculate the line currents I_a, I_b, and I_c.

Project activity

4.33 Based on your experience with the project activity in exercise 2.22, develop an Octave program that will analyze an AC circuit. The SPICE netlist [5] uses the following statement form to specify an AC source: `VXXX 1 0 ac 1`. The statement starts with a `V` to indicate a voltage source, then come the

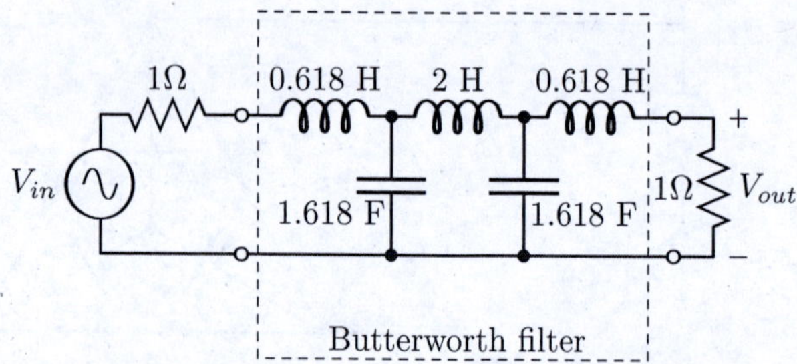

Figure 4.23: Schematic for project exercise 4.34.

connections, and the value is specified with the word `ac`, to indicate that it is a sinusoidally oscillating source. Modify your work in exercise 2.22 to incorporate inductors and capacitors.

4.34 A fifth order low-pass Butterworth filter is shown in Fig. 4.23. Set up the mesh or node equations to analyze the circuit. Apply V_{in} of $1\angle 0$ V and evaluate V_{out}. Analyze the circuit using Octave at different frequencies. *Hint: In Octave,* `w=[0:0.001:10]` *will generate an array for* `w`*, from 0 to 10, in steps of 0.001. Use the array form of* `w` *for your analysis.* Plot V_{out} as a function of the angular frequency, ω. The Butterworth filter was designed to have a 3-dB attenuation at 1 rad/s.

Know more

Historical profiles

The history of AC circuits and electrical technology is intriguing and exciting. On one side eminent scientists like Hertz, Faraday and Maxwell were experimenting and building the foundations of electromagnetism, and on the other side there were the genius minds of Edison and Tesla who were engineering the future.

Some of the prominent engineers who helped in the development of sinusoidal steady state analysis of electrical circuits were Oliver Heaviside [7, 8], Arthur Kennelly [9, 10], and Charles Steinmetz [11]. Oliver Heaviside was a recluse self-taught British physicist and electrical engineer who contributed substantially to the use of complex algebra to analyze electrical networks. Heaviside's attraction to electrical circuits came through his uncle, Wheatstone.

Oliver Heaviside[‡] Arthur Kennelly[‡] Charles Steinmetz[‡]

Kennelly, born to Irish parents in India (Mumbai), who later migrated to the United States, developed the concept of impedance. Steinmetz, a Polish engineer who later migrated to the United States, brought in the modern representation and conventions for phasors. Independently, Steinmetz, Kennelly and Heaviside contributed significantly to transmission lines, distributed networks, long distance telephony, radio, and the ionosphere.

A special mention must be made of Sir Jagadish Chandra Bose (1858-1937). Sir J.C. Bose was performing experiments at frequencies up to 60 GHz in 1897. Even though the circuit understanding and analysis techniques were not developed at that time, he could demonstrate radio transmitters and receivers at millimeter-wave frequencies. Sir J.C. Bose developed the world's first wireless communication link.[12], [13]. Read more about the life and times of Sir J.C. Bose in his modern biography [14].

Three-phase circuits were advocated first by the Westinghouse Corporation. George Westinghouse, the founder of the Westinghouse Corporation saw the promise in Nikola Tesla's proposal for three phase alternating current systems and took it up. The collaboration placed Westinghouse and Tesla in competition with Edison, as Edison was a strong proponent of direct current. Tesla, originally from Serbia, emigrated to the United States and first worked for Edison for a short time before starting on his own.

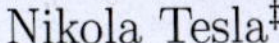

Nikola Tesla[‡] George Westinghouse[‡] Charles Fortescue[‡]

‡: Photographs were taken from the public domain.

Charles Fortescue, in 1918, while working for the Westinghouse Corporation, had proved [15] that an arbitrary set of three phasors can always be described as the sum of a positive, a negative, and a zero sequence. Fortescue's paper has been referred to as the single most important contribution in power engineering in the entire 20th century.

The first street to have electric lighting in India was Harrison road (now Mahatma Gandhi road) in Calcutta (Kolkata), in 1889. Calcutta Electric Supply Company (CESC) electrified Calcutta ten years after electric power was introduced in London. The first hydroelectric power station was the Sidrabong power station in Darjeeling, that generated AC power at 83.3 Hz, with a capacity of 130 kW. From roughly 1.3 GW of installed capacity in 1947 with electricity only reaching urban areas, India has roughly 400 GW of installed capacity as of March 2022, with less than 20 thousand households without access to electricity in 2019.[16]

Understand in depth

For more practice:

- Engineering Circuit Analysis, by William H. Hayt and Jack E. Kemmerley, McGraw-Hill, chapters 9, 10, 11 and most of 12.
- Schaum's Outline of Theory and Problems of Basic Circuit Analysis, by John O'Malley, McGraw-Hill, chapters 10, 11, 12, 13, 14, 15, 17, and relevant parts of 16.

Unit 5

Laplace transforms

Unit specifics

In this unit we have discussed the following:

- Laplace transforms, circuit analysis using Laplace transforms
- The convolution integral and the inverse Laplace transform
- Transformed network with initial conditions
- Transfer function representation, poles and zeros
- Frequency response, magnitude and phase plots
- Series and parallel resonances

Rationale

We have studied circuit analysis at DC, circuit analysis with differential equations, and circuit analysis at sinusoidal steady state. We have worked with differential equations only for first and second order networks, not for a general network.

In this unit we will study a mathematical technique, namely the Laplace transform, that aids us in solving differential equations rapidly. The techniques shown in this unit will enable the student to solve an arbitrarily complex linear time-invariant circuit (with L, C, R, M, and linear dependent sources) rapidly, for any general input. The Laplace transform is also used to directly solve a circuit with initial conditions.

The circuit may also be viewed as a system with an impulse response. In this unit we will learn about the system transfer function and appreciate the roles of poles and zeros of a system.

Pre-requisites

- Mathematics: Differential and integral calculus. Complex analysis and residue integration [17] is required for section 5.2.1.
- Circuit Theory: Thoroughness with Unit-3 of this book.
- Software: We will continue using Octave.

Unit outcomes

The list of outcomes of this unit are as follows.

U5-O1: Be able to convert a signal to its Laplace transform, and invert it back to time domain.

U5-O2: Be able to transform circuits with initial conditions into the Laplace domain.

U5-O3: Obtain the time domain response of circuits for any input.

U5-O4: Appreciate the concept of the transfer function.

Unit-5 outcomes	**Expected mapping with course outcomes** *(1: Weak, 2: medium, and 3: strong correlation)*					
	CO-1	**CO-2**	**CO-3**	**CO-4**	**CO-5**	**CO-6**
U5-O1	-	1	3	-	-	-
U5-O2	2	-	3	-	-	-
U5-O3	3	-	3	-	-	2
U5-O4	-	3	-	-	3	2

5.1 The Laplace transform

The Laplace transform is an essential tool in the analysis and synthesis of LTI circuits. In this section, we will learn about the Laplace transform.

Definition 5.1 (Laplace transform): The Laplace transform of a function $f(t)$ that exists for $t \in [0, \infty)$, $\mathcal{L}[f(t)]$, is defined as:

$$F(s) = \mathcal{L}[f(t)] = \int_{0-}^{\infty} f(t)e^{-st}dt \tag{5.1}$$

s can be a complex number, $s = \sigma + j\omega$, with σ as the real part of s and ω the imaginary part. [Note: We have specifically chosen s as the symbol for a complex number, just like in Unit-3.]

Definition 5.2 (RoC, the region of convergence): The Laplace transform integral is not convergent for all values of s. For a given function $f(t)$, the RoC is the set of all s such that $F(s) = \mathcal{L}[f(t)]$ is convergent.

If $f(t)$ is a rapidly increasing function, the real part of s, σ, has to be large enough to make the integral of (5.1) converge. The RoC of the function will accordingly be defined.

Example 5.1. What is the Laplace transform of $f(t) = 2e^{5t}$? What is the corresponding RoC?

$$\begin{aligned}
\mathcal{L}[f(t)] &= \int_{0-}^{\infty} (2e^{5t}e^{-st})\,dt = 2\int_{0-}^{\infty} e^{(5-s)t}\,dt \\
&= \frac{2}{5-s}\left[e^{(5-s)t}\right]_{0-}^{\infty} \\
&= \frac{2}{5-s}\Big(0-1\Big) \;\forall\, s \text{ such that } \mathrm{Re}[s] > 5 \\
&= \frac{2}{s-5} \;\forall\, s \text{ s.t. } \mathrm{Re}[s] > 5
\end{aligned}$$

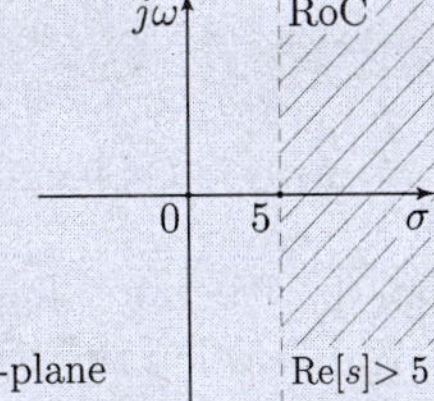

The RoC, Re[s] > 5, is indicated in the shaded region of the sketch above.

5.1.1 Convergence of the Laplace transform

The Laplace transform integral is absolutely convergent if the integral of the absolute value of the integrand does not blow up. That is, $\mathcal{L}[f(t)]$ exists if for some value of s,

$$\begin{aligned} \int_{0-}^{\infty} |f(t)e^{-st}|\,dt &< \infty \\ \text{or, } \int_{0-}^{\infty} |f(t)||e^{-\sigma t}|e^{-j\omega t}|\,dt &< \infty, \text{ where } s = \sigma + j\omega \\ \text{or, } \int_{0-}^{\infty} |f(t)|e^{-\sigma t}\,dt &< \infty, \because |e^{-j\omega t}| = 1 \end{aligned} \quad (5.2)$$

The Laplace transform of a signal exists as long as there is some real σ, for which the absolute integral in (5.2) does not blow up. Automatically, the Laplace transform will converge for all values of s that have a real part greater than the above σ.

The condition in (5.2) is a very powerful criterion. Almost all useful signals (e.g. polynomial, exponential) have a Laplace transform that converges. One can construct a signal such as $e^{(t^2)}$ that does not satisfy (5.2) for any value of σ. However, such rapidly diverging signals are not of practical importance, because every practical signal will be bounded.

5.1.2 Important properties of the Laplace transform

Theorem 5.1 (Linearity of the Laplace transform): The Laplacian operator, $\mathcal{L}$, is linear.

Proof.

$$\begin{aligned} \mathcal{L}[a_1 f_1(t) + a_2 f_2(t)] &= \int_{0-}^{\infty} (a_1 f_1(t) + a_2 f_2(t))e^{-st}dt \\ &= a_1 \int_{0-}^{\infty} f_1(t)e^{-st}dt + a_2 \int_{0-}^{\infty} f_2(t)e^{-st}dt \\ &= a_1\mathcal{L}[f_1(t)] + a_2\mathcal{L}[f_2(t)] \end{aligned}$$

The above demonstrates superposition and homogeneity at the same time, and proves that the Laplace transform of the scaled sum of two signals is the sum of the scaled Laplace transforms of the two signals.

With both superposition and homogeneity properties, we have proved that the Laplace transform is a linear operation. ■

Convention. It is common convention to use lowercase letters for a function of time and corresponding uppercase letters for its Laplace transform. For example, $H(s)$ is the Laplace transform of $h(t)$, $Y(s)$ is the Laplace transform of $y(t)$.

Theorem 5.2 (Laplace transform of a derivative): If $g(t) = \dot{f}(t)$, then

$$G(s) = \mathcal{L}[g(t)] = sF(s) - f(0^-), \text{ where } F(s) = \mathcal{L}[f(t)]$$

Proof.

$$G(s) = \mathcal{L}[g(t)] = \int_{0^-}^{\infty} \frac{df(t)}{dt} e^{-st} dt$$

Now let us consider the derivative of the function $f(t)e^{-st}$.

$$\begin{aligned}
\frac{d}{dt}(f(t)e^{-st}) &= \frac{df(t)}{dt}e^{-st} - sf(t)e^{-st}, \\
\text{or, } \int_{0^-}^{\infty} \frac{d}{dt}(f(t)e^{-st})dt &= \int_{0^-}^{\infty} \frac{df(t)}{dt}e^{-st}dt - \int_{0^-}^{\infty} sf(t)e^{-st}dt, \\
\text{or, } \left[f(t)e^{-st}\right]_{0^-}^{\infty} &= G(s) - sF(s) \\
\therefore G(s) &= sF(s) - f(0^-),
\end{aligned}$$

provided we choose s within the RoC of $F(s)$. If s is within the RoC of $F(s)$, $\lim_{t\to\infty}(f(t)e^{-st})$ will be 0. ∎

Theorem 5.3 (Integration with the Laplace transform): $g(t)$ is $\int_{0^-}^{t} f(\tau)d\tau$. Then,

$$G(s) = \mathcal{L}[g(t)] = F(s)/s, \text{ where } F(s) = \mathcal{L}[f(t)]$$

Proof. Consider the function $g(t)e^{-st}/(-s)$.

$$\begin{aligned}
\frac{d}{dt}\left\{g(t)\frac{e^{-st}}{-s}\right\} &= \frac{dg(t)}{dt} \cdot \frac{e^{-st}}{-s} + g(t)e^{-st} \\
&= f(t)\frac{e^{-st}}{-s} + g(t)e^{-st}, \\
\text{or, } \int_{0^-}^{\infty} \frac{d}{dt}\left\{g(t)\frac{e^{-st}}{-s}\right\}dt &= \int_{0^-}^{\infty} f(t)\frac{e^{-st}}{-s}dt + \int_{0^-}^{\infty} g(t)e^{-st}dt \\
\text{or, } \left[g(t)\frac{e^{-st}}{-s}\right]_{0^-}^{\infty} &= -\frac{F(s)}{s} + G(s) \\
\text{or, } G(s) &= \frac{F(s)}{s} + 0 + \frac{g(0^-)}{s}
\end{aligned}$$

However, $g(0^-)$ is 0 because $g(t)$ is the integral of $f(t)$ starting from 0^-. So at 0^-, $g(t)$ is 0.

$$\therefore G(s) = F(s)/s$$

■

Theorem 5.4 (The derivative of the Laplace transform): If the Laplace transform of $f(t)$ is $F(s)$, then the Laplace transform of $tf(t)$ is $-dF(s)/ds$.

Proof.

$$\begin{aligned}
F(s) &= \int_{0^-}^{\infty} f(t)e^{-st}\,dt \\
\therefore \frac{dF(s)}{ds} &= \int_{0^-}^{\infty} \frac{d(f(t)e^{-st})}{ds}\,dt \\
&= \int_{0^-}^{\infty} (-tf(t)e^{-st})\,dt \\
&= -\mathcal{L}[tf(t)]
\end{aligned}$$

■

Theorem 5.5 (Laplace transform of a delayed signal): If $g(t) = f(t - t_0)$, where t_0 is a real and positive delay, then $G(s) = e^{-st_0}F(s)$.

Proof. $g(t)$ is the same as $f(t)$, but delayed by t_0. While $f(t)$ may start at $t = 0$, $g(t)$ will correspondingly start at $t = t_0$.

$$\begin{aligned}
F(s) = \mathcal{L}[f(t)] &= \int_{0^-}^{\infty} f(t)e^{-st}\,dt \\
G(s) = \mathcal{L}[g(t)] &= \int_{0^-}^{\infty} g(t)e^{-st}\,dt \\
&= \int_{t_0^-}^{\infty} f(\tau - t_0)e^{-s\tau}\,d\tau \\
\text{(substituting } t = \tau - t_0\text{)} &= \int_{0^-}^{\infty} f(t)e^{-s(t+t_0)}\,dt \\
&= e^{-st_0}F(s)
\end{aligned}$$

■

5.1.3 A few important Laplace transforms

We will derive a few important Laplace transforms.

Impulse function

$$\mathcal{L}[\delta(t)] = \int_{0^-}^{\infty} \delta(t)e^{-st}dt = \int_{0^-}^{0^+} \delta(t) \cdot 1dt = 1$$

Step function

$$\begin{aligned}\mathcal{L}[u(t)] &= \int_{0^-}^{\infty} u(t)e^{-st}dt = \int_{0^+}^{\infty} e^{-st}dt \\ &= \left[\frac{e^{-st}}{-s}\right]_{0^+}^{\infty} = \frac{1}{s}\end{aligned}$$

Notice that the result is consistent with the fact that $u(t)$ is the integral of $\delta(t)$, and $\delta(t)$ is the derivative of $u(t)$.

- Because $u(t)$ is the integral of $\delta(t)$, by theorem 5.3, $\mathcal{L}[u(t)] = \mathcal{L}[\delta(t)]/s$. $\mathcal{L}[\delta(t)]$ is 1, and therefore $\mathcal{L}[u(t)]$ is $1/s$.
- Conversely, $\delta(t)$ is the derivative of $u(t)$. Then, by theorem 5.2, $\mathcal{L}[\delta(t)]$ is $s\mathcal{L}[u(t)] - u(0^-)$. But $\mathcal{L}[u(t)]$ is $1/s$ and $u(0^-)$ is 0. This gives us $\mathcal{L}[\delta(t)]$ as 1.

Ramp function

The unit ramp function, $\rho(t)$, is defined as:

$$\rho(t) = \begin{cases} 0, & \text{for } t < 0 \\ t & \text{for } t \geq 0. \end{cases}$$

The unit ramp function is the definite integral of $u(t)$. As such, $\mathcal{L}[\rho(t)] = 1/s \cdot \mathcal{L}[u(t)]$. $\therefore \mathcal{L}[\rho(t)] = 1/s^2$.

The exponential

We will now derive the Laplace transform of $f(t) = e^{at}u(t)$.

$$\begin{aligned}\mathcal{L}[e^{at}u(t)] &= \int_{0^-}^{\infty} e^{at}u(t)e^{-st}dt \\ &= \int_{0}^{\infty} e^{-(s-a)t}dt \\ &= -\frac{1}{s-a} \cdot \left[e^{-(s-a)t}\right]_0^{\infty} \\ &= \frac{1}{s-a}\end{aligned}$$

We could derive this using a second, more elegant method without computing the integral. Let us assume $\mathcal{L}[e^{at}u(t)]$ is $F(s)$.

$$\begin{aligned}
\mathcal{L}[\frac{d}{dt}(e^{at}u(t))] &= \mathcal{L}[ae^{at}u(t)] + \mathcal{L}[\delta(t)] \\
\text{or, } sF(s) &= aF(s) + 1, \ \because e^{at}u(t) = 0 \text{ at } t = 0^- \\
\text{or, } F(s) &= \frac{1}{s-a}
\end{aligned}$$

Note that the constant a does not have to be real. Since we have allowed complex a, a large class of signals are covered (refer back to section 3.1.1). The RoC of the Laplace transform of this function is the set of all s such that $\text{Re}[s] > \text{Re}[a]$.

Sine and cosine

We can express sine and cosine functions as the superposition of complex exponentials.

$$\begin{aligned}
\cos(\omega_0 t)u(t) &= \frac{e^{j\omega_0 t} + e^{-j\omega_0 t}}{2}u(t) \\
\therefore \mathcal{L}[\cos(\omega_0 t)] &= \frac{1}{2}\mathcal{L}[e^{j\omega_0 t}u(t)] + \frac{1}{2}\mathcal{L}[e^{-j\omega_0 t}u(t)] \\
&= \frac{1}{2}(\frac{1}{s - j\omega_0} + \frac{1}{s + j\omega_0}) \\
&= \frac{s}{s^2 + \omega_0^2}
\end{aligned}$$

$$\begin{aligned}
\sin(\omega_0 t)u(t) &= \frac{e^{j\omega_0 t} - e^{-j\omega_0 t}}{2j}u(t) \\
\therefore \mathcal{L}[\sin(\omega_0 t)] &= \frac{1}{2j}\mathcal{L}[e^{j\omega_0 t}u(t)] - \frac{1}{2}\mathcal{L}[e^{-j\omega_0 t}u(t)] \\
&= \frac{1}{2j}(\frac{1}{s - j\omega_0} - \frac{1}{s + j\omega_0}) \\
&= \frac{\omega_0}{s^2 + \omega_0^2}
\end{aligned}$$

The regions of convergence for both the sine and cosine functions are the set of all s such that $\text{Re}[s] > 0$. This is also known as the right half s-plane (RHP).

Other signals

In most cases, direct evaluation of the Laplace transform integral will not be necessary. With the help of the linearity principle (theorem 5.1), we will break up the signal into exponentials, sines and cosines. Obtaining the Laplace transform will then be a matter of looking up Tab. 5.1 and writing out the result.

Occasionally, we will use the integral, differential, and derivative properties (theorems 5.2, 5.3, 5.4) to deduce the transforms of functions that we are not directly familiar with. The strategy will always be to reduce the function to be transformed into a linear combination of familiar functions that we are comfortable with.

- $\mathcal{L}[te^{at}u(t)]$: There is a multiplication with t. We know $\mathcal{L}[e^{at}u(t)]$. We can use theorem 5.4.

$$\therefore \mathcal{L}[te^{at}u(t)] = -\frac{d}{ds}\left(\frac{1}{s-a}\right) = \frac{1}{(s-a)^2}$$

- $\mathcal{L}[t^2e^{at}u(t)]$: There is a further multiplication with t. We can use theorem 5.4 again.

$$\therefore \mathcal{L}[t^2e^{at}u(t)] = -\frac{d}{ds}\left(\frac{1}{(s-a)^2}\right) = \frac{2}{(s-a)^3}$$

- $\mathcal{L}[t^ne^{at}u(t)]$: From the previous two steps, by induction, we will obtain:

$$\mathcal{L}[t^ne^{at}u(t)] = \frac{n!}{(s-a)^{n+1}}$$

Example 5.2. Find the Laplace transforms of:

$$\begin{aligned}
\text{(a) } f_1(t) &= (6e^{-5t} + e^{3t} + 2t^2 - 9)u(t) \\
\text{(b) } f_2(t) &= (2e^{2t} + 3\cos(6t + \pi/3) + e^{-3t}\cos(6t))u(t) \\
\text{(c) } f_3(t) &= (3\sinh t + 2\cosh t)u(t)
\end{aligned}$$

(a) The transforms are mostly derived with the help of Tab. 5.1. For the third term, t^2 is $2\int t\,dt$. Using the integral property of the Laplace transform in theorem 5.3, the Laplace transform of t^2 is $2/s^3$.

Table 5.1: Laplace transform pairs

Name	$f(t) = \mathcal{L}^{-1}[F(s)]$	$F(s) = \mathcal{L}[f(t)]$
Impulse	$\delta(t)$	1
Step	$u(t)$	$1/s$
Ramp	$\rho(t)$	$1/s^2$
Exponential	$e^{at}u(t)$	$\frac{1}{s-a}$
Cosine	$\cos(\omega_0 t)u(t)$	$\frac{s}{s^2+\omega_0^2}$
Sine	$\sin(\omega_0 t)u(t)$	$\frac{\omega_0}{s^2+\omega_0^2}$
Cosine with exponential	$e^{-at}\cos(\omega_0 t)u(t)$	$\frac{s+a}{(s+a)^2+\omega_0^2}$
Sine with exponential	$e^{-at}\sin(\omega_0 t)u(t)$	$\frac{\omega_0}{(s+a)^2+\omega_0^2}$
t^n times exponential	$t^n e^{at}u(t)$	$\frac{n!}{(s-a)^{n+1}}$

$$\begin{aligned} F_1(s) &= 6\frac{1}{s-(-5)} + \frac{1}{s-3} + 2\frac{2}{s^3} - 9\frac{1}{s} \\ &= \frac{6}{s+5} + \frac{1}{s-3} + \frac{4}{s^3} - \frac{9}{s} \end{aligned}$$

(b) The second term can be broken up into sine and cosine terms using trigonometry. It can not be expressed as a delay.

$$\cos(6t + \pi/3) = \cos(\pi/3)\cos(6t) - \sin(\pi/3)\sin(6t)$$

Now we can use Tab. 5.1 to arrive at the transform.

$$\begin{aligned} F_2(s) &= \frac{2}{s-2} + \cos(\pi/3)\frac{3s}{s^2+6^2} - \sin(\pi/3)\frac{18}{s^2+6^2} + \frac{s+3}{(s+3)^2+6^2} \\ &= \frac{2}{s-2} + \frac{9s/2 - 27\sqrt{3}}{s^2+36} + \frac{s+3}{s^2+6s+45} \end{aligned}$$

(c) Hyperbolic functions can be broken up in terms of exponentials.

$$\begin{aligned}
\sinh t &= \frac{e^t - e^{-t}}{2} \text{ and } \cosh t = \frac{e^t + e^{-t}}{2} \\
\therefore F_3(s) &= \frac{3/2}{s-1} - \frac{3/2}{s+1} + \frac{2/2}{s-1} + \frac{2/2}{s+1} \\
&= \frac{5/2}{s-1} - \frac{1/2}{s+1} = \frac{2s+3}{s^2-1}
\end{aligned}$$

It turns out we can also use Octave to perform Laplace transforms.

```
octave:1> pkg load symbolic
octave:2> syms t
Symbolic pkg v3.0.0: Python communication link active, SymPy v1.9.
octave:3> f = 3*sinh(t) + 2*cosh(t);
octave:4> F = laplace(f)
F = (sym)
  2·s + 3
  -------
    2
   s  - 1
```

5.1.4 Initial and final value theorems

Theorem 5.6 (Initial value theorem): If $f(t)$ has a Laplace transform $F(s)$,

$$f(0^+) = \lim_{s\to\infty} sF(s)$$

Proof. From Theorem 5.2, we have:

$$\int_{0-}^{\infty} \frac{df(t)}{dt} e^{-st}\, dt = sF(s) - f(0^-)$$

$$\begin{aligned}
\therefore \lim_{s\to\infty} sF(s) &= \lim_{s\to\infty} \int_{0-}^{\infty} \dot{f} e^{-st} dt + f(0^-) \\
&= \lim_{s\to\infty} \int_{0-}^{0^+} \dot{f} e^{-st} dt + \lim_{s\to\infty} \int_{0+}^{\infty} \dot{f} e^{-st} dt + f(0^-) \\
&= \lim_{s\to\infty} \int_{0-}^{0^+} \dot{f} e^{-st} dt + f(0^-) \\
&= f(0^+) - f(0^-) + f(0^-) = f(0^+)
\end{aligned}$$

■

Theorem 5.7 (Final value theorem): If $f(t)$ has a Laplace transform $F(s)$,

$$\lim_{t\to\infty} f(t) = \lim_{s\to 0} sF(s)$$

Proof. Once again, from Theorem 5.2, we have:

$$\int_{0^-}^{\infty} \frac{df(t)}{dt} e^{-st}\, dt = sF(s) - f(0^-)$$

$$\begin{aligned}
\therefore \lim_{s\to 0} sF(s) &= \lim_{s\to 0} \int_{0^-}^{\infty} \dot{f} e^{-st} dt + f(0^-) \\
&= \int_{0^-}^{\infty} \dot{f} dt + f(0^-) \\
&= \lim_{t\to\infty} f(t)
\end{aligned}$$

■

The initial and final value theorems can be used to rapidly ascertain the values of $f(t)$ at $t = 0^+$ and as $t \to \infty$. The initial and final values are useful in many ways. They can be used to verify the Laplace transform.

5.2 The inverse Laplace transform

The inverse Laplace transform, $\mathcal{L}^{-1}$, is most often performed with the help of a table look-up. A table of Laplace transform pairs is shown in Tab. 5.1. From the table, if the Laplace transform of $\rho(t)$ is $1/s^2$, then the inverse Laplace transform of $1/s^2$ must be $\rho(t)$.

One would argue that such a technique for inverse Laplace transforms must be severely limiting. However, for almost all practical functions that **we need to invert**, it is possible to use the look-up table.

5.2.1 Inversion integral

For completeness, in this section we will briefly present the inversion integral that one can use to work out the inverse Laplace transform.

For a Laplace-domain function $F(s)$, the following expression gives its inverse Laplace transform.

$$f(t) = \mathcal{L}^{-1}[F(s)] = \frac{1}{j2\pi} \int_{-j\infty+\sigma}^{+j\infty+\sigma} e^{st} F(s)\, ds \cdot u(t) \tag{5.3}$$

In (5.3), σ is a real number such that all the points of singularity (**known as poles**) of $F(s)$ are to the left of the contour of integration. For instance, if all the poles of $F(s)$ are in the left-half s-plane (LHP), one may choose σ to be 0.

No proof of the inverse Laplace transform is offered in this text. The mathematically inclined inquisitive reader may refer to [18] for the proof. We can use residue integration [17] to work out the integral in specific cases.

Example 5.3. Use the inverse Laplace transform formula to obtain

$$\mathcal{L}^{-1}[\frac{1}{s+a}]$$

The only **pole** of the expression is at $s = -a$. As such we can select a contour for integration along a path that encloses the pole. The integration contour is shown in the sketch below. σ is greater than $-a$.

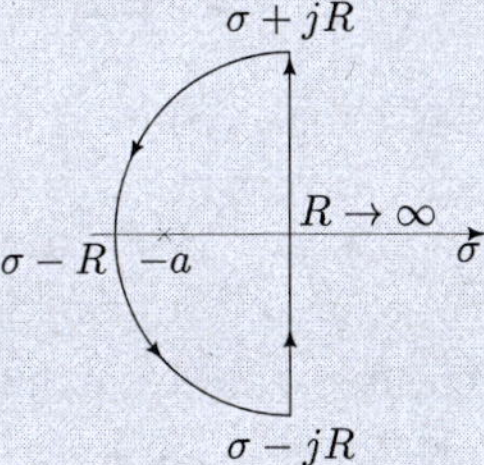

In the limit as $R \to \infty$, the contour will enclose the entire LHP. The half-circular part of the contour will have a zero integral because the integrand will vanish as $R \to \infty$ along the half circle. What we will be left with is the contour of integration desired for the inverse Laplace transform in (5.3). Now we can perform residue integration. There is a single pole enclosed within the contour.

$$\begin{aligned}
\therefore \int_{-j\infty+\sigma}^{+j\infty+\sigma} \frac{e^{st}}{s+a}\, ds &= j2\pi \operatorname*{Res}_{s=-a}\left(\frac{e^{st}}{s+a}\right) \\
&= j2\pi \lim_{s\to -a}(s+a)\frac{e^{st}}{s+a} \\
&= j2\pi e^{-at} \\
\therefore \frac{1}{j2\pi}\int_{-j\infty+\sigma}^{+j\infty+\sigma} \frac{e^{st}}{s+a}\, ds \cdot u(t) &= e^{-at}u(t)
\end{aligned}$$

The inverse Laplace transform is not difficult as long as one uses residue integration on the complex plane.

The closed-form expression for the inverse Laplace transform tells us that the inverse Laplace transform is unique. No two functions have the same Laplace transform. This is the basis for our next, and more popular, technique of inverting the Laplace transform, based on table look-up.

5.2.2 Partial fraction expansions for inverse Laplace transforms

We can break up any rational polynomial function, $F(s) = N(s)/D(s)$, where both $N(s)$ and $D(s)$ are polynomials in s, with the help of partial fractions. If $D(s)$ is a polynomial of order n, if the order of $N(s)$ is less than the order of $D(s)$, and $\{s_k\}_{k=1,2\ldots n}$ are non-repeated roots of $D(s)$, then $F(s)$ can be expressed as:

$$F(s) = \sum_{k=1}^{n} \frac{r_k}{s - s_k} \tag{5.4}$$

If $D(s)$ has repeated roots, say q repeated roots of s_p, then the related terms will be:

$$\frac{r_{p1}}{s - s_p} + \frac{r_{p2}}{(s - s_p)^2} + \cdots \frac{r_{pq}}{(s - s_p)^q}$$

When the roots are non-repeated (most common), the evaluation of the residues ($\{r_k\}$) is very straightforward. Note, these residues are the same as the residues discussed in section 5.2.1.

$$r_k = \lim_{s \to s_k} (s - s_k)F(s)$$

In other words, multiply both sides of (5.4) by $(s - s_k)$ and plug-in $s = s_k$. All the terms on the right evaluate to 0, except the term r_k. Therefore, r_k is the value of $(s - s_k)F(s_k)$.

When the roots are repeated (not common, common in difficult examination questions), we may use one of two methods to break up into partial fractions. The first method involves repeated differentiation, while the second method is more direct. We will discuss these methods in the following section 5.2.3.

Popularized by Oliver Heaviside, partial fraction expansions of rational polynomial functions are very useful for evaluating inverse Laplace transforms. Once expanded into partial fractions, the inverse Laplace transform is straightforward, with the help of Tab. 5.1. We will now demonstrate with a few examples. Examples with repeated roots will be taken up in the following section 5.2.3.

Example 5.4. Let us find $\mathcal{L}^{-1}[\frac{2s+3}{s^2+3s+2}]$.

$$\begin{aligned}
\frac{2s+3}{s^2+3s+2} &= \frac{2s+3}{(s+1)(s+2)} \\
&= \frac{r_1}{s+1} + \frac{r_2}{s+2} \\
r_1 &= \lim_{s \to -1} \frac{2s+3}{s+2} = 1. \\
r_2 &= \lim_{s \to -2} \frac{2s+3}{s+1} = 1. \\
\therefore \frac{2s+3}{s^2+3s+2} &= \frac{1}{s+1} + \frac{1}{s+2} \\
\therefore \mathcal{L}^{-1}[\frac{2s+3}{s^2+3s+2}] &= e^{-t}u(t) + e^{-2t}u(t)
\end{aligned}$$

Example 5.5. Let us find the inverse Laplace transform of

$$F(s) = \frac{s+4}{s^2+2s+10}$$

$$\begin{aligned}
F(s) &= \frac{s+4}{(s+(1+3j))(s+(1-3j))} \\
&= \frac{r_1}{s+(1+3j)} + \frac{r_2}{s+(1-3j)} \\
r_1 &= \lim_{s \to -(1+3j)} \frac{s+4}{s+(1-3j)} \\
&= 1/2 \cdot (1+j) \\
r_2 &= \lim_{s \to -(1-3j)} \frac{s+4}{s+(1+3j)} \\
&= 1/2 \cdot (1-j) \\
F(s) &= \frac{1/2 \cdot (1+j)}{s+(1+3j)} + \frac{1/2 \cdot (1-j)}{s+(1-3j)} \\
\therefore f(t) &= \{1/2 \cdot (1+j) \cdot e^{-t}e^{-3jt} + 1/2 \cdot (1-j) \cdot e^{-t}e^{3jt}\}u(t) \\
&= e^{-t}u(t)\{\cos(3t) + \sin(3t)\}
\end{aligned}$$

We can work this out efficiently using the lookup tables (Table 5.1) a little more effectively. The last two rows of the lookup table reveal that for our function $F(s)$, which has a denominator of $s^2 + 2s + 10$, $a = 1$. We can break up the denominator as $(s+1)^2 + 3^2$, and obtain $a = 1$, $\omega_0 = 3$. The numerator, $s + 4$, can now be broken up as $(s + a) + \omega_0$. We obtain the inverse Laplace transform as $f(t) = e^{-t}u(t)\{\cos(3t) + \sin(3t)\}$.

5.2.3 Partial fraction expansion with repeated poles

In the presence of repeated roots, the partial fraction expansion technique is more involved. In this text, we will demonstrate only through examples. There are two general methods. The first method, developed by Oliver Heaviside, uses Heaviside's expansion theorem.

Theorem 5.8 (Heaviside expansion theorem): Consider a rational polynomial function with q repeated poles at s_p.

$$F(s) = \frac{P(s)}{(s - s_p)^q}$$

$F(s)$ can be broken up into k partial fractions as follows:

$$\frac{r_{p_1}}{s - s_p} + \frac{r_{p_2}}{(s - s_p)^2} + \cdots \frac{r_{p_q}}{(s - s_p)^q}$$

where:

$$r_{p_n} = \frac{1}{(q-n)!} \frac{d^{q-n} P(s)}{ds^{q-n}}\bigg|_{s=s_p}$$

Proof. First we can construct $P(s)$ as $F(s)(s - s_p)^q$.

$$P(s) = F(s)(s - s_p)^q = r_{p_1}(s - s_p)^{q-1} + r_{p_2}(s - s_p)^{q-2} + \cdots + r_{p_q}$$

If we now set $s = s_p$, then all the terms except the last one vanish.

$$r_{p_q} = P(s)|_{s=s_p}$$

Next, we can differentiate $P(s)$. All terms will reduce by one order and the last one will vanish. If we now set $s = s_p$, we will obtain $r_{p_{q-1}}$.

$$r_{p_{q-1}} = P'(s)|_{s=s_p}$$

The next time we differentiate and plug in $s = s_p$, we will obtain $2r_{p_{q-2}}$. The process can be continued till we have all the partial fraction coefficients. ∎

The second technique is much simpler in practice and involves a modified power series expansion [19]. We will demonstrate with the help of examples.

Example 5.6. Find the inverse Laplace transform of:

$$F(s) = \frac{s-2}{s(s+1)^3}$$

We have repeated poles at $s = -1$. Let us change the variable and substitute $p = s + 1$. Then,

$$\begin{aligned} F(p-1) &= \frac{p-3}{(p-1)p^3} \\ \text{or, } p^3F(p-1) &= \frac{p-3}{p-1} \end{aligned}$$

Now we will expand $(p-3)/(p-1)$ using long division, in **increasing powers of** p.

$$\begin{array}{r|l} & 3+2p+2p^2 \\ \hline 1-p & 3-\ p \\ & \underline{3-3p} \\ & \quad 2p \\ & \quad \underline{2p-2p^2} \\ & \qquad 2p^2 \\ & \qquad \underline{2p^2-2p^3} \\ & \qquad\quad 2p^3 \end{array}$$

We stop at three terms in the long division, because we need only three partial fraction coefficients. We now have:

$$\begin{aligned} p^3F(p-1) = \frac{p-3}{p-1} &= 3 + 2p + 2p^2 + \frac{2p^3}{1-p} \\ \text{or, } F(p-1) &= \frac{3}{p^3} + \frac{2}{p^2} + \frac{2}{p} + \frac{2}{1-p} \\ \text{or, } F(s) &= \frac{3}{(s+1)^3} + \frac{2}{(s+1)^2} + \frac{2}{s+1} - \frac{2}{s} \end{aligned}$$

Now we will evaluate $f(t)$ based on Tab. 5.1.

$$f(t) = (3t^2e^{-t}/2 + 2te^{-t} + 2e^{-t} - 2)u(t)$$

Let us now demonstrate a slightly more difficult example with two roots, each repeated twice.

Example 5.7. Find the inverse Laplace transform of:

$$F(s) = \frac{5s^2 - 4}{s^2(s+2)^2}$$

Let us first expand for the pole at $s = 0$.

$$s^2F(s) = \frac{-4 + 5s^2}{4 + 4s + s^2}$$

We have organized the expression in **increasing powers of** s. Now using long division,

$$\begin{array}{r l}
 & \underline{-1 + s \qquad} \\
4 + 4s + s^2 \big) & -4 \qquad +5s^2 \\
 & \underline{-4 - 4s - s^2} \\
 & \qquad 4s + 6s^2 \\
 & \qquad \underline{4s + 4s^2 + s^3} \\
 & \qquad\quad 2s^2 - s^3
\end{array}$$

$$\therefore F(s) = -\frac{1}{s^2} + \frac{1}{s} + \frac{2 - s}{4 + 4s + s^2}$$

Now we will expand the fractional term around the pole at $s = -2$. In this case, we will use a new variable, $p = s + 2$, or, $s = p - 2$.

$$(s+2)^2 \frac{2 - s}{4 + 4s + s^2} = 2 - s = 4 - p$$

Further long division is not necessary, since there is no denominator. Dividing both sides by $(s + 2)^2$ and replacing p thereafter, we obtain the following.

$$\begin{aligned}
\frac{2 - s}{4 + 4s + s^2} &= \frac{4}{(s+2)^2} - \frac{1}{s+2} \\
\therefore F(s) &= -\frac{1}{s^2} + \frac{1}{s} + \frac{4}{(s+2)^2} - \frac{1}{s+2} \\
\therefore f(t) = \mathcal{L}^{-1}[F(s)] &= (-t + 1 + 4te^{-2t} - e^{-2t})u(t)
\end{aligned}$$

The technique happens to be simpler than the first method that uses repeated differentiation.

5.3 Solving circuits with Laplace transforms

5.3.1 Solving a differential equation

We will now try using the Laplace transform to solve a differential equation. As an example, let us try to work out the differential equation in example 3.9.

Example 5.8. Solve the below differential equation from example 3.9.

$$\ddot{v_C} + \dot{v_C} + v_C = 1, \text{ with } v_C(0) = 0.5, \text{ and, } \dot{v_C}(0) = 0.5.$$

We are going to take the Laplace transform of both sides of the equation. If we assume $\mathcal{L}[v_C(t)] = V_C(s)$, then $\mathcal{L}[\dot{v_C}] = sV_C(s) - 0.5$. Further, $\mathcal{L}[\ddot{v_C}] = s(s(V_C(s) - 0.5) - 0.5$. On the right side, the 1 translates to $1/s$ because it is the same as $u(t)$ for $t > 0$.

$$\begin{aligned} s(s(V_C(s) - 0.5) - 0.5 + sV_C(s) - 0.5 + V_C(s) &= 1/s \\ \text{or, } V_C(s)(s^2 + s + 1) &= \frac{s^2 + 2s + 2}{2s} \\ \text{or, } V_C(s) &= \frac{s^2 + 2s + 2}{2s(s^2 + s + 1)} \end{aligned}$$

Breaking this up into partial fractions, we obtain:

$$V_C(s) = \frac{1}{s} + \frac{-s/2}{s^2 + s + 1}$$

Now we can take the inverse Laplace transform of $V_C(s)$ to obtain $v_C(t)$ using Tab. 5.1. The first term gives us $u(t)$. The denominator of the second term gives us $a = 1/2$ and $\omega_0 = \sqrt{3}/2$. We expect a factor of $-1/2$ to satisfy the s term in the numerator with the cosine. To satisfy the units term of the numerator, we need the sine term to have a factor of $+1/\sqrt{12}$.

$$\therefore v_C(t) = u(t)\left\{1 + e^{-t/2}\left(-\frac{1}{2}\cos(\frac{\sqrt{3}t}{2}) + \frac{1}{\sqrt{12}}\sin(\frac{\sqrt{3}t}{2})\right)\right\}$$

The expression is in a different form of the result obtained in example 3.9.

With the Laplace transform, solving the linear differential equation reduces to algebra, manipulation, and a lookup table. Zero-input and zero-state responses are dealt with at the same time. The solution technique implicitly takes care of initial conditions.

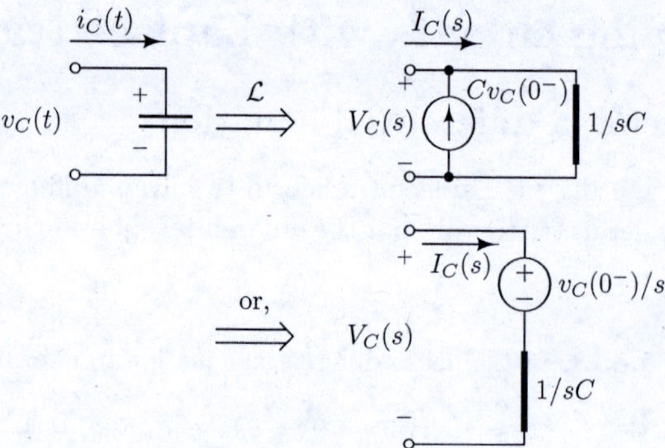

Figure 5.1: Laplace domain model for a capacitor

5.3.2 Directly solving a circuit

Laplace transforms of the voltages in a loop will obey KVL because the Laplace transform is a linear operation. i.e., the Laplace transform of the sum of two or more voltages is the sum of the Laplace transforms of the voltages, as proved in Theorem 5.1. Similarly, the Laplace transforms of the branch currents of a circuit will obey KCL. We are going to see how we can solve a circuit directly in the Laplace domain.

The capacitor

If the current through a capacitor, C, is $i_C(t)$ and the voltage across the capacitor is $v_C(t)$, then:

$$\begin{aligned} i_C(t) &= C\frac{dv_C(t)}{dt} \\ \text{or, } \mathcal{L}[i_C] &= C\mathcal{L}[\frac{dv_C}{dt}] \\ \text{or, } I_C(s) &= sCV_C(s) - Cv_C(0^-) \end{aligned}$$

We have shown the model for a capacitor in Fig 5.1. One can transform the current source $Cv_C(0^-)$ in shunt with the impedance $1/sC$ into a voltage source $v_C(0^-)/s$ in series with the impedance $1/sC$. Recall that the model for a capacitor C in the sinusoidal steady state is $1/j\omega C$. In the Laplace domain, $j\omega$ is replaced with s, and the model for the capacitor is $1/sC$. To account for initial conditions, there is an additional series $v_C(0^-)/s$ in the Laplace domain.

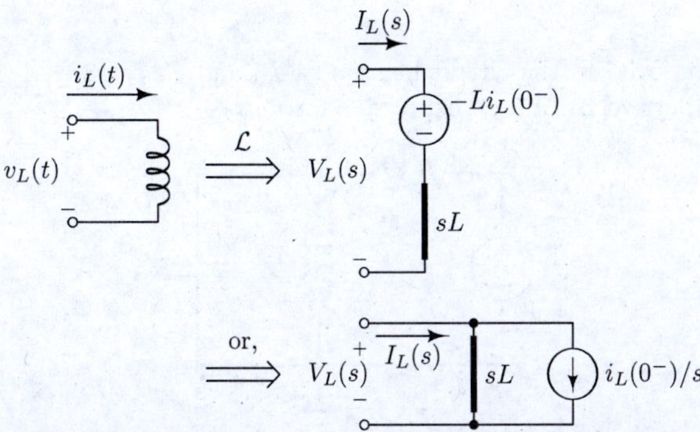

Figure 5.2: Laplace domain model for an inductor

The inductor

If the current through an inductor, L, is $i_L(t)$ and the voltage across the inductor is $v_L(t)$, then:

$$\begin{aligned} v_L(t) &= L\frac{di_L(t)}{dt} \\ \text{or, } \mathcal{L}[v_L] &= L\mathcal{L}[\frac{di_L}{dt}] \\ \text{or, } V_L(s) &= sLI_L(s) - Li_L(0^-) \end{aligned}$$

Just like the capacitor, we have shown the model for an inductor in Fig. 5.2. The inductor is modeled as an impedance of sL in series with a voltage source of $-Li_L(0^-)$. One can transform the model into an impedance sL in shunt with a current source, $i_L(0^-)/s$. Recall that the model for an inductor L in the sinusoidal steady state is $j\omega L$. In the Laplace domain, $j\omega$ is replaced with s, and the model for the inductor is sL. To account for initial conditions, there is an additional shunt $i_L(0^-)/s$ in the Laplace domain.

In any given linear time-invariant circuit, we can replace the input source with its Laplace transform, the inductors with their Laplace-domain models, the capacitors with their Laplace-domain models. Following the replacement step, we can perform a Laplace-domain analysis. We can use any method, e.g., tableau, node-voltage, mesh, Thévenin, Norton, superposition, and reciprocity. We can evaluate the output quantity of interest and then perform an inverse Laplace transform.

Typically, the input sources will be of the form $u(t)$ or $\cos(\omega_0 t)u(t)$, and the Laplace transforms of these will be available. The inverse Laplace transforms will have to be worked out with the help of partial-fraction expansion.

Example 5.9. In the circuit below, R_2, L_3 and C_4 are 2 Ω, 1/3 H and 1/4 F respectively. Find an expression for $v_C(t)$.

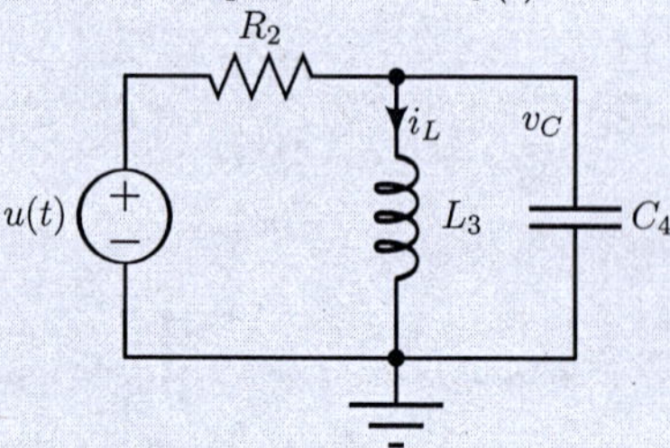

The circuit has been set up a long time before $t = 0$. As such, at $t = 0^-$, the voltage across the capacitor must be 0, and the current through the inductor must be 0. We will replace the capacitor, the inductor, and the voltage source with their Laplace-domain models. The model for the inductor is only an impedance of sL_3 and the model for the capacitor is only an admittance of sC_4.

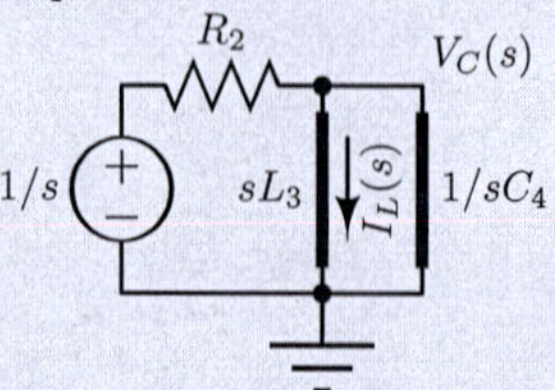

From the figure, $V_C(s)$ is given by:

$$
\begin{aligned}
V_C(s) &= \frac{1}{s} \cdot \frac{sL_3}{R_2 + sL_3 + s^2 R_2 L_3 C_4} = \frac{2}{s^2 + 2s + 12} \\
&= \frac{2}{\sqrt{11}} \cdot \frac{\sqrt{11}}{(s+1)^2 + (\sqrt{11})^2} \\
\therefore v_C(t) &= \frac{2}{\sqrt{11}} \cdot e^{-t} \sin(\sqrt{11}t)\, u(t)
\end{aligned}
$$

The two-capacitor first-order circuit of example 3.4 was probably a little disturbing, because it demonstrated an instantaneous voltage change across a capacitor. We will now verify the same problem using the Laplace transform technique. Laplace transforms and inverse Laplace transforms can also be handled by the Octave symbolic package.

Example 5.10. In the circuit below evaluate $v_{C1}(t)$ and $v_{C2}(t)$.

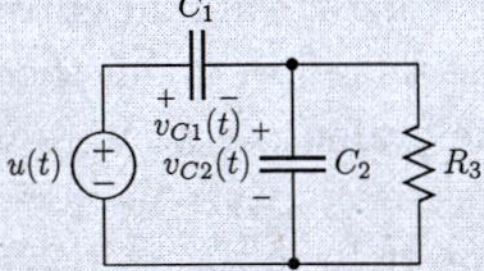

$v_{C1}(0^-)$ and $v_{C2}(0^-)$ are both 0.

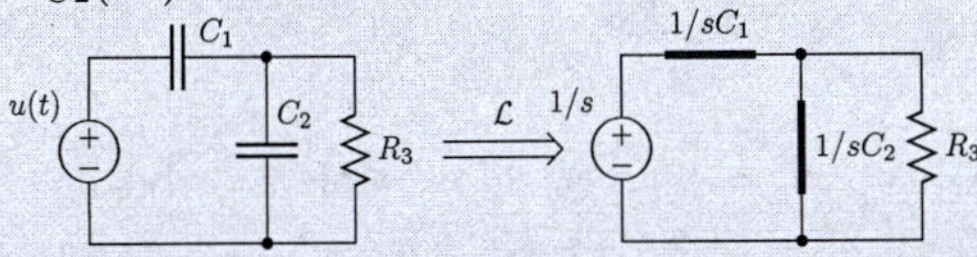

We can evaluate the voltage across C_2 using potential division.

$$
\begin{aligned}
V_{C2}(s) &= \frac{1}{s} \cdot \frac{(1/sC_2) \parallel R_3}{1/sC_1 + ((1/sC_2) \parallel R_3)} \\
&= \frac{C_1/(C_1 + C_2)}{s + 1/(C_1R_3 + C_2R_3)} \\
\therefore v_{C2}(t) &= \frac{C_1}{C_1 + C_2} \cdot e^{-\frac{t}{R_3(C_1+C_2)}} \cdot u(t)
\end{aligned}
$$

Inverse Laplace transforms can also be done using Octave.

Program 5.1. rcc.m: Inverse Laplace transforms with Octave.

```
pkg load symbolic % Load symbolic toolbox
syms s; % Declare s
C1 = 1/2; C2 = 1/2; R3 = 3; % Values of C1, C2, R3
VC1 = (1/s)*(1/(s*C1))/(1/(s*C1)+1/(s*C2+1/R3)); % V_C1(s)
VC2 = (1/s)*(1/(s*C2+1/R3))/(1/(s*C1)+1/(s*C2+1/R3)); % V_C2(s)
vc1t = ilaplace(VC1); % v_C1(t) expression
vc2t = ilaplace(VC2); % v_C2(t) expression
fh1 = function_handle(vc1t); % Function based on the expression
fh2 = function_handle(vc2t); % Function based on the expression
t = [0:2e-2:10]; % Initialize t
vc1 = fh1(t); vc2 = fh2(t); % Evaluate v_C1(t) and v_C2(t)
plot(t, vc1); hold on; % Plot, do not erase
plot(t, vc2); % Plot
```

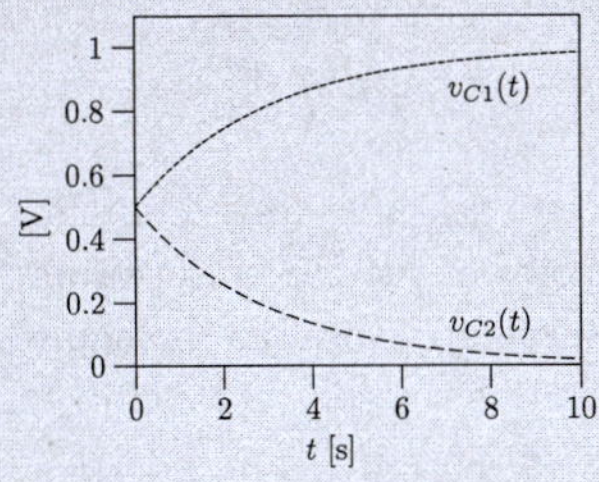

In the last two examples, there were no switches and the initial conditions were zero. In the next two examples we will have switches; naturally the initial voltages across capacitors and initial currents through inductors are not expected to be zero. One should work out the initial conditions mentally.

In unit 3, if v_x had to be evaluated, we were working out $v_x(0^-)$, $\dot{v}_x(0^-)$ (if second order), and further derivatives in case of higher order. With the Laplace transform technique, only initial voltages across all capacitors and initial currents through all inductors are required.

Example 5.11. This is a repeat of example 3.9. In the circuit below, after being closed for a long time, the switch is opened at $t = 0$. Obtain the complete expression for $v_C(t)$ for $t \geq 0$.

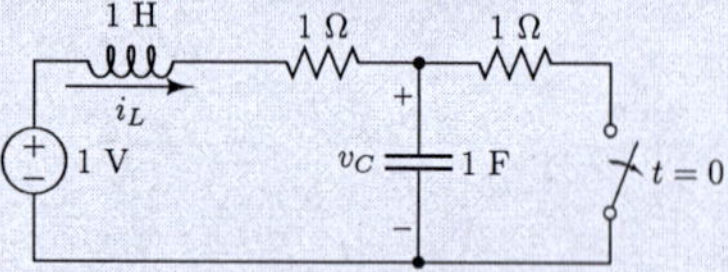

The initial voltage, $v_C(0^-)$ is 0.5 V. The initial current, $i_L(0^-)$ is 0.5 A. We have drawn the Laplace domain model of the circuit below.

The mesh equation is:

$$-1/s - 0.5 + sI_L + I_L + 0.5/s + I_L/s = 0$$

$$\begin{aligned} \text{The above results in } I_L(s + 1 + 1/s) &= 0.5/s + 0.5 \\ \text{or, } I_L &= \frac{0.5(1+s)}{s^2+s+1} \\ \text{Now, } V_C = I_L/s + 0.5/s &= \frac{s^2+2s+2}{2s(s^2+s+1)} \end{aligned}$$

The inverse Laplace transform was worked out in example 5.8. Note the use of $1/s$ as the transform of the DC voltage source. The voltage source did not have a step function in the time domain. However, the Laplace transform is only for $t \geq 0$ and a constant source behaves like $u(t)$.

Example 5.12. Analyze the circuit below and evaluate $v(t)$, the voltage across the capacitor.

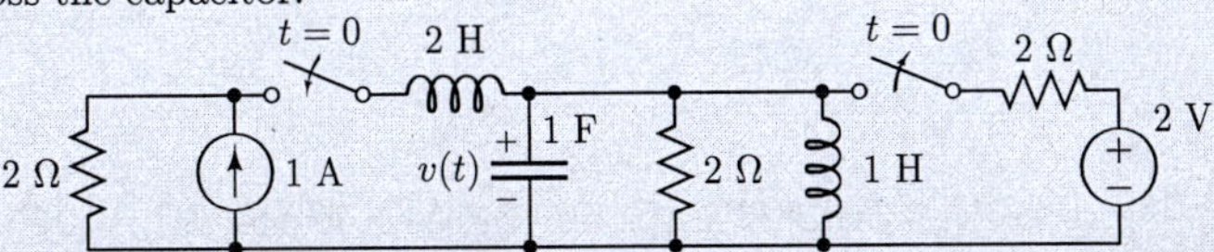

At $t = 0^-$, the circuit is in the steady-state with the 2 V voltage source in the circuit. The current in the 2 H inductor is 0, the current in the 1 H inductor is 1 A, the voltage $v(0^-)$ across the 1 F capacitor is 0. We have shown the Laplace model of the circuit for $t \geq 0$ below.

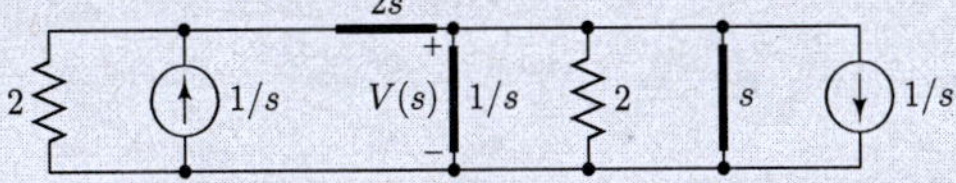

The circuit can be solved using superposition. We switch the current sources on, one at a time, and write out $V(s)$. Accordingly,

$$\begin{aligned} V(s) &= 2/s \cdot \frac{(2 \parallel s \parallel (1/s))}{2 + 2s + (2 \parallel s \parallel (1/s))} - 1/s \cdot (2 \parallel s \parallel (1/s) \parallel (2s+2)) \\ &= -\frac{2s}{2s^3 + 3s^2 + 4s + 2} \\ \therefore v(t) &= 0.51e^{-0.69t} - e^{-0.40t}\{0.75\sin(1.13t) + 0.51\cos(1.13t)\} \end{aligned}$$

The inverse Laplace transform may fail when you use the Octave symbolic package. This is because the denominator polynomial is cubic. We can solve this numerically using a second technique with the `residue` function.

Program 5.2. res.m: Inverse Laplace transforms using residues at poles.

```
B=[-2 0]; A=[2 3 4 2]; % Numerator=B, denominator=A, in decreasing orders of s
[R, P, K, E] = residue(B, A); % R is the residue at each pole P.
% We expect three poles because A is of order 3. The length of A is 4.
t = [0:1e-2:20]; vt = zeros(size(t)); % Initialize t and v(t)
for m=1:(length(A)-1) % For each pole P_m,
  vt = vt + R(m)*e.^(P(m)*t); % add the corresponding R_m e^(P_m t)
end;
```

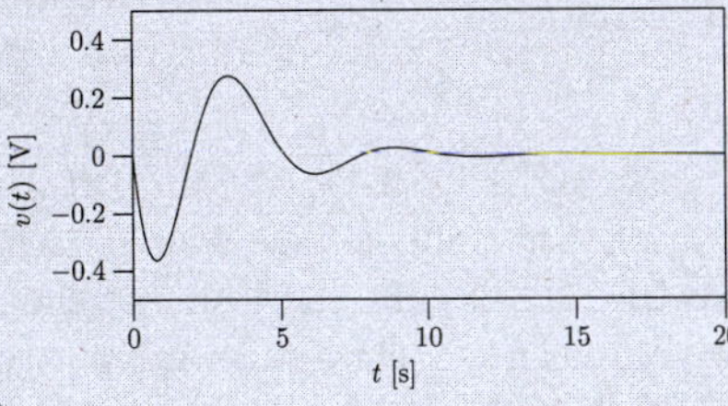

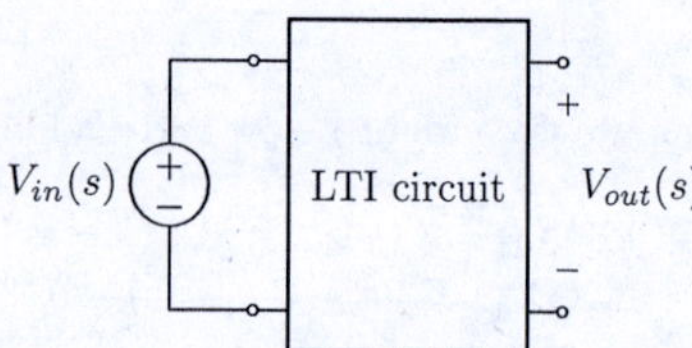

Figure 5.3: A circuit as a system. An independent voltage or current source is applied at the input. The output is any node voltage, or branch voltage, or branch current.

5.4 The circuit as a system

5.4.1 Convolution

In unit 4, we had studied the impulse response of a linear time-invariant system. The response to an input applied to a system is given by the convolution integral that we had studied in (4.2) and (4.3).

As a refresher, an input of $\delta(t)$ may be applied to a linear time-invariant circuit. The corresponding output of the system is the impulse response. Now visualize a circuit as a system, with its input as an independent voltage or current source. The output of this system may be any node voltage or branch current. Fig. 5.3 depicts our setup.

If the input is $\delta(t)$, $V_{in}(s)$ is 1. The Laplace transform of the corresponding output is the **transfer function** of the circuit, $H(s)$. The inverse Laplace transform of $H(s)$ is $h(t)$, the impulse response of the circuit.

For some other input applied, say $x(t)$, because the circuit is LTI, the output, $y(t)$ is the convolution of $x(t)$ and $h(t)$, given by (4.2) and (4.3), and is reproduced below.

$$y(t) = \int_{-\infty}^{\infty} h(\tau)x(t-\tau)d\tau = \int_{-\infty}^{\infty} x(\tau)h(t-\tau)d\tau$$

The process of convolution is represented in short-hand with a $*$. That is, in short, the above equation is expressed as:

$$y(t) = h(t) * x(t) = x(t) * h(t)$$

We know that if we apply an input voltage source of $x(t)$, with a Laplace transform of $X(s)$, to the LTI circuit, the response should correspondingly be $X(s)H(s)$ (because of linearity). Does this work out to convolution in the time domain? Let us check.

Theorem 5.9: Convolution in the time domain is equivalent to multiplication in the Laplace domain.

Proof.

$$\begin{aligned}\mathcal{L}[y(t)] &= \mathcal{L}[\int_{-\infty}^{\infty} x(\tau)h(t-\tau)\,d\tau] \\ &= \int_{t=0^-}^{\infty}\int_{\tau=-\infty}^{\infty} x(\tau)h(t-\tau)\,d\tau\; e^{-st}\,dt\end{aligned}$$

We will now interchange the order of integration. First we will integrate with respect to t, and we will integrate with respect to τ after that.

$$\begin{aligned}\therefore \mathcal{L}[y(t)] &= \int_{\tau=-\infty}^{\infty} x(\tau)\left(\int_{t=0^-}^{\infty} h(t-\tau)e^{-st}\,dt\right)d\tau \\ &= \int_{\tau=-\infty}^{\infty} x(\tau)\left(\int_{t=0^-}^{\infty} h(t-\tau)e^{-s(t-\tau)}dt\right)e^{-s\tau}d\tau\end{aligned}$$

Now we will replace $t-\tau$ with a new variable, t'.

$$\therefore \mathcal{L}[y(t)] = \int_{\tau=-\infty}^{\infty} x(\tau)\left(\int_{t'=-\tau}^{\infty} h(t')e^{-st'}dt'\right)e^{-s\tau}d\tau$$

Now both $h(t')$ and $x(\tau)$ are 0 for all $t' < 0$ and $\tau < 0$, respectively.

$$\begin{aligned}\therefore \mathcal{L}[y(t)] &= \int_{\tau=-\infty}^{\infty} \left(\int_{t'=0}^{\infty} h(t')e^{-st'}dt'\right)e^{-s\tau}d\tau \\ &= H(s)\int_{-\infty}^{\infty} x(\tau)e^{-s\tau}d\tau \\ &= H(s)\int_{0}^{\infty} x(\tau)e^{-s\tau}d\tau = H(s)X(s)\end{aligned}$$

■

The overall procedure is shown in Fig. 5.4. In the Laplace domain, each block processes the signal by multiplying its transfer function. From an input independent source to the output voltage or current, a circuit behaves just like any other system, with a transfer function in the Laplace domain.

5.4.2 Transfer function representation

An LTI circuit will comprise of L, C, R, M, and linear dependent sources. We will assume the circuit has a single independent source, which is considered as

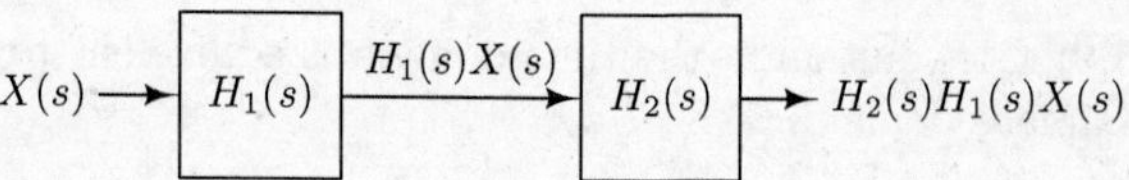

Figure 5.4: A system processes an input by multiplying its transfer function in the Laplace domain.

the input. To work out the impulse response of the circuit, the independent source is set to $\delta(t)$, or 1 in the Laplace domain. The inductors, capacitors, and mutual inductances are replaced with their respective Laplace-domain models.

Now consider working out the output with the help of the mesh-current method. The impedances of each branch are tabulated in a matrix with correct signs, and this matrix is multiplied by the unknown mesh currents to obtain a vector comprising of functions of the single independent voltage source. The mesh currents will be obtained by inverting the matrix of impedances. As such, every output of the circuit, and **the impulse response of the circuit will be a rational polynomial in s, i.e., of the form $H(s) = N(s)/D(s)$, where $N(s)$ and $D(s)$ are polynomials.**

5.4.3 Poles and zeros

Definition 5.3 (Poles): The roots of $D(s)$, the denominator of the transfer function, are the poles.

Definition 5.4 (Zeros): The roots of $N(s)$, the numerator of the transfer function, are the zeros.

The units for both pole and zero locations are the same as the units of s, that is, radians per second.

Example 5.13. In example 5.9, the output is v_C. What are the poles and zeros of the system?

The expression for $V_C(s)$, if the input voltage source is $\delta(t)$, is:

$$V_C(s) = \frac{sL_3}{R_2 + sL_3 + s^2 R_2 L_3 C_4} = \frac{2s}{s^2 + 2s + 12}$$

The numerator of the system transfer function is $2s$. The system transfer function has a zero at $s = 0$ rad/s.

The denominator of the system transfer function is $s^2 + 2s + 12$. The poles of the system are at $s = (-1 + j\sqrt{11})$ rad/s and $s = (-1 - j\sqrt{11})$ rad/s.

The poles and zeros of a system are denoted pictorially on the s-plane. The poles are shown as crosses, whereas the zeros are located with circles.

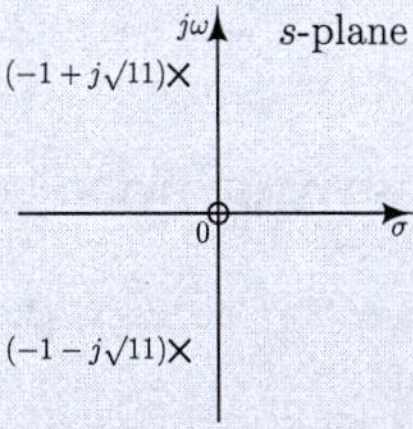

We have defined poles twice, first in Definition 3.3 as the roots of the characteristic equation, and now in Definition 5.3 as the roots of the denominator polynomial of the impulse response. Is there any discrepancy?

It can be proved that the roots of the characteristic equation are identical to the roots of the denominator polynomial. Regardless of the chosen output quantity, the poles of the system will be identical.

Scan the QR-code, and learn more about why the roots of the characteristic equation are identical to the roots of the denominator of the system transfer function.

The response of a system to any input can be broken up with the help of partial fractions. For each pole s_k with a residue r_k, the impulse response will be $r_k e^{s_k t}$ in the time domain. In case the input is not an impulse, the residues may change, but the evaluation technique remains the same. If the pole is in the LHP, i.e., if $\text{Re}[s_k] < 0$, then $e^{s_k t}$ reduces to 0 as $t \to \infty$. Therefore, the impulse response of a system reduces to 0 as $t \to \infty$ if the poles of the system are in the LHP. As $t \to \infty$, in such a system, the response of the system is only the forced response and the free response dies out. This is true even when the poles are repeated and the impulse response has terms like $t^n e^{s_k t}$. The negative exponential decays faster than any polynomial.

5.4.4 Stability of a system

A system is said to be stable if for bounded inputs the output is bounded. In other words, if the output does not blow up for any reasonable input, the system is stable.

If all the poles of the system are in the LHP, the free response decays to 0 as $t \to 0$. If we apply a bounded input to such a system, the forced response will remain bounded. A system with all its poles in the LHP is stable.

5.5 A system in sinusoidal steady state

Let us now relate the Laplace transform with the sinusoidal steady state analysis that we learned in unit 4. The Laplace transform of a cosine input is $s/(s^2 + \omega_0^2)$. The system transfer function will have multiple poles; the poles that are not in the left half s-plane will lead to unstable behavior. As such, we will consider a system that has poles only in the left half s-plane (LHP).

Let us consider a system with a transfer function of the following form:

$$H(s) = K \cdot \frac{(s - s_{z_1})(s - s_{z_2}) \cdots (s - s_{z_n})}{(s - s_{p_1})(s - s_{p_2}) \cdots (s - s_{p_m})}$$

The system has m poles and n zeros. The input to the system is a cosine at ω_0, with a Laplace transform of $s/(s^2 + \omega_0^2)$. Then, the output of the system, $Y(s)$, is given by:

$$Y(s) = K \cdot \frac{s}{(s^2 + \omega_0^2)} \frac{(s - s_{z_1})(s - s_{z_2}) \cdots (s - s_{z_n})}{(s - s_{p_1})(s - s_{p_2}) \cdots (s - s_{p_m})}$$

We should be able to transform $Y(s)$ into the time domain with the help of the partial fraction method. Each pole of the system will have a complex residue. Let r_k be the residue for the pole s_{p_k}. Let us assume the residues at $s = \pm j\omega_0$ are r_a and r_b. Then, we have $y(t)$ as:

$$y(t) = r_a e^{j\omega_0 t} + r_b e^{-j\omega_0 t} + \underbrace{r_1 e^{s_{p_1} t} + r_2 e^{s_{p_2} t} + \cdots + r_m e^{s_{p_m} t}}_{\text{exponentially decaying to 0}}$$

All the poles of the system are in the LHP. As such, a term $r_k e^{s_{p_k} t}$ will exponentially decay to zero with time. In the **steady state**, a long time after the experiment was started, all such terms will contribute to 0. Then, in the steady state, we will be left with only the first two terms.

To compute the residue r_a, we multiply $Y(s)$ by $(s - j\omega_0)$ and plug in $s = j\omega_0$.

$$\begin{aligned}
\therefore r_a = \lim_{s \to j\omega_0} (s - j\omega_0) Y(s) &= \frac{s}{s + j\omega_0} \cdot K \frac{(s - s_{z_1}) \cdots}{(s - s_{p_1}) \cdots} \\
= \frac{j\omega_0}{2j\omega_0} \cdot H(j\omega_0) &= \frac{1}{2} H(j\omega_0)
\end{aligned}$$

Similarly, for r_b, we multiply $Y(s)$ by $(s + j\omega_0)$ and plug in $s = -j\omega_0$.

$$\begin{aligned} r_b = \lim_{s \to -j\omega_0} (s + j\omega_0)Y(s) &= \frac{s}{s - j\omega_0} \cdot K\frac{(s - s_{z_1})\cdots}{(s - s_{p_1})\cdots} \\ = \frac{-j\omega_0}{-2j\omega_0} \cdot H(-j\omega_0) &= \frac{1}{2}H(-j\omega_0) \end{aligned}$$

Finally, in the steady state, we have the response of the system as:

$$y(t) = \frac{H(j\omega_0)e^{j\omega_0 t} + H(-j\omega_0)e^{-j\omega_0 t}}{2}$$

The input to our system was real. It is imperative that $y(t)$, the output, is also real. We can break up $y(t)$ into real and imaginary parts; the imaginary part necessarily needs to be zero.

$$\begin{aligned} \therefore 2\,\mathrm{Re}[y(t)] &= \mathrm{Re}[H(j\omega_0)]\cos(\omega_0 t) - \mathrm{Im}[H(j\omega_0)]\sin(\omega_0 t) \\ &+ \mathrm{Re}[H(-j\omega_0)]\cos(\omega_0 t) + \mathrm{Im}[H(-j\omega_0)]\sin(\omega_0 t) \quad (5.5) \\ \text{and } 2\,\mathrm{Im}[y(t)] &= \mathrm{Re}[H(j\omega_0)]\sin(\omega_0 t) + \mathrm{Im}[H(j\omega_0)]\cos(\omega_0 t) \\ &- \mathrm{Re}[H(-j\omega_0)]\sin(\omega_0 t) + \mathrm{Im}[H(-j\omega_0)]\cos(\omega_0 t) \quad (5.6) \end{aligned}$$

(5.6) is zero at all times. The cosine terms must add up to zero, and the sine terms must add up to zero separately. This means the real parts of $H(j\omega_0)$ and $H(-j\omega_0)$ are equal, and the imaginary parts of $H(j\omega_0)$ and $H(-j\omega_0)$ add up to zero. To summarize the two, $H(-j\omega_0)$ is the conjugate of $H(j\omega_0)$.

(5.5) can now be simplified as follows.

$$\begin{aligned} 2y(t) &= 2\,\mathrm{Re}[H(j\omega_0)]\cos(\omega_0 t) - 2\,\mathrm{Im}[H(j\omega_0)]\sin(\omega_0 t) \\ \text{or, } y(t) &= |H(j\omega_0)|\cos(\omega_0 t + \phi) \quad (5.7) \\ \text{where } \phi &= \angle H(j\omega_0) \end{aligned}$$

Equation (5.7) implies that the sinusoidal steady state response of a system can be quickly computed by plugging in $s = j\omega_0$ and evaluating the magnitude and the phase. The input amplitude is scaled by the magnitude and modified in phase by the phase of $H(j\omega_0)$. This discussion reconciles the Laplace transform method with our development of phasors in unit 4.

5.5.1 Significance of poles and zeros

The poles and zeros of a transfer function are significant in the following ways.

1. If the transfer function has a zero at a frequency s_z, and we apply an input at the complex frequency s_z, i.e., the input is $e^{s_z t}$, then the forced response is zero. If the system has a pair of zeros at $\pm j\omega_0$ and we apply $\cos\omega_0 t$ as the input, the forced response will be 0.

2. A similar statement cannot be made for the poles. Poles of the transfer function express the nature of the free (homogeneous, zero-input) response of the system. Given a system with poles at $s_{p_1}, s_{p_2}, \ldots$, for any initial conditions, if we release the system with zero input, the response will be of $A_1 e^{s_{p_1} t} + A_2 e^{s_{p_2} t} + \cdots$, where A_1, A_2 are coefficients that can be determined by obtaining the residues.

3. Poles in the RHP will result in an unstable system because its response to initial conditions will rise exponentially. Poles on the $j\omega$ axis will result in a marginally stable system; its response to initial conditions will not decay. A stable system necessarily requires all its poles in the LHP.

4. The frequency response of a system can be constructed from its poles and zeros.

5.5.2 Frequency response

While using a circuit as a system, it is of great importance to understand the response of the circuit to different frequencies, in magnitude and in phase. For a system $H(s)$, the magnitude, $|H(j\omega)|$ and the phase $\angle H(j\omega)$ together comprise the frequency response.

Example 5.14. In the circuit shown below, evaluate the frequency response of V_{out} with respect to the input V_{in}. Assume R_0 is 1 Ω, L and C are 1/2 H and 1/3 F respectively.

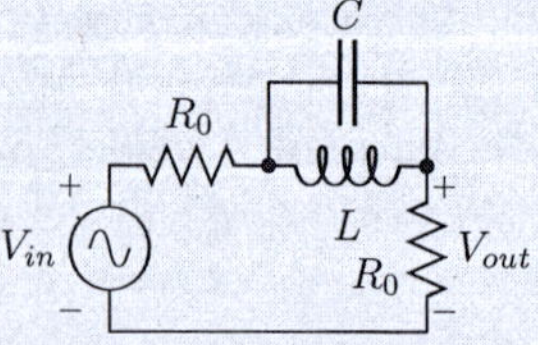

The inductor has an impedance of sL, the capacitor has an impedance of $1/(sC)$. The two components in parallel have an impedance of $sL/(1 + s^2LC)$. We can obtain V_{out} through potential division.

$$V_{out}(s) = \frac{R_0}{R_0 + \frac{sL}{1+s^2LC} + R_0} \cdot V_{in}(s)$$

The system transfer function can be simplified as:

$$H(s) = \frac{V_{out}(s)}{V_{in}(s)} = \frac{R_0(1+s^2LC)}{2R_0(1+s^2LC)+sL} = \frac{1}{2} \cdot \frac{1+s^2LC}{s^2LC + \frac{sL}{2R_0} + 1}$$

$$\therefore H(j\omega) = \frac{1-\omega^2LC}{2(1-\omega^2LC) + \frac{j\omega L}{R_0}} = \frac{6-\omega^2}{12-2\omega^2+3j\omega}$$

We can now obtain the magnitude and phase responses.

$$|H(j\omega)|^2 = \frac{(6-\omega^2)^2}{(12-2\omega^2)^2+9\omega^2}$$

$$\angle H(j\omega) = -\tan^{-1}\frac{3\omega}{12-2\omega^2}$$

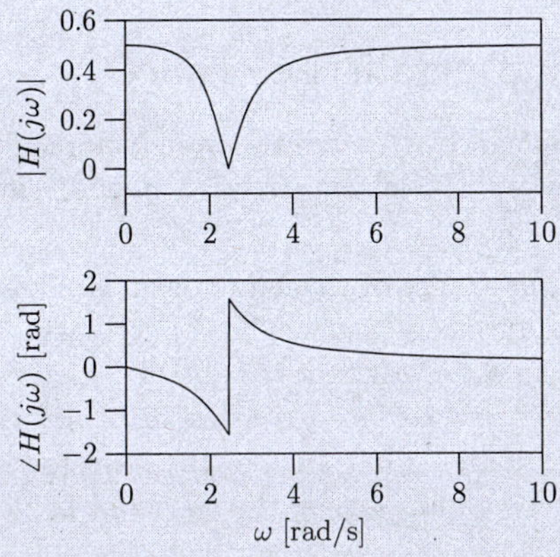

The frequency response of a system is the value of $H(s)$ as we sweep s along the $j\omega$ axis. The system may have multiple poles in the LHP and multiple zeros. There is no restriction on the location of zeros. Complex poles and zeros necessarily have to come in pairs with their complex conjugates. The complex conjugate of every complex pole and zero is required to ensure the overall transfer function remains real. This is visualized in Fig. 5.5.

The frequency response is a complex number and can be visualized either in Cartesian coordinates (as $H_r + jH_i$) or in polar coordinates (as $|H|e^{j\phi}$). A polar coordinate plot is generally preferred, because the magnitude plot directly tells us how much the input amplitude will be amplified or attenu-

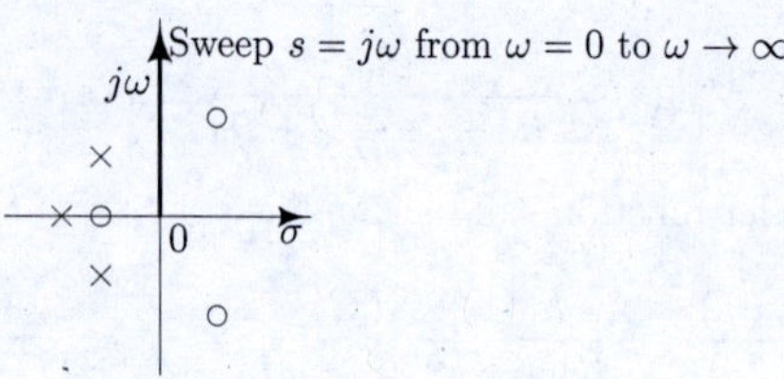

Figure 5.5: The frequency response of $H(s)$ is its value as we sweep s along the positive $j\omega$ axis.

ated.

The plots of the frequency response give the engineer intuition about the circuit. For instance, the frequency response plots in example 5.14 reveal that the output is zero for $\sqrt{6}$ rad/s. For frequencies sufficiently far from this spot frequency, the magnitude gain is 0.5.

5.5.3 Series and parallel resonance

Definition 5.5 (Resonance): A circuit with inductors and capacitors is said to be in resonance when at a specific frequency its driving point impedance is purely real.

For $s = j\omega$, the impedance of the inductor is purely imaginary and positive, while the impedance of the capacitor is purely imaginary but negative. If we combine them in series, then at $\omega = 1/\sqrt{LC}$ the combined impedance is 0.

On the other hand, the admittance of the inductor is purely imaginary and negative, while the admittance of the capacitor is purely imaginary and positive. If we combine them in parallel, then at $\omega = 1/\sqrt{LC}$ the combined admittance is 0.

Corollary 5.1. An inductor L and a capacitor C resonate in series at an angular frequency of $1/\sqrt{LC}$ to form a short circuit (zero impedance).

Corollary 5.2. An inductor L and a capacitor C resonate in shunt at an angular frequency of $1/\sqrt{LC}$ to form an open circuit (zero admittance).

In example 5.14, the inductor and capacitor resonate at $\omega = \sqrt{6}$ rad/s. At this frequency, the parallel-LC combination behaves like an open circuit, and there is no connection between V_{in} and V_{out}. As such, the frequency response at this frequency is 0, as one will observe in the graph for $|H(j\omega)|$.

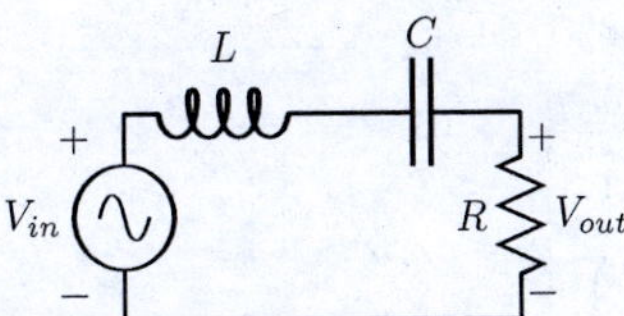

Figure 5.6: A circuit with series LC resonance

Series-LC resonance

In the series RLC network of Fig. 5.6,

$$H(s) = \frac{V_{out}(s)}{V_{in}(s)} = \frac{R}{R + sL + \frac{1}{sC}} = \frac{sCR}{1 + sCR + s^2LC}$$

For $s = j\omega$, we can obtain the magnitude and phase of the transfer function as follows.

$$|H(j\omega)|^2 = \frac{\omega^2R^2C^2}{(1-\omega^2LC)^2 + \omega^2R^2C^2}$$

$$\angle H(j\omega) = \frac{\pi}{2} - \tan^{-1}\frac{\omega RC}{1-\omega^2LC}$$

The magnitude of $H(j\omega)$ is of particular interest and is plotted in the graph of Fig. 5.7. For the graph, L, C and R are chosen as 1 H, 1 F and 1 Ω, without any loss of generality. It is customary to show the magnitude plot in decibels. In most cases, showing the magnitude plot with the ω axis in the log-scale reveals more information.

A signal exactly at $\omega = 1/\sqrt{LC}$ will have no attenuation, and a magnitude response of 1. Other frequencies will be attenuated by the circuit. The response of the circuit is known as **band-pass**, because a band of frequencies around $\omega_0 = 1/\sqrt{LC}$ pass through, while all other frequencies are stopped. The graph in Fig. 5.7 is in dB; -40 dB on the y-axis means an attenuation of 100 times.[1]

The band of frequencies that are declared to have passed are arbitrarily chosen as the frequencies that are attenuated by less than 3 dB, or in amplitude terms, are at least $1/\sqrt{2}$ times the input amplitude. Similarly, the frequencies that have not passed through this "filter" are such that they are

[1]A ratio can be expressed in decibels (dB). If a ratio x is of two amplitudes, then x is expressed in dB as $20 \times \log_{10} x$. If x is the ratio of two powers, it is expressed in dB as $10 \times \log_{10} x$.

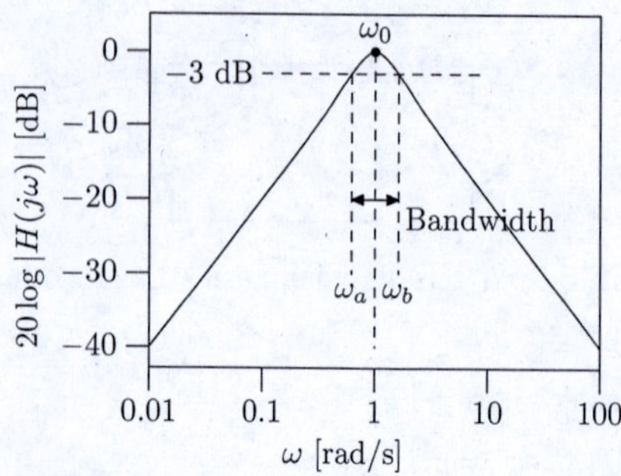

Figure 5.7: $|H(j\omega)|$ as a function of ω for the circuit in Fig. 5.6. R, L, C were chosen as 1 Ω, 1 H, and 1 F, respectively.

attenuated by more than 3 dB. ω_a and ω_b in Fig. 5.7 are frequencies such that $|H(j\omega_{a,b})|^2 = 1/2$. Let us solve.

$$
\begin{aligned}
|H(j\omega_{a,b})|^2 = \frac{1}{2} &= \frac{\omega_{a,b}^2 R^2 C^2}{(1-\omega_{a,b}^2 LC)^2 + \omega_{a,b}^2 R^2 C^2} \\
\therefore \omega_{a,b}^2 R^2 C^2 &= (1-\omega_{a,b}^2 LC)^2 \\
\text{or, } \pm\, \omega_{a,b} RC &= 1 - \omega_{a,b}^2 LC \\
\text{or, } \omega_{a,b}^2 LC \pm \omega_{a,b} RC - 1 &= 0
\end{aligned}
$$

It is clear from the above two ($\pm$) quadratic equations that the product of the roots is $-1/(LC)$. This reveals that if one root is positive, the other will be negative. Since ω_a and ω_b are both positive, they need to be roots of two different ($\pm$) equations. Alternately, if for one equation ω_a is a root, then $-\omega_b$ is the other root. For the second equation ω_b and $-\omega_a$ are the roots. The sum of roots in a quadratic equation is the negative ratio of the first and second coefficients. In other words, $\omega_b - \omega_a$ is $RC/LC = R/L$.

We define the **quality factor**, Q, of the network as the ratio of the center frequency to the bandwidth. In this case, Q is $\omega_0 L/R$, or $\sqrt{L/C}/R$, or $1/(\omega_0 RC)$.

The resistance in the series-RLC network of Fig. 5.6 is changed to vary the Q, without changing the resonant frequency. In Fig. 5.8, we have shown the frequency response of the network for different quality factors, as we have changed the value of R. The bandwidth becomes narrower as we increase Q; the band-pass nature of the circuit becomes sharper as we increase Q.

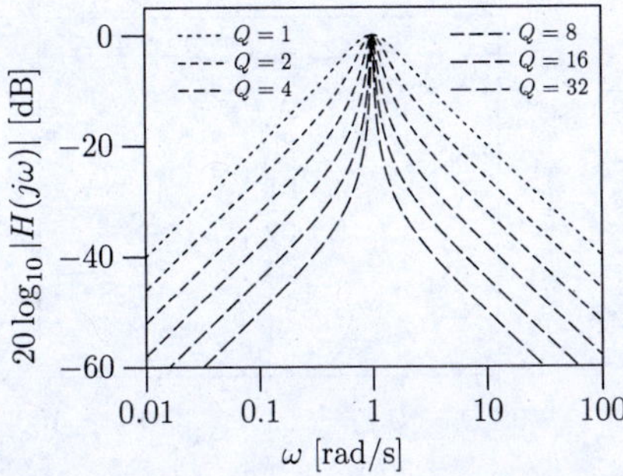

Figure 5.8: Frequency response of the series RLC network as Q is increased from 1 to 32.

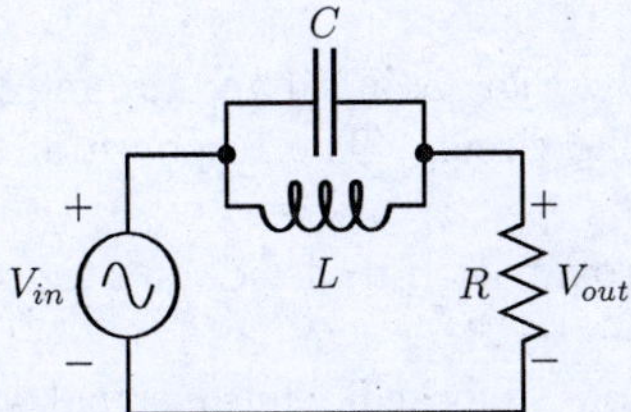

Figure 5.9: A circuit with shunt LC resonance

Shunt LC resonance

The circuit in Fig. 5.9 has an inductor and capacitor in shunt, that resonate at $\omega_0 = 1/\sqrt{LC}$. At the resonance frequency the shunt combination behaves as an open circuit. It is expected that at the resonant frequency, the magnitude of the transfer function will be 0. At DC the inductor, and at infinitely high frequencies, the capacitor will behave as a short circuits.

$$
\begin{aligned}
H(s) &= \frac{R}{R + \frac{sL}{1+s^2LC}} = \frac{1 + s^2LC}{s^2LC + sL/R + 1} \\
H(j\omega) &= \frac{1 - \omega^2 LC}{(1 - \omega^2 LC) + j\omega L/R} \\
|H(j\omega)|^2 &= \frac{(1 - \omega^2 LC)^2}{(1 - \omega^2 LC)^2 + \omega^2 L^2/R^2}
\end{aligned}
$$

Once again, $|H(j\omega)|$ is of interest, and is plotted in Fig. 5.10 for R, L, C of 1 Ω, 1 H and 1 F, respectively. A signal at $\omega_0 = 1/\sqrt{LC}$ will be blocked by

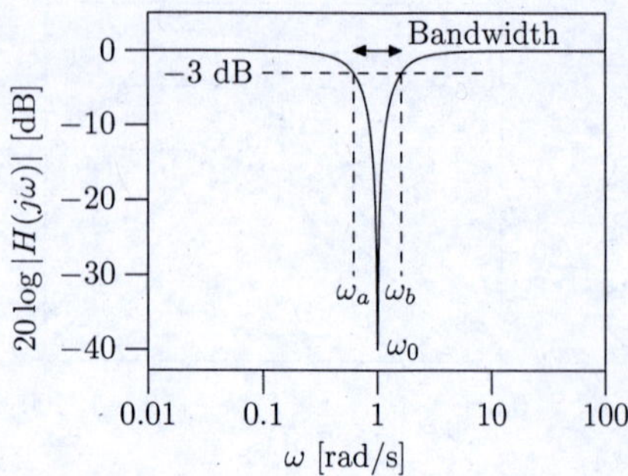

Figure 5.10: $|H(j\omega)|$ as a function of ω for the shunt-LC resonant circuit in Fig. 5.9. R, L, C were chosen as 1 Ω, 1 H, and 1F, respectively.

the LC-shunt impedance. Signals in a band around this resonant frequency will be attenuated by the circuit. The response of the circuit is known as **band-stop**.

An engineering approximation to decide if a signal is passed or not is to check for an attenuation of 3 dB, or a gain of $1/\sqrt{2}$ in the magnitude. A signal that has more attenuation is blocked, while a signal with less attenuation has passed.

In the case of the LC-shunt resonant circuit, frequencies between ω_a and ω_b in Fig. 5.10 have been blocked, while all others pass. ω_a and ω_b are the roots of the expression:

$$\frac{1}{2} = \frac{(1 - \omega_{a,b}^2 LC)^2}{(1 - \omega_{a,b}^2 LC)^2 + \omega_{a,b}^2 L^2/R^2}$$

This simplifies to:

$$\omega_{a,b}^2 LC + \pm\omega_{a,b} L/R - 1 = 0$$

The −3 dB cut-off frequencies can be worked out. More importantly, the bandwidth may be expressed as:

$$\omega_b - \omega_a = \frac{1}{RC} = \frac{\omega_0}{\omega_0 RC} = \frac{\omega_0}{Q}$$

The Q of the circuit is $\omega_0 RC$, or $R/\sqrt{L/C}$, or $R/(\omega_0 L)$. In the case of the shunt-LC resonant circuit, the quality factor is proportional to R, whereas, in the case of the series-LC resonant circuit, the quality factor is inversely proportional to R. The bandwidth of the band-stop circuit can be made narrower and narrower by increasing the values of R.

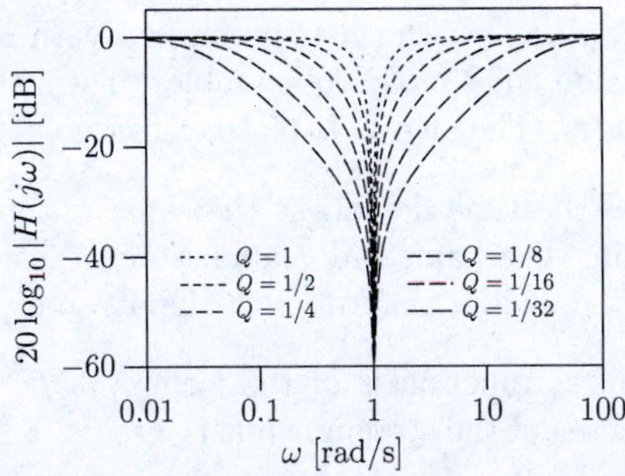

Figure 5.11: The bandwidth of the shunt-LC resonating circuit increases as Q is decreased.

Fig. 5.11 shows the magnitude response of the LC-shunt resonant circuit of Fig. 5.9, with L and C set to 1 H and 1 F, respectively. The value of Q is changed from 1 to (1/32) by decreasing the value of R from 1 Ω to (1/32) Ω. As Q decreases, the bandwidth of the band-stop response increases. An Octave program to generate the plot in Fig. 5.11 is given in Prog. 5.3.

Program 5.3. qshuntLC.m: Evaluation and plotting

```
w = 10.^[-2:0.01:2]; % Set of values of ω
jw = j*w;
tosave = transpose(w);
for R = 1./[1 2 4 8 16 32] % Set of resistor values
   B = [1 0 1]; % Numerator as a polynomial
   A = [1 1/R 1]; % Denominator as a polynomial
   num = B*[jw.^2; jw; ones(size(jw))]; % N(jω)
   den = A*[jw.^2; jw; ones(size(jw))]; % D(jω)
   Hjw = num./den; % H(jω)
   semilogx(w, 20*log10(abs(Hjw))); hold on;
end;
```

5.6 Unit summary

- The Laplace transform can be used to work out the transient response rapidly. Every capacitor is replaced by an impedance of $1/sC$ in series with a voltage source $v_C(0^-)/s$. Every inductor is replaced by an impedance of sL in shunt with a current source $i_L(0^-)/s$. The rest of the analysis is performed using ordinary DC analysis, without solving any differential equation.

- The inverse Laplace transform is performed with the help of partial fraction expansion and simple lookup tables. The inverse Laplace transform of $1/s$ is $u(t)$. The inverse Laplace transform of $1/(s-a)$ is $e^{at}u(t)$.
- Convolution in the time domain is the same as multiplication in the Laplace domain. The Laplace-domain system transfer function simply multiplies the Laplace-transform of the input, to give the output.
- A system transfer function is of the form $N(s)/D(s)$. The roots of $N(s)$ are the zeros of the system, while the roots of $D(s)$ are the poles.
- A system is stable if all its poles are in the LHP (left half s-plane). There are no constraints on the locations of the zeros for the stability of a system.
- To obtain the sinusoidal steady-state response of a system, s is to be substituted with $j\omega$. The magnitude of $H(j\omega)$ scales the input amplitude. The phase of $H(j\omega)$ modifies the phase of the input sinusoid.
- The magnitude and phase of $H(j\omega)$ are plotted to represent complete information about the system. The magnitude is usually plotted in decibels. Both the magnitude and phase are usually **plotted with the ω axis in the log scale**.
- An inductor in series with a capacitor resonate at $1/\sqrt{LC}$ to form a short-circuit. An inductor in shunt with a capacitor resonate at $1/\sqrt{LC}$ to form an open circuit.
- The quality factor or Q of a series-LC circuit is inversely proportional to R. The Q of a shunt-LC circuit is proportional to R. Q is the ratio of the center-frequency and the bandwidth.

5.7 Exercises

Multiple choice type questions

5.1 The Laplace transform of $u(t)$ is:
(a) s (b) 1 (c) $1/s$ (d) $1/s^2$

5.2 The inverse Laplace transform of $1/(s^2+1)$ is:
(a) $u(t)\cos t$ (b) $u(t)\sin t$ (c) $e^{-t}u(t)\cos t$ (d) $e^{-t}u(t)\sin t$

5.3 The operation of convolution is:
(a) Commutative (b) Distributive (c) Associative (d) All of these

5.4 Which one of the following systems **is not** stable?
(a) $\dfrac{s-1}{s+1}$ (b) $\dfrac{s+1}{s-1}$ (c) $\dfrac{s+1}{s+2}$ (d) $\dfrac{s+2}{s+1}$

5.5 Which one of the following systems **is** stable?
(a) $\dfrac{s-1}{s^2+2s+2}$ (b) $\dfrac{s+1}{s^2-2s+2}$ (c) $\dfrac{s-1}{s^2+1}$ (d) $\dfrac{s+1}{s^2}$

5.6 A pulse is given by the function $u(t)-u(t-1)$. What is its Laplace transform?
(a) $\dfrac{1}{se^s}\cdot(e^s-1)$ (b) $\dfrac{1}{s}\cdot(1-e^s)$ (c) $\dfrac{1}{se^s}\cdot(1-e^s)$ (d) $\dfrac{1}{s}\cdot(e^s-1)$

5.7 A series-RLC circuit is designed with $R = 10\ \Omega$, $L = 100$ nH, $C = 10$ pF. The output of the circuit is the voltage across R. The center frequency is:
(a) 10 MHz (b) 159.1 MHz (c) 1 GHz (d) 6.28 GHz

5.8 A series-RLC circuit is designed with $R = 10\ \Omega$, $L = 100$ nH, $C = 10$ pF. The output of the circuit is the voltage across R. The bandwidth of the circuit is:
(a) 1 MHz (b) 15.9 MHz (c) 100 MHz (d) 628 MHz

5.9 A signal $v(t)$ is $e^{5t}u(t)$. What is the region-of-convergence (RoC) for its Laplace transform $V(s)$?
(a) $s > 5$ (b) $s < 5$ (c) $s > -5$ (d) $s < -5$

5.10 A signal $v(t)$ is $e^{5t}(u(t)-u(t-1))$. What is the RoC of its Laplace transform $V(s)$?
(a) $s > 5$ (b) Entire RHP (c) Entire LHP (d) All s

Short answer type questions

5.11 In the Laplace transform domain, an inductor of value L with an initial current of i_0 is represented by a ______ source of value $-Li_0$ in ______ with an impedance sL, or a ______ source of value i_0/s in ______ with an impedance sL.

5.12 A voltage $v(t)$ has units of volts. What are the units of its Laplace transform, $V(s)$? What is the unit of s? The Laplace transform of the current through a capacitor is $I_C(s) = sCV_C(s) - Cv_C(0^-)$. Justify the dimensions of the equation. How do you reconcile this result with the dimensions of a system transfer function, $H(j\omega)$? Also, reconcile the dimensions of all the entries in Tab. 5.1.

5.13 The current through a capacitor is taken as the input to a system. The output of the system is the voltage across the same capacitor. Is this a stable system? Try to construct unstable systems using only L, C, R, M. Can you make any conclusions about systems designed only with L, C, R, M?

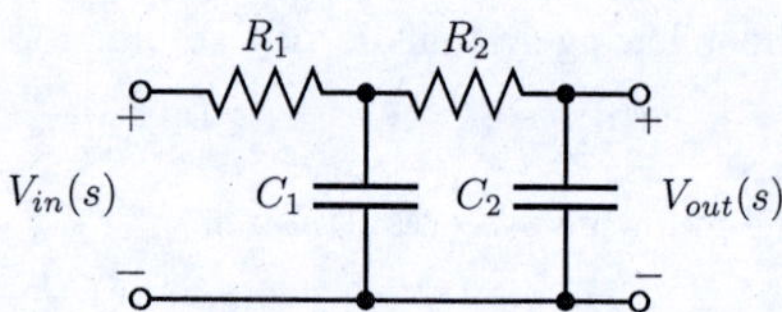

Figure 5.12: Schematic for exercise 5.18

5.14 If we ignore the initial conditions, the derivative in the time domain implies multiplying by s in the Laplace domain. A derivative in the Laplace domain implies multiplying by t in the time domain. Find other properties of the Laplace transform that behave in this dual manner.

5.15 The impulse response of a system is $h(t)$. Prove that if an input $u(t)$ is applied to the system, the output will be $\int_{0^-}^{t} h(\tau)\, d\tau$.

Numericals

5.16 Find the time domain impulse responses corresponding to the following system transfer functions.

(a) $\dfrac{3s^2 + 20s + 31}{s^3 + 10s^2 + 31s + 30}$ (b) $\dfrac{s^2 + 5s + 9}{s^3 + 9s^2 + 24s + 20}$

(c) $\dfrac{4s + 4}{s^2 + 4s + 5}$ (d) $\dfrac{3s^2 + 12s + 13}{s^3 + 6s^2 + 13s + 10}$

(e) $\dfrac{-2s^3 - 3s^2 + 1}{s^4 + 6s^3 + 17s^2 + 28s + 20}$ (f) $\dfrac{2s^3 + 18s^2 + 54s + 53}{s^4 + 11s^3 + 45s^2 + 81s + 54}$

(g) $\dfrac{3s^3 + 4s^2 + 5s + 6}{(s+2)^2(s^2 + 2s + 10)}$ (h) $\dfrac{10s^3 + 38s + 127s}{(s^2 + 2s + 10)(s^2 + 4s + 13)}$

5.17 Solve exercises 3.21, 3.22, 3.23, 3.26, 3.28, 3.29, 3.30, 3.31 again, now using Laplace transforms.

5.18 In the schematic of Fig. 5.12, assume all resistors are equal to 1 Ω and all capacitors are equal to 1 F. A unit step, $u(t)$, is applied as $v_{in}(t)$. Work out the rise time at the output, $v_{out}(t)$.

5.19 In the schematic of Fig. 5.13, once again assume all resistors are 1 Ω and all capacitors are 1 F. Evaluate the rise time of the unit step response of the circuit.

5.20 In the schematic of Fig. 5.14, the values of L and C are such that $\sqrt{L/C}$ is R_0 and $\omega_0^2 LC = 1$. Evaluate $H(j\omega)$. Keep R_0 and ω_0 as variables in your expression. Evaluate the phase response and the magnitude response of the system. Plot the phase and magnitude responses.

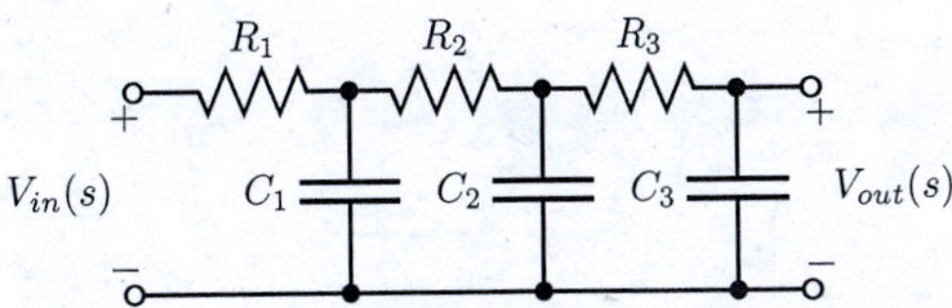

Figure 5.13: Schematic for exercise 5.19

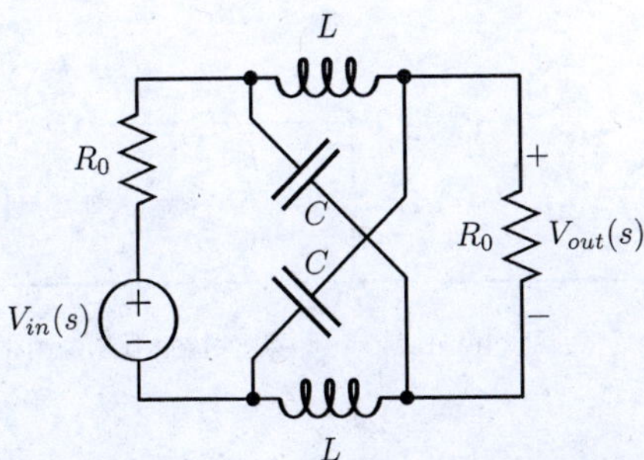

Figure 5.14: Schematic for exercise 5.20.

5.21 In the circuit of Fig. 5.15, R is 1 Ω, L is 2 H, C is 1 F. $H(s)$ is $V_{out}(s)/V_{in}(s)$. Find an expression for $|H(j\omega)|^2$.

5.22 In the circuit of Fig. 5.16, R is 1 Ω, C_1 is $2\cos(3\pi/8)$ F, L_2 is $2\cos(\pi/8)$ H, C_3 is $2\cos(\pi/8)$ F, L_4 is $2\cos(3\pi/8)$ H. $H(s)$ is $V_{out}(s)/V_{in}(s)$. Find an expression for $|H(j\omega)|^2$.

5.23 The circuit shown in Fig. 5.17 is excited by a voltage source $v_{in}(t)$ given by $e^{-\sigma_0 t}\cos(\omega_0 t)u(t)$. Express the voltage $v_{out}(t)$ as a function of time. σ_0 is 1 rad/s, ω_0 is 1 rad/s. Plot $v_{out}(t)$.

5.24 $v_{in}(t)$ in Fig. 5.17 is given by $e^{+\sigma_0 t}\cos(\omega_0 t)$. Express $v_{out}(t)$ as a function of time. σ_0 and ω_0 are both 1 rad/s. Plot $v_{out}(t)$.

5.25 In the Wien bridge circuit shown in Fig. 5.18, $v_{in}(t)$ is $u(t)$. Assume R_2 and R_3 to be 2 Ω and 3 Ω, respectively. Assume C_2 is 1/2 F. Evaluate and plot

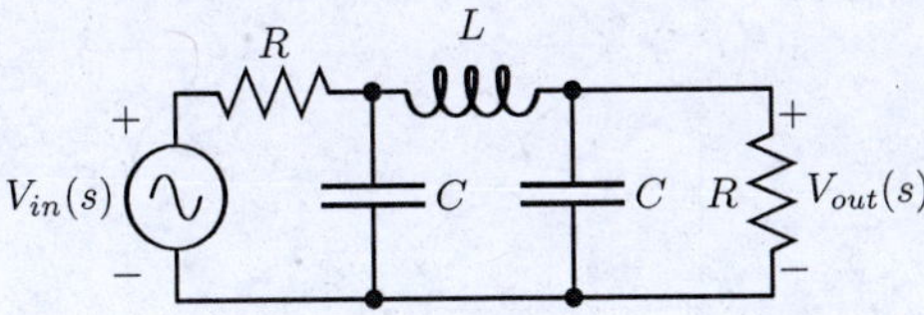

Figure 5.15: Schematic for exercise 5.21

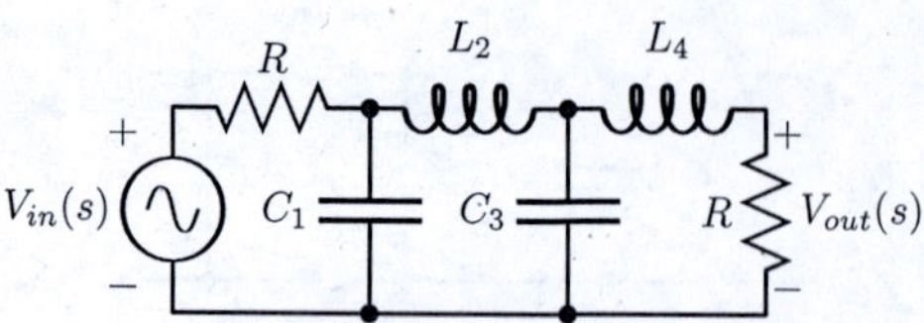

Figure 5.16: Schematic for exercise 5.22

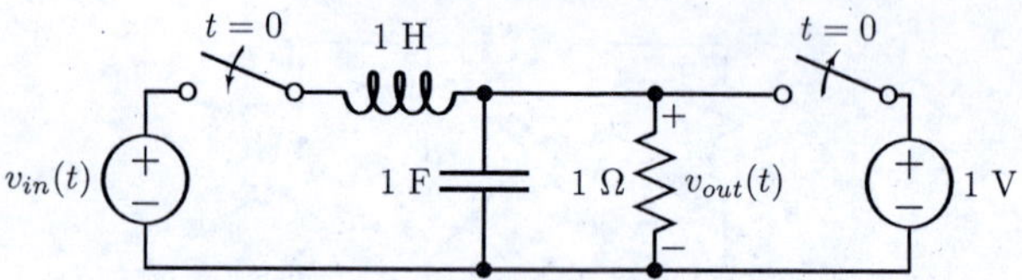

Figure 5.17: Schematic for exercises 5.23 and 5.24.

$v_{out}(t)$. What will be $v_{out}(t)$ if $v_{in}(t)$ is $\cos t$? Also find out $v_{out}(t)$ if $v_{in}(t)$ is $e^{-t}\cos t$. Relate your results to the frequency response, $|H(j\omega)|$, of the circuit.

5.26 Consider a shunt-LC network, where an inductor of value 1 nH is placed in shunt with a capacitor of value 1 nF. Such a network is known as a tank circuit. The inductor is not ideal, and a resistance of 100 Ω in shunt with the inductor represents the non-ideality. At $t = 0^-$, the voltage across the capacitor is 1 V. Evaluate the voltage across the capacitor with time. Come up with a value of Q, the quality factor, for the tank. Prove that the number of times the voltage across the capacitor oscillates is roughly equal to Q. (Assume that you can detect an oscillation as long as the amplitude has not decayed $e^{\pi} \approx 23$ times.) This technique is often used in the laboratory to measure the Q of a circuit.

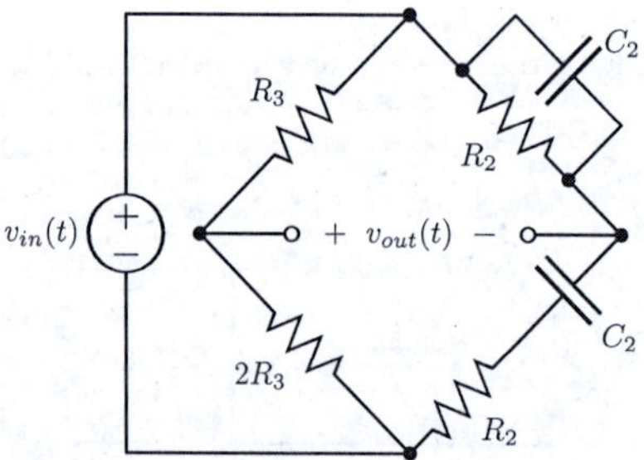

Figure 5.18: Wien bridge circuit for exercise 5.25.

Know more

Historical profiles

Pierre-Simon Laplace (1749 to 1827) was a French mathematician and astronomer. The Laplace transform is named after him. With the origin of the Laplace transform, classical mechanics was transformed from a geometrical study to a study based on calculus. Known as the Newton of France, Laplace's contributions ranged from the hypothesis of the black hole, an accurate model of the tides, to the formulation of Bayesian probability.

The Laplacian operator, ∇^2, is named after Laplace. Laplace showed that the gradient of the divergence of a potential will always be zero.

Pierre-Simon Laplace‡

‡: Photograph taken from the public domain.

Laplace's name is engraved on the northwest side of the Eiffel tower, along with many other French scientists and engineers. At École Militaire, Laplace was the teacher of Siméon Denis Poisson, whose contributions to mathematics and physics are also well recognized. Laplace was also the teacher of Napoleon Bonaparte. For a short time, after the French revolution, Napoleon had appointed Laplace as his minister of the interior. Napoleon had to dismiss him soon, as he realized that Laplace was not a good administrator.

Laplace transforms are ubiquitous in Electrical Engineering, and are frequently encountered in control engineering, electronic circuit design and power electronics.

Understand in depth

Laplace transforms have wide-ranging applications in physics, in structural engineering, mechanical engineering, and in EE. In EE, the main applications of Laplace transforms are in control engineering and robotics, circuit design, power systems, and power electronics. The Laplace transform is a tool to rapidly solve high order linear differential equations. Mastering the technique comes with practice.

For more practice:

- Network Analysis and Synthesis, by Franklin F. Kuo, John Wiley & Sons Inc., chapters 6 and 7.
- Engineering Circuit Analysis, by William H. Hayt and Jack E. Kemmerley, McGraw-Hill, chapter 13.

Unit 6

Fourier series and transforms

Unit specifics

In this unit we have discussed the following:

- Trigonometric and exponential Fourier series
- Discrete spectra and waveform symmetries
- Steady state response of a network to a periodic signal
- The Fourier transform and continuous spectra

Rationale

We have studied circuit analysis for DC inputs, steady-state sinusoidal inputs, and transient inputs that can be transformed into the Laplace domain. If the input is not a sinusoid, but periodic, circuit analysis with Laplace transforms or with phasors will not be useful. To analyze the steady state response of circuits to periodic but non-sinusoidal signals (for example, square waves, triangular waves), we will learn techniques that use the Fourier series and Fourier transform.

With the Fourier series, we will break up any periodic signal into a sum of sinusoidal waves comprising of a fundamental frequency and its harmonics. Once broken up into sinusoids, we will apply superposition along with phasor techniques to analyze the circuit completely.

Pre-requisites

- Mathematics: Differential and integral calculus.
- Physics: Familiarity with simple harmonic motion and periodicity.
- Circuit theory: Familiarity with Unit-3 and thoroughness with Unit-4 of this book.
- Software: We will continue using Octave.

Unit outcomes

The list of outcomes of this unit are as follows.

U6-O1: Be able to decompose a periodic signal into its Fourier series.

U6-O2: Be able to visually identify the presence or absence of harmonics in the spectrum.

U6-O3: Obtain the steady state response of a network for any periodic non-sinusoidal input.

U6-O4: Relate the Fourier series and Fourier transform with the Laplace transform.

Unit-6 outcomes	**Expected mapping with course outcomes** *(1: Weak, 2: medium, and 3: strong correlation)*					
	CO-1	**CO-2**	**CO-3**	**CO-4**	**CO-5**	**CO-6**
U6-O1	3	-	3	-	1	1
U6-O2	-	2	3	-	-	-
U6-O3	3	-	1	-	-	3
U6-O4	-	3	3	-	-	-

6.1 Preliminaries

6.1.1 Orthogonality of the trigonometric system

The sets of functions $\{\cos(kx)\}$ and $\{\sin(kx)\}$ where $k \in \mathbb{Z}$, is known as the **trigonometric system**. ($\mathbb{Z}$ is the set of all integers.) 1, $\cos x$, $\sin x$, $\cos(2x)$, $\sin(2x)$... are the functions in the trigonometric system.

Theorem 6.1: The trigonometric system forms a set of orthogonal functions. That is, the integral of the product of any two functions in the trigonometric system over the interval from 0 to 2π is 0. (One may also choose $-\pi$ to π for the integral, or any other interval of 2π.)

$$\int_0^{2\pi} \cos(ax)\cos(bx)\,dx = 0, \text{ where } a \neq b \text{ and } a, b \in \mathbb{Z} \quad (6.1)$$

$$\int_0^{2\pi} \sin(ax)\sin(bx)\,dx = 0, \text{ where } a \neq b \text{ and } a, b \in \mathbb{Z} \quad (6.2)$$

$$\int_0^{2\pi} \cos(ax)\sin(bx)\,dx = 0, \text{ where } a \neq b \text{ and } a, b \in \mathbb{Z} \quad (6.3)$$

Proof. We can use well-known trigonometric identities.

$$\cos(ax)\cos(bx) = \frac{1}{2}[\cos(a+b)x + \cos(a-b)x]$$

$$\sin(ax)\sin(bx) = \frac{1}{2}[\cos(a-b)x - \cos(a+b)x]$$

$$\cos(ax)\sin(bx) = \frac{1}{2}[\sin(a+b)x - \sin(a-b)x]$$

As long as $a \neq b$, the integrals of all the terms in the RHS from 0 to 2π are 0. ■

6.1.2 Periodic signals

The Fourier series is a convenient way to understand and represent periodic signals.

Definition 6.1 (Periodic signal): A signal $f(t)$ is periodic and has a period T if $f(t+T) = f(t)$ for all t.

In other words, if $f(t)$ repeats itself after every T period of time, it is periodic. A sine wave, $\sin(\omega t)$ is periodic with a period $2\pi/\omega$. There are other common periodic signals, for example a square wave, or a triangular wave, or the output of a half-wave or full-wave rectifier.

If a signal has a period of T, that is, if it repeats itself after T, then it will repeat itself after $2T$ as well, it will repeat itself after any nT where n is an integer. A signal with period T, also has a period of $2T$, $3T$, and so on. The **fundamental period** is the lowest value of T, such that $f(t)$ repeats itself.

Corollary 6.1. If $f(t)$ and $g(t)$ are both periodic functions with a period of T, then any linear combination of $f(t)$ and $g(t)$ will also be periodic with a period of T.

Proof.

$$\begin{aligned} f(t+T) &= f(t) \\ g(t+T) &= g(t) \\ \therefore af(t+T)+bg(t+T) &= af(t)+bg(t) \end{aligned}$$

■

The functions $\cos(k \cdot 2\pi/T \cdot t)$ and $\sin(k \cdot 2\pi/T \cdot t)$ are both periodic with a period of T for any integral value of k. A linear combination of these functions is also periodic with a period of T.

Corollary 6.2. If $f(t)$ is given by:

$$f(t) = a_0 + \sum_{k=1}^{\infty} a_k \cos\left(\frac{2\pi kt}{T}\right) + \sum_{k=1}^{\infty} b_k \sin\left(\frac{2\pi kt}{T}\right),$$

then $f(t)$ is periodic with a period T.

Proof. The proof follows from Corollary 6.1. Each function in the sets of functions $\{\cos(2\pi kt/T)\}$ and $\{\sin(2\pi kt/T)\}$ is periodic with a period of T. Their linear combination is also periodic with a period of T. ■

Joseph Fourier had proved the logical reverse of Corollary 6.2. He had proved that if **any** function is periodic with a period of T, then it can be broken up into a sum of sines and cosines from the trigonometric system, like $f(t)$. The required frequencies comprise a fundamental frequency and its harmonics (or multiples).

6.2 Trigonometric Fourier series

If $f(t)$ is periodic with T, and can be broken up into a sum of sines and cosines like in Corollary 6.2, then the expanded form of $f(t)$ is known as its **trigonometric Fourier series**, $\{a_k\}$ and $\{b_k\}$ are the Fourier coefficients of $f(t)$. There are a few very reasonable restrictions on $f(t)$ for which its Fourier series is valid.

Theorem 6.2: If $f(t)$ is periodic with T, and has **finite** number of discontinuities over its period, a finite number of maxima and minima over its period, and is absolutely integrable over its period, then:

$$f(t) = a_0 + \sum_{k=1}^{\infty} a_k \cos\left(\frac{2\pi kt}{T}\right) + \sum_{k=1}^{\infty} b_k \sin\left(\frac{2\pi kt}{T}\right),$$

where:

$$a_0 = \frac{1}{T}\int_0^T f(t)\,dt \tag{6.4}$$

$$a_k = \frac{2}{T}\int_0^T f(t)\cos\left(\frac{2\pi kt}{T}\right)\,dt \tag{6.5}$$

$$b_k = \frac{2}{T}\int_0^T f(t)\sin\left(\frac{2\pi kt}{T}\right)\,dt \tag{6.6}$$

Partial proof. We will only prove that the coefficients are as stated in (6.4), (6.5), (6.6). We will not prove equality of $f(t)$ with its Fourier series representation. We will not prove the conditions of validity (also known as the Dirichlet conditions) of the Fourier series. For a detailed proof, refer to [20].

To prove (6.4), let us find the average of $f(t)$.

$$\begin{aligned}
\int_0^T f(t)\,dt &= \int_0^T a_0\,dt + \sum_{k=1}^{\infty} a_k \underbrace{\int_0^T \cos\frac{2\pi kt}{T}\,dt}_{0} + \sum_{k=1}^{\infty} b_k \underbrace{\int_0^T \sin\frac{2\pi kt}{T}\,dt}_{0} \\
&= a_0 T \\
\therefore a_0 &= \frac{1}{T}\int_0^T f(t)\,dt
\end{aligned}$$

Next, we will multiply $f(t)$ by $\cos(2\pi nt/T)$ and integrate to prove (6.5).

$$\begin{aligned}
\int_0^T f(t)\cos\frac{2\pi nt}{T}\,dt &= \int_0^T a_0 \cos\frac{2\pi nt}{T}\,dt + \int_0^T \sum_{k=1}^{\infty} a_k \cos\frac{2\pi kt}{T}\cos\frac{2\pi nt}{T}\,dt \\
&+ \int_0^T \sum_{k=1}^{\infty} b_k \sin\frac{2\pi kt}{T}\cos\frac{2\pi nt}{T}\,dt
\end{aligned}$$

The first integral evaluates to 0. In the second integral, by Theorem 6.1, all the terms integrate to 0, except when $k = n$. In the third integral, by Theorem 6.1, all the terms integrate to 0. Finally, we have:

$$\begin{aligned}
\therefore \int_0^T f(t)\cos\frac{2\pi nt}{T}\,dt &= \int_0^T a_n \cos\frac{2\pi nt}{T}\cos\frac{2\pi nt}{T}\,dt \\
&= \int_0^T \frac{a_n}{2}(1+\cos(2\frac{2\pi nt}{T}))\,dt \\
&= \frac{a_n T}{2} \\
\therefore a_n &= \frac{2}{T}\int_0^T f(t)\cos\frac{2\pi nt}{T}\,dt
\end{aligned}$$

Lastly, we will multiply $f(t)$ by $\sin(2\pi nt/T)$ and integrate to prove (6.6).

$$\begin{aligned}
\int_0^T f(t)\sin\frac{2\pi nt}{T}\,dt &= \int_0^T a_0 \sin\frac{2\pi nt}{T}\,dt + \int_0^T \sum_{k=1}^{\infty} a_k \cos\frac{2\pi kt}{T}\sin\frac{2\pi nt}{T}\,dt \\
&+ \int_0^T \sum_{k=1}^{\infty} b_k \sin\frac{2\pi kt}{T}\sin\frac{2\pi nt}{T}\,dt
\end{aligned}$$

The first integral evaluates to 0. In the second integral, by Theorem 6.1, all the terms integrate to 0. In the third integral, by Theorem 6.1, all the terms integrate to 0, except when $k = n$.

$$\begin{aligned}
\therefore \int_0^T f(t)\sin\frac{2\pi nt}{T}\,dt &= \int_0^T b_n \sin\frac{2\pi nt}{T}\sin\frac{2\pi nt}{T}\,dt \\
&= \int_0^T \frac{b_n}{2}(1-\cos(2\frac{2\pi nt}{T}))\,dt \\
&= \frac{b_n T}{2} \\
\therefore b_n &= \frac{2}{T}\int_0^T f(t)\sin\frac{2\pi nt}{T}\,dt
\end{aligned}$$

Notice that we have shown that if $f(t)$ can be broken up into a Fourier series, then the coefficients are given by (6.4), (6.5), (6.6). We have not proved that the sequences of $\{a_k\}$ and $\{b_k\}$ converge. We have also not proved that $f(t)$ is indeed equal to its Fourier series expansion. ■

The Dirichlet conditions are the conditions under which the Fourier series converges. These conditions are of little interest to an engineering student.

All practical functions will have a finite number of discontinuities in every period, a finite number of maxima and minima, and will be absolutely integrable in every period.

Example 6.1. In the graph below $x(t)$ is a triangular waveform repeating with a period of 2 s. Find the Fourier series expansion of $x(t)$.

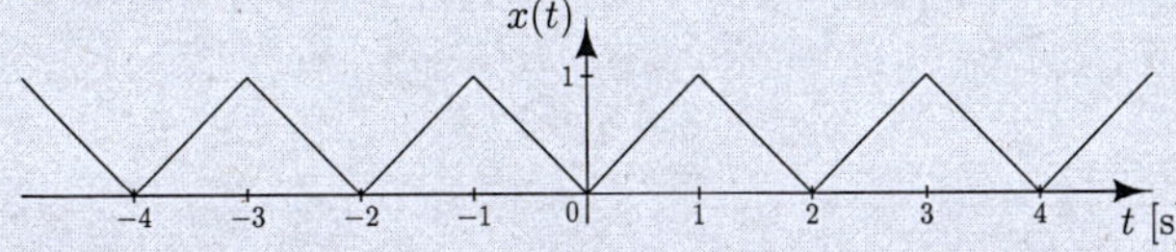

We will analyze $x(t)$ for its Fourier series coefficients in the period from 0 to 2 s.

$$a_0 = \frac{1}{2}\int_0^2 x(t)\ dt = \frac{1}{2}$$

Observe that the coefficient a_0 is the **average** of the signal over its period.

$$\begin{aligned}
a_n &= \frac{2}{2}\int_0^2 x(t)\cos(2n\pi t/2)\ dt \\
&= \int_0^1 t\cos(n\pi t)\ dt + \int_1^2 (2-t)\cos(n\pi t)\ dt \\
&= \frac{1}{n^2\pi^2}\Big[x\sin x + \cos x\Big]_0^{n\pi} + \frac{1}{n^2\pi^2}\Big[2n\pi\sin x - x\sin x - \cos x\Big]_{n\pi}^{2n\pi} \\
&= 2\frac{(-1)^n - 1}{n^2\pi^2} \\
&= \begin{cases} 0 & \text{for even } n \\ -\frac{4}{n^2\pi^2} & \text{for odd } n \end{cases}
\end{aligned}$$

Next we can obtain b_n from (6.6).

$$\begin{aligned}
b_n &= \int_0^1 t\sin(n\pi t)\ dt + \int_1^2 (2-t)\sin(n\pi t)\ dt \\
&= \frac{1}{n^2\pi^2}\Big[\sin x - x\cos x\Big]_0^{n\pi} + \frac{1}{n^2\pi^2}\Big[-2n\pi\cos x - \sin x + x\cos x\Big]_{n\pi}^{2n\pi} \\
&= \frac{-n\pi + 2n\pi - n\pi}{n^2\pi^2}(-1)^n = 0
\end{aligned}$$

For the $\{b_n\}$ coefficients, all the integrals evaluate to 0. Finally we have:

$$x(t) = \frac{1}{2} - \frac{4}{\pi^2}\cos(\pi t) - \frac{4}{9\pi^2}\cos(3\pi t) - \frac{4}{25\pi^2}\cos(5\pi t)\cdots$$

Definite and indefinite integration, performed with the help of Octave, can significantly reduce the effort of deducing the Fourier coefficients.

```
octave:1> pkg load symbolic
octave:2> syms t n
octave:3> assume n positive integer
octave:4> ft1 = t*cos(n*pi*t); ft2 = (2-t)*cos(n*pi*t);
octave:5> an = int(ft1, t, 0, 1) + int(ft2, t, 1, 2)
an = (sym)
          n
  2·(-1)       2
  ------- - -----
    2  2     2  2
   π ·n     π ·n
octave:6> ft1 = t*sin(n*pi*t); ft2 = (2-t)*sin(n*pi*t);
octave:7> bn = int(ft1, t, 0, 1) + int(ft2, t, 1, 2)
bn = (sym) 0
```

The Fourier series expansion can be confirmed with the help of Octave.

Program 6.1. fourierseries.m: Confirm that the Fourier series expansion in the limit converges to $x(t)$. As the number of terms is increased, the Fourier series will converge to the triangular $x(t)$.

```
figure(1); clf; hold on; grid on;
a0 = 0.5; %a0
t = [0:1e-4:2]; %t
x = [[0:1e-4:1], [1-1e-4:-1e-4:0]]; %x(t)
plot(t, x);
f = a0; %f is the Fourier estimate
for k = 1:7 % Number of terms to be used
  ak = 2*((-1)^k-1)/k^2/pi^2; %ak
  f = f + ak*cos(k*pi*t); % Update the estimate by ak cos k2πt/2
  plot(t, f); % Plot the Fourier-series estimate
end
```

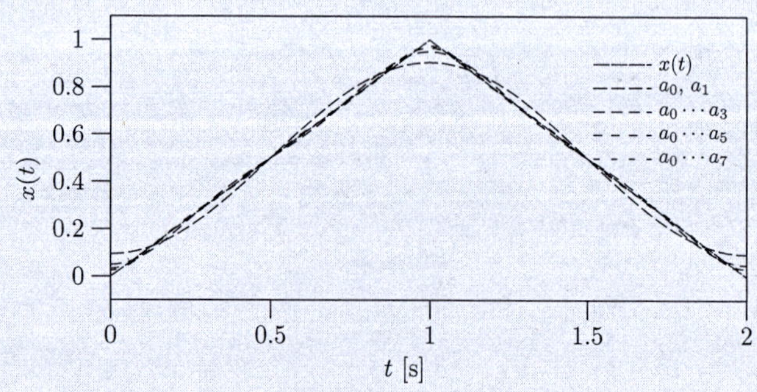

The Fourier series works for discontinuous waveforms. This came as a surprise to the mathematics community in Fourier's time, and delayed the acceptance of Fourier's work. In the next example, we will obtain the Fourier series expansion of a waveform with discontinuities, such as a square wave. Note that to compute the Fourier series coefficients, integration can be performed over any period of $f(t)$. The period chosen may be from 0 to T, or if convenient, from $-T/2$ to $+T/2$, or can be any other convenient whole period of $f(t)$.

Example 6.2. In the graph below $x(t)$ is a square pulse repeating with a period of 2 s. Find the Fourier series expansion of $x(t)$. Assume that the duty cycle of the wave is 50%.

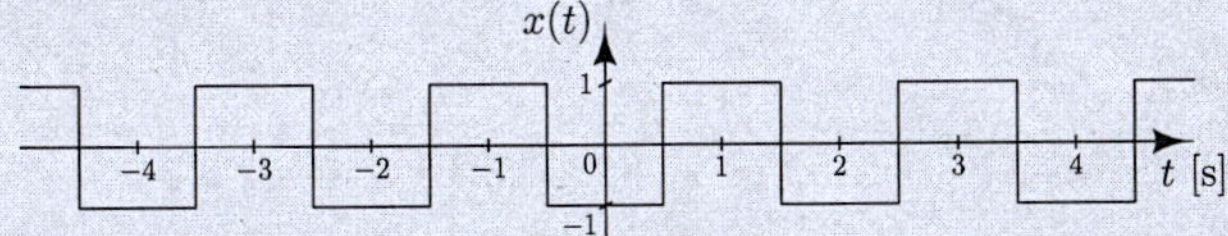

The first coefficient, a_0, is the average of $x(t)$ over a period. In this case, the average is 0. $a_0 = 0$.

$$\begin{aligned} a_n &= \frac{2}{2}\int_0^2 x(t)\cos(n2\pi t/2)\ dt \\ &= \int_{0.5}^{1.5} \cos(n\pi t)\ dt + \int_{1.5}^{2.5} (-1)\cos(n\pi t)\ dt \\ &= \int_{0.5}^{1.5} \cos(n\pi t)\ dt - \int_{1.5}^{2.5} \cos(n\pi t)\ dt \\ &= \begin{cases} \dfrac{4\cdot(-1)^{(n+1)/2}}{n\pi} & \text{for odd } n \\ 0 & \text{for even } n \end{cases} \end{aligned}$$

Similarly, for b_n, we obtain:

$$\begin{aligned} b_n &= \frac{2}{2}\int_0^2 x(t)\sin(n2\pi t/2)\ dt \\ &= \int_{0.5}^{1.5} \sin(n\pi t)\ dt - \int_{1.5}^{2.5} \sin(n\pi t)\ dt \\ &= 0 \end{aligned}$$

Finally we have:

$$x(t) = -\frac{4}{\pi}\cos(\pi t) + \frac{4}{3\pi}\cos(3\pi t) - \frac{4}{5\pi}\cos(5\pi t) + \frac{4}{7\pi}\cos(7\pi t) - \cdots$$

Let us now confirm the Fourier series expansion with the help of Octave.

Program 6.2. fourierseriessqwave.m: Confirm that the Fourier series expansion in the limit converges to $x(t)$. The error at the discontinuity will decrease as we increase the number of terms, and will only disappear in the limit as $n \to \infty$.

```
figure(1); clf; hold on; grid on;
a0 = 0; % a0
t = [0:1e-4:2]; % t
f = a0; % f is the Fourier estimate
to_plot = [1 3 5 7 99];
for k = 1:99 % Number of terms to be used
  if(mod(k, 2))
    ak = 4*(-1)^((k+1)/2)/k/pi;
  else
    ak = 0;
  end; % a_k
  f = f + ak*cos(k*pi*t);
  if(ismember(k,to_plot)) plot(t, f); endif % Plot the Fourier-series estimate
end
```

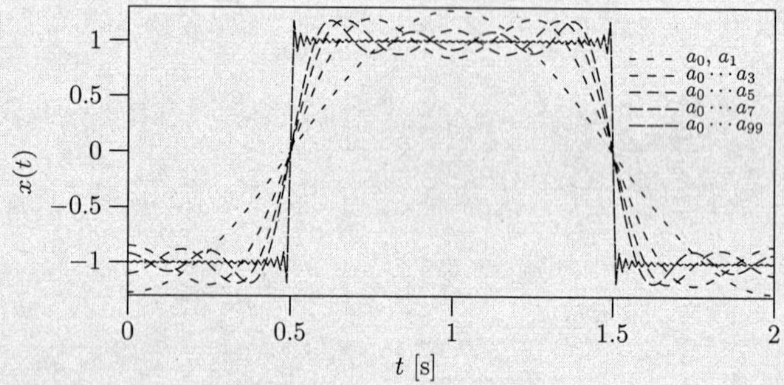

In the plot, we have shown the fundamental frequency component, the sum of the terms after including a_3, a_5, a_7, and all terms till a_{99}.

As we increase the number of terms, the sum of the Fourier terms converges to the desired profile. However, an error persists at discontinuities. The error persists at the edges of discontinuity, till the number of terms becomes so large that the width of the error is negligible. For example, in Example 6.2, the error at the edge is noticeable even when 99 Fourier terms have been included. It turns out that the error remains in height, but the width of the error becomes narrower and narrower as the number of terms increases. The error at the edges is known as **Gibb's phenomenon**, as described in [21] by Willard Gibbs in 1899, and in the limit approaches 9% of the jump discontinuity. In Example 6.2, the jump discontinuity has a height of 2 units; Gibb's error is approximately 0.18 units.

6.3 Waveform symmetries

In Example 6.1 and Example 6.2, notice that all the sine terms, b_n, worked out to zero. Also notice that a_n also worked out to zero for even values of n. There are a few short-cuts, and it is often inefficient to work out integrals.

6.3.1 Even functions

In the case of an even function, $x(-t) = x(t)$.

Corollary 6.3. If $f(t)$ is an even function, all Fourier coefficients for sine components, $\{b_n\}$, are 0.

Proof. The $\{b_n\}$ coefficients are given by (6.6). Since for an even function, $f(t) = f(-t)$, let us integrate from $-T/2$ to $+T/2$, instead of integrating from 0 to T.

$$\begin{aligned}
\therefore b_n \cdot \frac{T}{2} &= \int_{-T/2}^{T/2} f(t)\sin\left(\frac{2\pi nt}{T}\right) dt \\
&= \int_{-T/2}^{0} f(t)\sin(2\pi nt/T)\,dt + \int_{0}^{T/2} f(t)\sin(2\pi nt/T)\,dt \\
&\quad \text{Replacing } t \text{ with } -t \text{ in the first integral} \\
&= -\int_{T/2}^{0} f(-t)\sin(-2\pi nt/T)\,dt + \int_{0}^{T/2} f(t)\sin(2\pi nt/T)\,dt \\
&= \int_{0}^{T/2} -f(-t)\sin(2\pi nt/T)\,dt + \int_{0}^{T/2} f(t)\sin(2\pi nt/T)\,dt \\
&= 0
\end{aligned}$$

In the above derivation we used the identity $\sin(-x) = -\sin(x)$, and applied $f(-t) = f(t)$. ■

Thumb rule. If a function is even, we only need to work out $\{a_n\}$ and all $\{b_n\}$ are 0.

Examples of even functions were shown in Example 6.1 and Example 6.2. A few more possible even functions are shown in Fig. 6.1. In Fig. 6.1, the axis of symmetry (that is the y-axis corresponding to $t = 0$) is implied. The functions have only even symmetry.

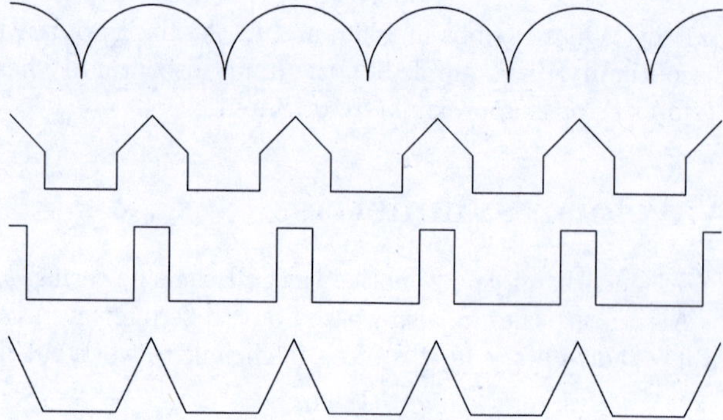

Figure 6.1: Functions with only even symmetry. The axis of symmetry corresponding to the y-axis or $t = 0$ axis is implied.

Figure 6.2: Functions with only odd symmetry. The axis of symmetry corresponding to the y-axis or $t = 0$ is implied.

6.3.2 Odd functions

In the case of an odd function, $x(-t) = -x(t)$. A few periodic odd functions are shown in Fig. 6.2. The axis of symmetry is implied in Fig. 6.2 and the functions are exclusively odd.

The triangular wave and square wave waveforms in Examples 6.1 and 6.2 are even as represented by their $t = 0$ axes. However, if we remove the average or DC to eliminate the a_0 component, and shift the y-axis to $t = 0.5$, the functions become odd.

Corollary 6.4. If $f(t)$ is an odd function, Fourier coefficients for cosine components, $\{a_n\}$, are 0. The constant term, a_0 is also 0.

Proof. First, we will work out the a_0 coefficient. The a_0 coefficient is given by (6.4). We will integrate from $-T/2$ to $+T/2$ to work out the average of $f(t)$. We will apply $f(-t) = -f(t)$, when required.

$$\begin{aligned}
T \cdot a_0 &= \int_{-T/2}^{T/2} f(t)\,dt \\
&= \int_{-T/2}^{0} f(t)\,dt + \int_{0}^{T/2} f(t)\,dt \\
&\quad \text{Replacing } t \text{ with } -t \text{ in the first integral} \\
&= -\int_{T/2}^{0} f(-t)\,dt + \int_{0}^{T/2} f(t)\,dt \\
&= \int_{0}^{T/2} (f(-t) + f(t))\,dt = 0.
\end{aligned}$$

Next, let us calculate a_n with the help of (6.5). We will integrate from $-T/2$ to $+T/2$.

$$\begin{aligned}
a_n \cdot \frac{T}{2} &= \int_{-T/2}^{T/2} f(t)\cos\left(\frac{2\pi nt}{T}\right) dt \\
&= \int_{-T/2}^{0} f(t)\cos(2\pi nt/T)\,dt + \int_{0}^{T/2} f(t)\cos(2\pi nt/T)\,dt \\
&\quad \text{Replacing } t \text{ with } -t \text{ in the first integral} \\
&= -\int_{T/2}^{0} f(-t)\cos(-2\pi nt/T)\,dt + \int_{0}^{T/2} f(t)\cos(2\pi nt/T)\,dt \\
&= \int_{0}^{T/2} \cos(2\pi nt/T)(f(t) + f(-t))\,dt = 0.
\end{aligned}$$

■

Thumb rule. In case of an odd function, only $\{b_n\}$ needs to be worked out. All $\{a_n\}$, including a_0, are 0.

Example 6.3. Evaluate the Fourier series expansion of the saw-tooth waveform shown in the following graph.

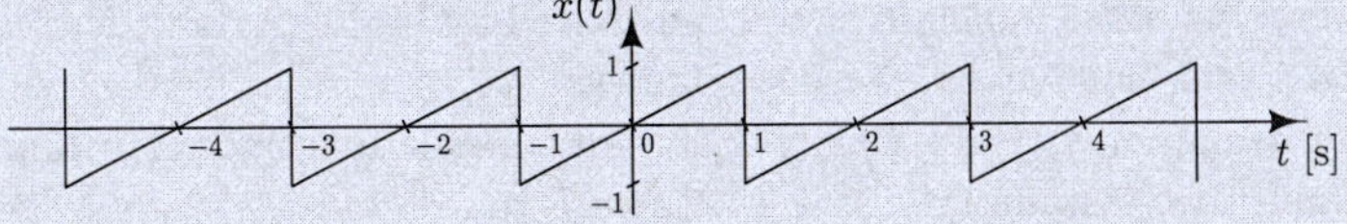

Figure 6.3: Functions with half-wave symmetry. When shifted by half of a period, the functions change in sign. The x-axis is implied.

The function is an odd function and $\{a_n\}$ are 0. We will need to evaluate $\{b_n\}$.

$$\begin{aligned} b_n &= \frac{2}{2}\int_{-1}^{1} t\sin(n\pi t)\,dt \\ &= -2\frac{(-1)^n}{n\pi} \\ \therefore x(t) &= \frac{2}{\pi}\sin(\pi t) - \frac{2}{2\pi}\sin(2\pi t) + \frac{2}{3\pi}\sin(3\pi t) - \frac{2}{4\pi}\sin(4\pi t) + \cdots \end{aligned}$$

The sum of the first 20 terms of the Fourier series is shown in the plot below. Notice the incidence of Gibb's phenomena at the jump discontinuity. Gibb's error is expected to be roughly 9% of the jump.

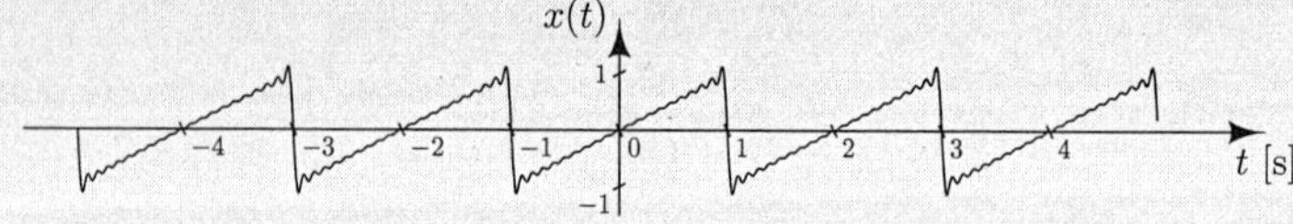

6.3.3 Functions with half-wave symmetry

A periodic function always remains the same when shifted by a period. Some periodic functions change in sign when shifted by half a period. These functions are said to have half-wave symmetry.

Mathematically, $f(x+T) = f(x)$ for all periodic functions. For functions with half-wave symmetry, $f(x+T/2) = -f(x)$.

The functions shown in Fig. 6.3 exhibit half-wave symmetry. One needs to first subtract out the average (a_0) component of the wave, before looking for half-wave symmetry. The portion of the wave in the first half of the period needs to be the opposite of the portion of the wave in the second half of the period.

Corollary 6.5. For a function with half-wave symmetry, Fourier coefficients for all even order harmonics ($\{a_0, a_2, a_4, a_6 \cdots\}$ and $\{b_2, b_4, b_6 \cdots\}$) are 0.

Proof. The second half-period of the wave is the opposite of the first half-period of the wave. The average of the waveform is 0. a_0 is 0. Let us now examine the $2k$th harmonic with coefficients a_{2k} and b_{2k}.

$$
\begin{aligned}
\frac{T}{2} \cdot a_{2k} &= \int_{-T/2}^{T/2} f(t) \cos(2k\frac{2\pi}{T}t)\, dt \\
&= \int_{-T/2}^{0} f(t) \cos(2k\frac{2\pi}{T}t)\, dt + \int_{0}^{T/2} f(t) \cos(2k\frac{2\pi}{T}t)\, dt
\end{aligned}
$$

In the integral for the second half of the period, let us create a new dummy variable, τ, with $\tau = t - T/2$. Then $d\tau = dt$. Just the integral for the second half of the period becomes:

$$
\begin{aligned}
\int_{0}^{T/2} f(t) \cos(2k\frac{2\pi}{T}t)\, dt &= \int_{-T/2}^{0} f(\tau + T/2) \cos(2k\frac{2\pi}{T}(\tau + T/2))\, d\tau \\
&= -\int_{-T/2}^{0} f(\tau) \cos(2k\frac{2\pi}{T}\tau + 2k\pi)\, d\tau \\
&= -\int_{-T/2}^{0} f(t) \cos(2k\frac{2\pi}{T}t)\, dt
\end{aligned}
$$

Going back to the integral over the entire period, the terms cancel out and we are left with 0. An identical derivation works for the b_{2k} coefficients, because $\cos(\theta+2k\pi)$ and $\sin(\theta+2k\pi)$ are the same as $\cos\theta$ and $\sin\theta$, respectively. ■

Coefficients of all even harmonics in the Fourier series are 0 if the waveform has half-wave symmetry. In addition, if the wave is either odd or even, for example in the illustrations of Examples 6.1 and 6.2, the b_n or a_n coefficients are 0. Waveforms which are both odd/even and half-wave symmetric are known as **quarter-wave symmetric**.

6.4 Exponential Fourier series

The trigonometric Fourier series sine and cosine terms can be combined together into an exponential series that uses the complex exponential. Coefficients of the exponential Fourier series are now complex.

Theorem 6.3: If $x(t)$ is periodic with a period T, then:

$$x(t) = \sum_{-\infty}^{\infty} C_n e^{jn\omega_0 t}, \tag{6.7}$$

where ω_0 is $2\pi/T$, and C_n is given as follows:

$$C_n = \frac{1}{T}\int_0^T x(t)e^{-jn\omega_0 t}\,dt \tag{6.8}$$

It is assumed that $x(t)$ complies with the Dirichlet conditions with finite number of maxima and minima in a period, finite number of discontinuities in a period, and is absolutely integrable over a period.

(6.7) can be elaborated, breaking up the complex coefficients and the complex exponentials. The nth term is $(\mathrm{Re}[C_n] + j\,\mathrm{Im}[C_n])(\cos(n\omega_0 t) + j\sin(n\omega_0 t))$, which gives $\mathrm{Re}[C_n]\cos(n\omega_0 t) - \mathrm{Im}[C_n]\sin(n\omega_0 t)$ as the real part. There is also a term for C_{-n}; notice that the summation runs from $-\infty$ to ∞. The $-n$th term will similarly give us a real and imaginary part. If the C_{-n} is the conjugate of C_n, then the real parts of the nth and $-n$th terms add up, while the imaginary parts cancel out.

If $x(t)$ is real, then C_{-n} is indeed the conjugate of C_n. This can be verified by observing (6.8). In the case of C_n, the complex exponential is $\cos - j\sin$, while in the case of C_{-n}, the complex exponential is $\cos + j\sin$.

Finally, we have the nth and $-n$th terms of (6.7) combining with each other as:

$$C_n e^{jn\omega_0 t} + C_n^* e^{-jn\omega_0 t} = 2\,\mathrm{Re}[C_n]\cos(n\omega_0 t) - 2\,\mathrm{Im}[C_n]\sin(n\omega_0 t)$$

Looking back at the expression for the trigonometric Fourier series, and comparing, we have the following relationships:

$$a_0 = C_0 \tag{6.9}$$

$$a_n = 2\,\mathrm{Re}[C_n] = C_n + C_n^* \tag{6.10}$$

$$b_n = -2\,\mathrm{Im}[C_n] = C_n^* - C_n \tag{6.11}$$

$$C_n = (a_n - jb_n)/2, \quad \text{for } n > 0, \text{ and } C_{-n} = C_n^* \tag{6.12}$$

No further proof of Theorem 6.3 is offered, since the results are identical to those of Theorem 6.2.

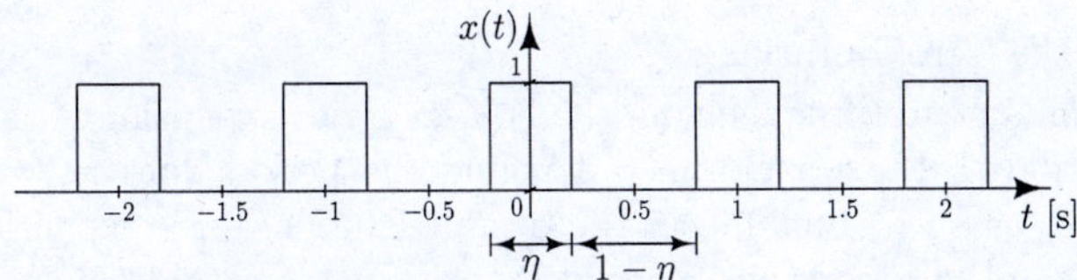

Figure 6.4: A square wave between 0 and 1, with a period of 1 s, and a duty cycle of η.

6.5 The square wave

The square wave is a very common periodic signal; often the Fourier series expansion of the square wave is required. In this section, we will analyze the square wave in detail.

$x(t)$, shown in Fig. 6.4, is a square wave with a period of 1 s, high and low levels of $+1$ and 0, respectively. $x(t)$ has a duty cycle of η. $x(t)$ is $+1$ from $t=-\eta/2$ to $\eta/2$, and 0 from $\eta/2$ to $1-\eta/2$. Let us obtain the exponential Fourier series for $x(t)$. Notice that $t=0$ has been adjusted such that the function is even.

A period of 1 s corresponds to ω_0 of 2π rad/s.

$$\begin{aligned} C_n &= \int_0^1 x(t)e^{-j2\pi nt}\,dt \\ &= \int_{-\eta/2}^{\eta/2} e^{-j2\pi nt}\,dt \\ &= \begin{cases} \dfrac{\sin(n\pi\eta)}{\pi n}, & \text{for } n \neq 0 \\ \eta, & \text{for } n = 0 \end{cases} \end{aligned} \tag{6.13}$$

Since the function is even, C_n has worked out to be purely real. If we had worked out the trigonometric Fourier series, a_n would be $2\,\mathrm{Re}[C_n]$, or $a_n = 2\sin(n\pi\eta)/(n\pi)$, and $a_0=\eta$, while $b_n=0$.

A second observation of our result in (6.13) is in the C_0 coefficient. If n is allowed to be a real number and not just an integer, then in the limit, as $n\to 0$, $\sin(n\pi\eta)/(\pi\eta)$ becomes nothing but η. This means that the two cases in (6.13) are, in the limit, the same.

6.5.1 Change in timing

The square wave can always be shifted to be an even function. The axis of symmetry can be placed either at the middle of the $+1$ duration, or at the middle of the 'zero' duration.

The fundamental frequency in the Fourier series expansion is the inverse of the period of the waveform. Changing the period does not affect the coefficients, as such, but affects the overall Fourier series expansion. The frequencies of the cosines and sines have their fundamental frequency as that of the waveform.

6.5.2 Change in values of states

If the square wave is not between 0 and +1, but between two other states (with a difference of 1), only the average of the waveform changes. As such, the value of $C_0 = a_0$ needs to be re-evaluated. For example, if the square wave is between −0.5 and +0.5, $C_0 = a_0$ will work out as the new average of the waveform; all other coefficients will remain the same.

If the two states of the square wave change such that the difference is not 1, the amplitudes of the coefficients will scale.

Example 6.4. Evaluate the Fourier series coefficients for a square wave between −0.1 and +2, with a duty cycle of 0.4. The period of the wave is 1 ms.

In one period, the square wave is at −0.1 for 0.6 times the period and at +2 for 0.4 times the period. The average of the square wave is therefore:

$$C_0 = a_0 = -0.1 \times 0.6 + 2 \times 0.4 = 0.74$$

Next, we can evaluate the other coefficients. For a square wave with duty cycle 0.4 and period of 1 s, the coefficients are $\sin(0.4n\pi)/(n\pi)$. The amplitude of the square wave has increased from 1 to 2.1. Therefore, the amplitudes of the coefficients will scale 2.1×. Finally, we have:

$$C_n = \begin{cases} 2.1\dfrac{\sin(0.4n\pi)}{n\pi}, & \text{for } n \neq 0 \\ 0.74 & \text{for } n = 0 \end{cases}$$

The Fourier series expansion is as follows:

$$x(t) = 0.74 + 4.2\sum_{n=1}^{\infty} \frac{\sin(0.4n\pi)}{n\pi} \cos(2\pi n \cdot 1000t)$$

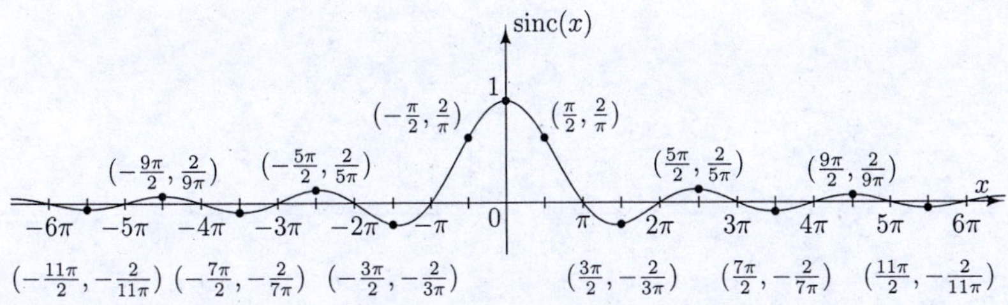

Figure 6.5: sinc(x)

6.5.3 The sinc function

Definition 6.2 (The sinc function):

$$\text{sinc}(x) = \begin{cases} \dfrac{\sin x}{x} & \text{for } x \neq 0 \\ 1 & \text{for } x = 0 \end{cases}$$

The above definition of the sinc function, used in this text, is the **unnormalized** sinc function used in Mathematics [22].

The Fourier series coefficients of a square wave are related to the sinc function. For example, in case of the waveform of Fig. 6.4, the coefficients are $\eta\,\text{sinc}(n\pi\eta)$ for all n including $n = 0$. A plot of sinc(x) is shown in Fig. 6.5.

When x is an integral multiple of π, (except at $x = 0$), the sinc function is 0 because $\sin(x)$ is 0. The sinc function has a central lobe, and several side lobes. The amplitude of the side lobes decreases as $1/|x|$, whereas the central lobe has an amplitude of 1. In Fig. 6.5, the values at odd multiples of $\pi/2$ are indicated. These points are the **approximate** maxima and minima of the function.

6.5.4 Discrete spectra

A plot of the Fourier coefficients of a waveform is known as its spectrum. The spectrum is often plotted for the coefficients of the exponential Fourier series, $|C_n|$. Only the magnitude is plotted, since C_n is otherwise a complex number. The x-axis is chosen as $\omega = n2\pi/T = n\omega_0$.

The spectrum of the waveform corresponding to Fig. 6.4, with $\eta = 0.4$ is shown in Fig. 6.6. The spectrum is the set of lines. In Octave, such a plot is generated with the `stem` command. It is possible to reconstruct the waveform from the spectrum of the waveform (as long as phase information is also known for each C_n). For each point in the spectrum, we use a cosine

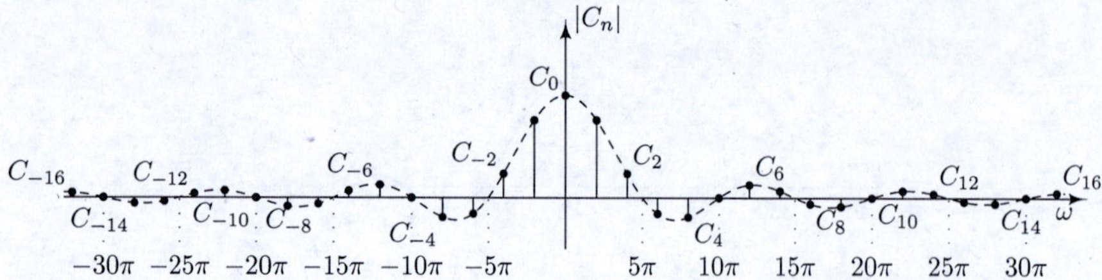

Figure 6.6: Spectrum of the square wave of Fig. 6.4. The duty cycle is 0.4.

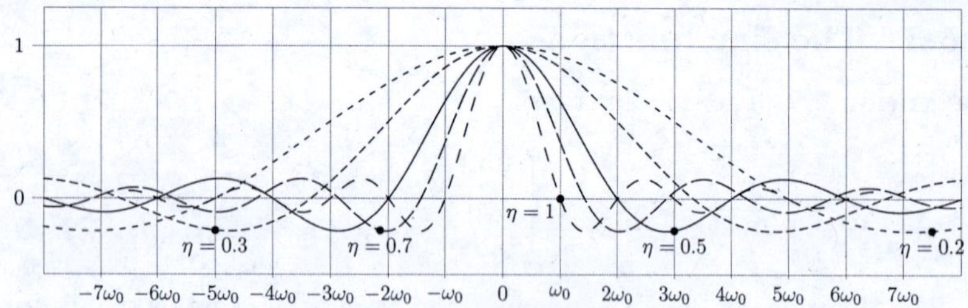

Figure 6.7: Expansion or compression of the sinc function with duty cycle, η. The coefficients are samples of the sinc function at ω_0, $2\omega_0$ etc. The entire set of coefficients need to be further scaled by a factor η, if the two states of the square wave are 0 and 1.

with the amplitude given by the height of the point and the angular frequency given by the value of ω. We add all these cosines to reconstruct the waveform from its spectrum.

The sinc function is shown in the background of Fig. 6.6. The Fourier series coefficients are samples of the scaled sinc function.

If we change the two states of the square wave, the C_0 coefficient is to be adjusted as the new average in one period, and all other coefficients are to be scaled by a factor that is the difference of the two states.

If the period of the waveform (or fundamental frequency of the square wave) changes, the entire ω-axis of the spectrum is scaled. The coefficients do not change; the shape of the sinc function remains the same, the amplitudes do not change; only the ω-axis is scaled by a factor that is the fundamental frequency of the square wave. The values of frequencies on the ω-axis in Fig. 6.6 are simply to be re-labeled.

The other factor that may change in the square wave is its duty cycle, η. A change in η stretches or compresses the sinc function in the background of Fig. 6.6. This will change the values of the coefficients. For example, when

$\eta = 0.5$, or the duty cycle is 50%, the waveform has quarter-wave symmetry. In such a case, the sinc function compresses from that of Fig. 6.6, and all the even harmonics, $C_{\pm 2}$, $C_{\pm 4}$, etc., fall on the zero crossings of the sinc function.

Fig. 6.7 shows the sinc function as it compresses and expands with different values of η. For a square wave between 0 and 1, the set of exponential Fourier series coefficients, $\{C_n\}$, are samples of this sinc function multiplied by η.

Observe the following in Fig. 6.7.

- When $\eta = 0.5$, the even harmonics are all zero. For a 50% duty cycle, the waveform has half-wave (and quarter-wave) symmetry. As such, all even harmonics become 0.
- In the limit, when η is increased to 1, that is, when the square wave becomes a constant 1 (and does not come back to 0 at all), the Fourier coefficients are 0 for all the harmonics. Only the C_0 coefficient remains.
- In the other limit, when η decreases to 0, the first lobe of the spectrum expands and covers all frequencies. However, remember that the entire spectrum also needs to be scaled by η. When η is 0, the entire spectrum will become a flat line at 0.
- When η is 1/3, the 3rd harmonic (and 6th, 9th, ...harmonics) are 0. When η is 1/4, the 4th (and 8th, 12th, ...harmonics) are 0. When η is 1/5, the 5th harmonic (and 10th ...) are 0.

6.6 Steady state response of a network

A periodic signal can be broken up into a fundamental frequency and its harmonics. For example, a periodic square wave voltage source can be dismantled into a voltage source at the fundamental frequency in series with voltage sources at each of the harmonics. The response of a linear network to the combination of these sources at the fundamental frequency and its harmonics can now be evaluated through superposition.

6.6.1 Case study: oscilloscope probes

Almost any Electrical Engineering laboratory, or any laboratory in which electrical signals are measured, has an oscilloscope. Along with an oscilloscope one uses oscilloscope probes to measure and display signals on a screen. You would have used one too. Oscilloscope probes are fairly expensive. Handle them gently. The circuit in Example 6.5 is the circuit inside an oscilloscope probe.

Example 6.5. The RC circuit shown below is excited by a square wave, $v_{in}(t)$, between 0 and 5 V, with 50% duty cycle and 1 kHz frequency. Find the steady state response of the circuit. R_a is 9 MΩ, R_b is 1 MΩ and C_b is 90 pF. Evaluate the response, $v_{out}(t)$, when (a) C_a is 5 pF, (b) C_a is 10 pF, and (c) C_a is 20 pF.

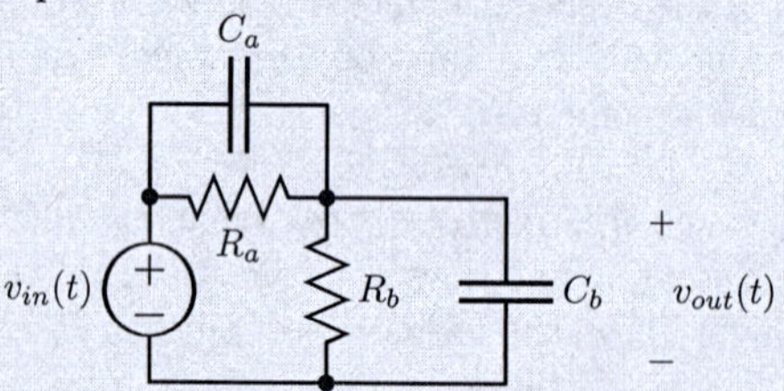

The fundamental frequency is 1 kHz which gives $\omega_0 = 2000\pi$ rad/s. $v_{in}(t)$ can be broken up as a Fourier series as follows:

$$v_{in}(t) = 2.5 + 10\sum_{1}^{\infty}\frac{\sin(n\pi/2)}{n\pi}\cos(n\omega_0 t) = 2.5\sum_{-\infty}^{\infty}\text{sinc}(n\pi/2)e^{jn\omega_0 t}$$

This corresponds to several voltage sources in series. We can now use superposition, evaluate the response to each of these voltage sources, and finally sum up the responses.

Let us work in the phasor domain to evaluate the response at any given frequency. The transfer function, $V_{out}(j\omega)/V_{in}(j\omega)$ is given as follows:

$$\begin{aligned}\frac{V_{out}(j\omega)}{V_{in}(j\omega)} &= \frac{R_b \parallel \frac{1}{j\omega C_b}}{R_a \parallel \frac{1}{j\omega C_a} + R_b \parallel \frac{1}{j\omega C_b}} \\ &= \frac{R_b}{R_a + R_b}\cdot\frac{1 + j\omega C_a R_a}{1 + j\omega(C_a + C_b)\frac{R_a R_b}{R_a + R_b}}\end{aligned}$$

Each Fourier series coefficient, C_n, will be amplified by the above transfer function to give us the corresponding Fourier series coefficient of $v_{out}(t)$.

$$\therefore v_{out}(t) = \frac{2.5R_b}{R_a + R_b}\cdot\sum_{-\infty}^{\infty}\text{sinc}(n\pi/2)\cdot\frac{1 + jn\omega_0 C_a R_a}{1 + jn\omega_0(C_a + C_b)\frac{R_a R_b}{R_a + R_b}}e^{jn\omega_0 t}$$

Program 6.3. rccsqwave.m: Evaluation of $v_{out}(t)$ using Octave.

```
Ra = 9e6; Rb = 1e6; Cb = 90e-12; % Component values
Ca1 = 5e-12; Ca2 = 10e-12; Ca3 = 20e-12; % Three cases
n = [-100:100]; % Index of cn; DC, fundamental and upto 100th harmonic.
cn = 2.5*sin(n*pi/2)./(n*pi/2); % cn coefficients for given vin(t)
cn(find(n==0)) = 2.5; % c0 for vin(t)
w0 = 2*pi*1e3; % Fundamental, ω0
w = n*w0; % ω
%stem(w, cn); Stem plot for line spectra
dn1=cn*(Rb/(Ra+Rb)).*(1+j*w*Ca1*Ra)./(1+j*w*(Ca1+Cb)*Ra*Rb/(Ra+Rb));
dn2=cn*(Rb/(Ra+Rb)).*(1+j*w*Ca2*Ra)./(1+j*w*(Ca2+Cb)*Ra*Rb/(Ra+Rb));
dn3=cn*(Rb/(Ra+Rb)).*(1+j*w*Ca3*Ra)./(1+j*w*(Ca3+Cb)*Ra*Rb/(Ra+Rb));
% dn1, dn2, dn3: Fourier coefficients for cases (a), (b), (c)
t = 0:2e-6:1e-3; % t. Step size to be at lower than 1/max(n)*T
vout1 = 0; vout2 = 0; vout3 = 0; % Initialize vout(t) in cases a,b,c
vin = 0; % Initialize vin(t)
for k=1:length(n) % For each index
  % Add corresponding cn e^jωt and dn e^jωt
  vin   = vin   + cn(k)*e.^(j*w(k)*t);
  vout1 = vout1 + dn1(k)*e.^(j*w(k)*t);
  vout2 = vout2 + dn2(k)*e.^(j*w(k)*t);
  vout3 = vout3 + dn3(k)*e.^(j*w(k)*t);
end;
figure(1); clf; % Create and clear a plot window
plot(t, vout1); hold on; % Plot vout for case (a). Do not erase.
plot(t, vout2); % Plot vout for case (b)
plot(t, vout3); % Plot vout for case (c)
```

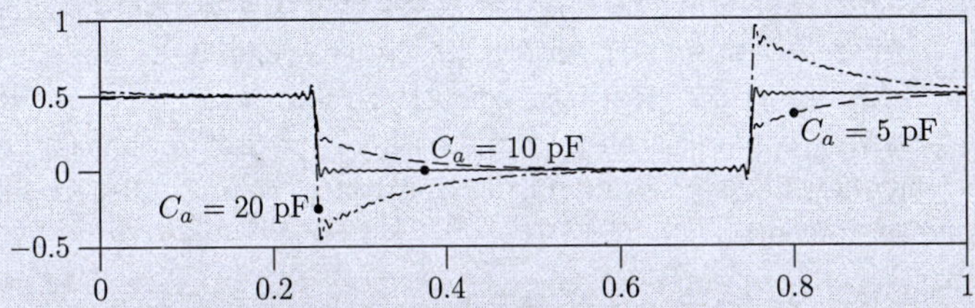

The square wave input gives a square wave output when C_a is 10 pF. When C_a is 20 pF, there is considerable overshoot and undershoot. When C_a is 5 pF, the response is damped. All three responses have half-wave symmetry, because all even-order harmonics in the input signal are zero.

There is a tiny screw on the oscilloscope probe that one can adjust to obtain a precise square-wave response to an input square wave. The oscilloscope itself provides a reference square-wave signal that is used to **compensate** the probe.

The values given in Example 6.5 are very realistic. The oscilloscope input impedance is usually 1 MΩ (R_b) in shunt with some capacitance (C_b) in the vicinity of 100 pF. A 10X probe will have a series 9 MΩ resistance (R_a) with

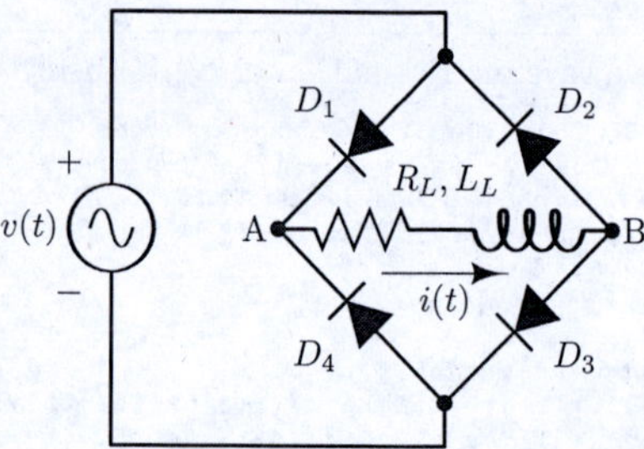

Figure 6.8: A full-wave rectifier with an R-L load. When $v(t)$ is positive, D_1 and D_3 conduct. When $v(t)$ is negative, D_4 and D_2 conduct.

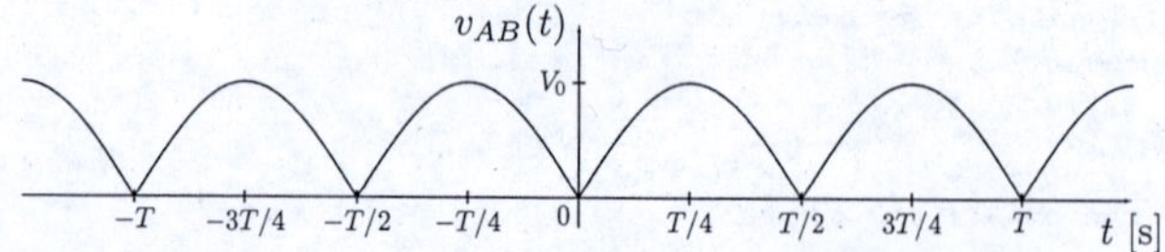

Figure 6.9: Full-wave rectified voltage. In the circuit of Fig. 6.8, $v_{AB}(t)$ is shown assuming ideal diodes.

a tun-able shunt capacitance (C_a). The test signal in a standard oscilloscope is a 0-5 V square wave ($v_{in}(t)$), with a 50% duty cycle, at 1 kHz.

Once the probe is compensated correctly, the measurement system will provide a 10 MΩ input resistance (along with a capacitance that is the series combination of C_a and C_b). A high input impedance will not disturb sensitive voltage measurements.

Did you know that the oscilloscope is used to measure only periodic signals? Scan the QR-code, and learn more about the oscilloscope and its probe.

6.6.2 Case study: full-wave rectifiers

A very common non-linear circuit is a voltage rectifier. Let us use the Fourier-series method to analyze the ripple content in the output current of a full-wave rectifier, shown in Fig. 6.8.

For a voltage amplitude that is much larger than the cut-in voltage of the diodes, we can assume that the diodes are ideal. The voltage profile of

$v_{AB}(t)$, the rectified voltage across the load, is shown in Fig. 6.9. First, we will expand the voltage $v_{AB}(t)$ as its Fourier series.

$$\begin{aligned}
C_n &= \frac{1}{T}\int_0^T v_{AB}(t)e^{-jn\omega_0 t}\,dt \\
&= \frac{1}{T}\int_0^{T/2} V_0\sin(\omega_0 t)e^{-jn\omega_0 t}\,dt + \frac{1}{T}\int_{T/2}^{T} V_0\sin(-\omega_0 t)e^{-jn\omega_0 t}\,dt \\
&= \begin{cases} \dfrac{2V_0}{\pi(1-n^2)} & \text{for even } n \\ 0 & \text{for odd } n \end{cases}
\end{aligned}$$

Notice that the Fourier coefficients for all the odd harmonics are zero because $v_{AB}(t)$ is periodic with $T/2$ and the fundamental frequency is $2\omega_0$. Also, notice that all coefficients are real because the function was even. The trigonometric Fourier series will only have cosine terms. The DC component, C_0 is $2V_0/\pi$.

$$\therefore v_{AB}(t) = \frac{2V_0}{\pi}\sum_{k=-\infty}^{\infty}\frac{1}{1-4k^2}\cdot e^{jk2\omega_0 t}$$

The voltage v_{AB} can be visualized as a series combination of several sinusoids at $2\omega_0$ and its harmonics, as given above. At an angular frequency of ω, the impedance provided by the load is $R_L + j\omega L_L$. At $2k\omega_0$, the impedance is $R_L + j2k\omega_0 L_L$. Therefore we have $i(t)$ as follows.

$$i(t) = \frac{2V_0}{\pi}\sum_{k=-\infty}^{\infty}\frac{1}{1-4k^2}\cdot\frac{1}{R_L + j2k\omega_0 L_L}\cdot e^{j2k\omega_0 t}$$

Program 6.4. fullwaverect.m: Evaluation of $i(t)$ in the full-wave rectifier using Octave.

```
VO = 110; w0 = 100*pi; % Input voltage v(t)
RL = 1; LL = 5e-3; % R_L, L_L
t = [0:5e-5:40e-3]; %t
n = [-100:100]; % Index
vABt = 0; it = 0; %Initialize v_AB(t) and i(t)
for k=1:length(n) % Sum the Fourier coefficients
  vABt = vABt + 2*V0/pi*(1/(1-4*n(k)^2))*e.^(2*n(k)*j*w0*t);
  it = it + 2*V0/pi/(1-4*n(k)^2)/(RL+2*j*n(k)*w0*LL)*e.^(2*n(k)*j*w0*t);
end;
figure(1); clf; plot(t, vABt); hold on; grid on;
plot(t, it);
```

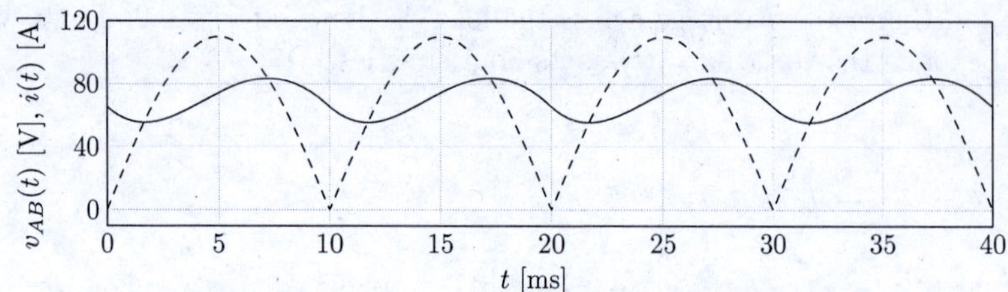

Figure 6.10: Full-wave rectified voltage and current through an RL load. The input voltage is of amplitude 110 V and frequency 50 Hz. R_L and L_L are 1 Ω and 5 mH respectively.

The waveforms from Prog. 6.4 are plotted in Fig. 6.10, for an input voltage of 110 V amplitude and 50 Hz frequency, and a load of 1 Ω in series with 5 mH. The voltage across the resistance is not a rectified sinusoid any more, but a somewhat skewed 100-Hz sinusoid with an average of $2/\pi$ times the input amplitude.

6.7 Power in a periodic signal

6.7.1 Series or shunt RC or RL loads

Power consumed in a load is always the effect of the resistive part of the load. The inductive or capacitive part of a load does not consume power. As an illustration, let us analyze the power dissipated in R_L of Fig. 6.8. The current, $i(t)$, through R_L is given in Fig. 6.10. The power consumed by the load is therefore:

$$P_L = (i(t))^2 R_L$$

To compute the average power in the load, it will therefore suffice to only study the mean-square current through a series R-X load, or the mean-square voltage across a shunt R-X load.

If $x(t)$ is periodic with a period T, then it may be represented with a trigonometric Fourier series, with coefficients $\{a_n\}$ and $\{b_n\}$.

$$\therefore x^2 = \left[a_0 + \sum_{n=1}^{\infty}\{a_n \cos(n\omega_0 t) + b_n \sin(n\omega_0 t)\}\right]^2$$

When we evaluate the **average** of $(x(t))^2$ over a period, all the cross terms integrate to zero, because of Theorem 6.1. Only the square terms need to be averaged over the period.

$$\therefore \overline{x^2} \quad = \quad a_0^2 + \sum_{n=1}^{\infty} \frac{1}{T} \int_0^T \{a_n^2 \cos^2(n\omega_0 t) + b_n^2 \sin^2(n\omega_0 t)\}\, dt$$

The square terms, each, can now be split into a component at DC and a cosine at $2n\omega_0$. The cosine at $2n\omega_0$ integrates to zero over one period and only the component at DC remains.

$$\therefore \overline{x^2} = a_0^2 + \sum_{n=1}^{\infty} (a_n^2 + b_n^2)/2$$

We can relate the above expression to the coefficients of the exponential Fourier series with the help of (6.9), (6.10), (6.11), (6.12). $(a_n^2 + b_n^2)/2$ is nothing but $|C_n|^2$. a_0^2 is C_0^2.

$$\therefore \overline{x^2} = \sum_{n=-\infty}^{\infty} |C_n|^2 \tag{6.14}$$

Corollary 6.6 (Parseval's equality). The mean square of a periodic signal over a period is the sum of the magnitude-squares of all its exponential Fourier series coefficients.

Proof. We have proved this in (6.14). ∎

The average power in a resistor R_L is therefore given as follows:

$$\overline{P_L} = \overline{(i(t))^2} R_L = R_L \sum_{n=-\infty}^{\infty} |i_n|^2,$$

where $\{i_n\}$ are the set of exponential Fourier series coefficients of the current through R_L.

Example 6.6. A 110 V 50 Hz sinusoid is rectified with a full wave rectifier. What is the power in a 50 Ω load?

The rectified voltage is of the form:

$$v_L(t) = \frac{220}{\pi} \sum_{k=-\infty}^{\infty} \frac{1}{1 - 4k^2} e^{jk2\omega_0 t}$$

Therefore, the mean-square voltage is:

$$\overline{v_L^2} = \frac{220^2}{\pi^2} \sum_{k=-\infty}^{\infty} \frac{1}{(1 - 4k^2)^2} = \frac{220^2}{\pi^2} \cdot \frac{\pi^2}{8}$$

Finally we obtain the average power as 11 W.

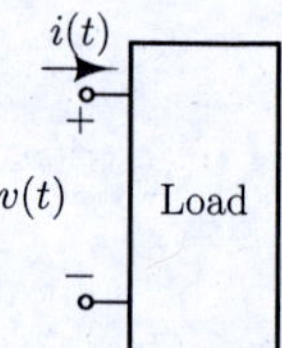

Figure 6.11: The voltage across the network is $v(t)$ and the current going into the network is $i(t)$. $i(t)$ and $v(t)$ have a period of T.

6.7.2 Any general load

Let us now analyze the power in a load which cannot be resolved as a simple series or shunt network. Let us assume $\{v_n\}$ are the coefficients of the exponential Fourier series of the voltage across the load and $\{i_n\}$ are the coefficients of the exponential Fourier series of the current through the load. Let us formalize our problem statement with the help of Fig. 6.11.

$$v(t) = \sum_{n=-\infty}^{\infty} v_n e^{jn\omega_0 t}$$

$$i(t) = \sum_{n=-\infty}^{\infty} i_n e^{jn\omega_0 t}$$

$$\therefore P = v(t)i(t) = \left(\sum_{n=-\infty}^{\infty} v_n e^{jn\omega_0 t}\right)\left(\sum_{n=-\infty}^{\infty} i_n e^{jn\omega_0 t}\right) \qquad (6.15)$$

The average power over a period is our main interest. The average of any term like $C_n e^{jn\omega_0 t}$ over a period of $T = 2\pi/\omega_0$ is zero, as long as $n \neq 0$. If $n = 0$, the average of $C_n e^{jn\omega_0 t}$ is C_n. The nth term in the first sum can combine with the $-n$th term in the second sum to give $v_n i_n$ and provide a non-zero average. All other product terms average out to zero. Refer to (6.12) for the relationship between the nth and $-n$th coefficients.

$$\therefore \overline{P} = v_0 i_0 + \sum_{n=1}^{\infty}(v_n i_{-n} + v_{-n} i_n) \qquad (6.16)$$

$$= v_0 i_0 + \sum_{n=1}^{\infty}(v_n i_n^* + v_n^* i_n)$$

$$= v_0 i_0 + \sum_{n=1}^{\infty} 2\,\mathrm{Re}[v_n i_n^*] \qquad (6.17)$$

The form of (6.16) is convenient if we have the exponential Fourier series for the voltage and the current. The form of (6.17) resolves further as the product of the rms voltage at each frequency, the rms current at each frequency, and $\cos\phi$ at each frequency (ϕ is the angle between the voltage and current phasors). Refer back to (4.6); at each harmonic the voltage and current phasors can be used to obtain the power.

Thumb rule. Break up the voltage and the current into different frequencies. The total power consumed is the sum of the powers consumed at each individual frequency. The cross terms do not contribute because of orthogonality.

Example 6.7. The voltage across a load and the current through the load are:

$$\begin{aligned} v(t) &= 110\cos(\omega_0 t) + 20\cos(3\omega_0 t - \pi/3) + 3\cos(5\omega_0 t + \pi/3) \\ i(t) &= 4\cos(\omega_0 t - \pi/6) + 0.1\cos(3\omega_0 t) + 0.01\cos(5\omega_0 t + \pi/4) \end{aligned}$$

What is the average power delivered to the load? What is the percentage harmonic content in the average power?

At the fundamental frequency the voltage is $110\cos(\omega_0 t)$ and current is $4\cos(\omega_0 t - \pi/6)$. Therefore the power is $110 \times 4 \times \cos(\pi/6)/2$ or 190.5 W.

At the third harmonic, the voltage amplitude is 20 V, the current amplitude is 0.1 A, the phase difference is $\pi/3$. The power in the third harmonic is $20 \times 0.1 \times cos(\pi/3)/2$ or 0.5 W.

At the fifth harmonic, the voltage amplitude is 3 V, the current amplitude is 0.01 V, the phase difference is $\pi/12$. The power in the fifth harmonic is $3 \times 0.01 \times \cos(\pi/12)/2$ or 0.014 W.

The total average power delivered to the load is 191.014 W. The percentage of the power in the harmonics is 0.27%.

6.8 The Fourier transform

As the period of the signal increases, in the limit, as $T \to \infty$, the Fourier series resolves into the Fourier transform. In Fig. 6.7, we saw how the Fourier coefficients evolve as we change the duty cycle. Let us now observe how the Fourier coefficients evolve as we change the period.

In Fig. 6.12(a) shows a 50% duty cycle square wave which is 1 from 0.5 s to 1 s and has a period of 1 s. As we increase the period, without changing

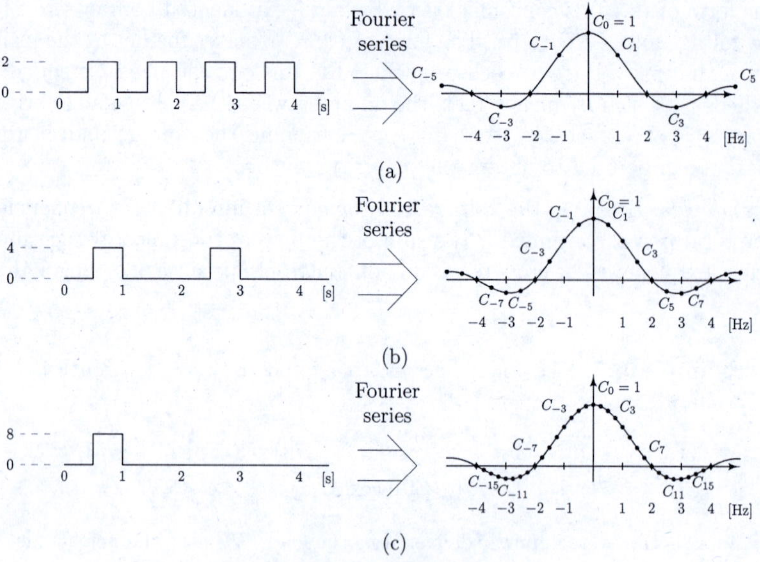

Figure 6.12: Fourier coefficients as the period changes. The waves are all of height $2T$ from 0.5 s to 1 s. (a) Fourier coefficients of a wave with a period of 1 s. (b) Fourier coefficients of a wave with a period of 2 s. (c) Fourier coefficients of a wave with a period of 4 s.

the shape of the wave, the fundamental frequency decreases and the Fourier coefficients become denser. In the limit, as the period becomes infinitely large, there are Fourier coefficients at every frequency.

The exponential Fourier series coefficients of a periodic wave are given as:

$$C_n = \frac{1}{T}\int_{-T/2}^{T/2} x(t)e^{-jn\omega_0 t}\, dt$$

Let us examine the quantity TC_n. In the limit, as $T \to \infty$, we have:

$$TC_n = \int_{-\infty}^{\infty} x(t)e^{-jn\omega_0 t}\, dt$$

In the above expression, let us define $n\omega_0$ as ω. Further, let us define TC_n as $X(j\omega)$. Then:

Definition 6.3 (Fourier transform): The Fourier transform of a signal $x(t)$ is written as $X(j\omega)$, where:

$$X(j\omega) = \int_{-\infty}^{\infty} x(t)e^{-j\omega t}\, dt \tag{6.18}$$

We can reconstruct the signal $x(t)$ from its Fourier coefficients. We have:

$$\begin{aligned} x(t) &= \sum_{n=-\infty}^{\infty} C_n e^{jn\omega_0 t} \\ &= \frac{1}{2\pi} \sum_{n=-\infty}^{\infty} TC_n e^{jn\omega_0 t} \cdot \frac{2\pi}{T} \end{aligned}$$

Now we replace TC_n as $X(j\omega)$, $n\omega_0$ as ω, and $2\pi/T$ as ω_0.

$$\therefore x(t) = \frac{1}{2\pi} \sum_{n=-\infty}^{\infty} X(j\omega) e^{j\omega t} \omega_0$$

In the limit, as $T \to \infty$, $\omega_0 = 2\pi/T$ will become infinitesimally small; let us rename this as $\Delta\omega$.

$$\therefore x(t) = \frac{1}{2\pi} \sum_{n=-\infty}^{\infty} X(j\omega) e^{j\omega t} \Delta\omega$$

In the limit the sum becomes an integral.

Definition 6.4 (Inverse Fourier transform): The inverse Fourier transform of $X(j\omega)$ is given as $x(t)$, with:

$$x(t) = \frac{1}{2\pi} \int_{-\infty}^{\infty} X(j\omega) e^{j\omega t} \, d\omega \tag{6.19}$$

The Fourier transform (6.18) and the inverse Fourier transform (6.19) form a consistent system. The Fourier transform for a signal converges if the signal is absolutely integrable (sufficient, but not necessary). The derivation given in this discussion is not rigorous; for a full derivation the reader is referred to [20].

6.8.1 Relationship with the Laplace transform

Notice the similarity between the Fourier transform of (6.18) and the Laplace transform of (5.1). In case of the Laplace transform, $j\omega$ is replaced by s. The integral is taken from 0 to ∞ in the Laplace transform, as opposed to $-\infty$ to ∞. However, in our discussion of Laplace transforms, we worked with signals that were always 0 for $t < 0$. The Fourier and Laplace transforms of such signals are identical, with only a replacement of the variable $j\omega$ with s.

The inverse Fourier transform given in (6.19) is also very similar to the inverse Laplace transform integral given in (5.3). In case of the Laplace transform, the signal is restricted to be 0 for $t < 0$. For the Fourier transform there is no such restriction. The factor $u(t)$ at the end of (5.3) is not required in the inverse Fourier transform. A simple replacement of s with $j\omega$, and an integration along the $j\omega$ axis with $\sigma = 0$ will change the inverse Laplace transform into an inverse Fourier transform.

The Fourier transforms of the functions in Tab. 5.1 are mostly identical, with a simple replacement of s with $j\omega$. Fourier transforms can be deduced for other functions for which Laplace transforms cannot be constructed, specifically functions which exist before $t = 0$. Periodic functions can only be treated with Fourier transforms (and Fourier series), but not with Laplace transforms. On the other hand, Laplace transforms can handle initial conditions, which Fourier transforms do not.

A circuit with initial conditions has to be analyzed with Laplace transforms. The homogeneous response and the steady-state response are both evaluated together with Laplace and inverse Laplace transforms. On the other hand, a circuit which is in steady state can be analyzed with the help of Fourier transforms. Only the steady-state response is evaluated with Fourier and inverse Fourier transforms; the homogeneous response cannot be evaluated with this technique.

6.8.2 Properties of Fourier transforms

A table of Fourier transform pairs that cannot be evaluated with Laplace transforms are given in Tab. 6.1. A few important properties of Fourier transforms are listed below. The reader is encouraged to prove all of these. In the interest of brevity, the proofs will be omitted from this text.

Fourier transform at DC: The Fourier transform of $x(t)$ at $\omega = 0$ is the total area under $x(t)$.

Linearity: Fourier transforms are linear. Both superposition and homogeneity are valid for Fourier transforms.

Convolution: Convolution in the time domain is equivalent to multiplication in the frequency domain. This allows us to use the frequency domain for working out responses of a system to different signals.

Multiplication: The Fourier integrals are duals of each other. As a result, multiplication in the time domain is equivalent to convolution in the frequency domain, with a factor of $1/(2\pi)$ in the frequency domain.

Time shifting: A time shift or delay by t_0 is the same as a convolution with $\delta(t-t_0)$. As such, the Fourier transform is to be multiplied with $e^{-j\omega t_0}$.

Table 6.1: Fourier transform pairs. These are in addition to the Laplace transform pairs of Tab. 5.1, where a replacement of s with $j\omega$ will give us a Fourier transform pair.

Name	$f(t)$	$F(j\omega)$
Delta	$\delta(t)$	1
Constant	1	$2\pi\delta(\omega)$
Delayed delta	$\delta(t-t_0)$	$e^{-j\omega t_0}$
Complex exponential	$e^{j\omega_0 t}$	$2\pi\delta(\omega-\omega_0)$
Cosine	$\cos(\omega_0 t)$	$\pi\{\delta(\omega-\omega_0)+\delta(\omega+\omega_0)\}$
Sine	$\sin(\omega_0 t)$	$-j\pi\{\delta(\omega-\omega_0)-\delta(\omega+\omega_0)\}$
Fourier series	$\sum_{n=-\infty}^{\infty} C_n e^{jn\omega_0 t}$	$2\pi\sum_{n=-\infty}^{\infty} C_n\delta(\omega-n\omega_0)$
Impulse train	$\sum_{k=-\infty}^{\infty}\delta(t-kT)$	$\frac{2\pi}{T}\sum_{n=-\infty}^{\infty}\delta(\omega-\frac{2\pi n}{T})$
Unit step	$u(t)$	$\frac{1}{j\omega}+\pi\delta(\omega)$
Rectangular pulse	$u(t+T)-u(t-T)$	$2T\,\text{sinc}(\omega T)$
Sinc	$\frac{W}{\pi}\,\text{sinc}(Wt)$	$u(\omega+W)-u(\omega-W)$

Differentiation in time: Differentiating in time is equivalent to a multiplication by $j\omega$ in the frequency domain.

Integration in time: The Fourier transform of $\int_{-\infty}^{t} x(t)dt$ is $\pi X(0)\delta(\omega) + X(j\omega)/(j\omega)$. In words, a delta at DC corresponding to the area under $x(t)$, and the Fourier transform of $x(t)$ divided by $j\omega$.

Energy: The energy in the signal, $\int_{|} x(t)|^2\,dt$, over the entire duration of the signal is the energy in the frequency domain scaled by $1/(2\pi)$. This is Parseval's relation.

$$\int_{-\infty}^{\infty} |x(t)|^2\,dt = \frac{1}{2\pi}\int_{-\infty}^{\infty} |X(j\omega)|^2\,d\omega$$

Scan the QR-code, and learn more about Fourier transforms, its inverse, derivations of the properties of Fourier transforms, and commonly used Fourier transform pairs.

Example 6.8. In the circuit shown below, the voltage source $v_{in}(t)$ is given. Determine the amplitude and phase spectra of the output, $v_{out}(t)$.

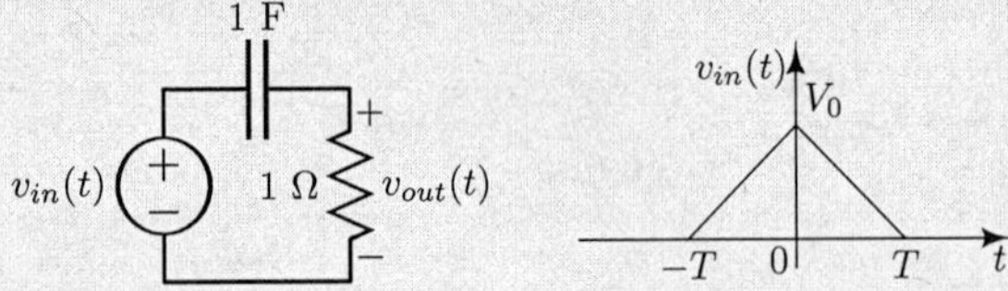

The Fourier transform of the input, $v_{in}(t)$, can be worked out as:

$$\begin{aligned} V_{in}(j\omega) &= V_0 \int_{-T}^{T} (1 - \frac{|t|}{T}) e^{-j\omega t} dt \\ \therefore V_{in}(j\omega) &= V_0 T \,\text{sinc}^2(\omega T/2) \end{aligned}$$

The transfer function of the circuit is:

$$\begin{aligned} \frac{V_{out}(j\omega)}{V_{in}(j\omega)} &= \frac{j\omega}{1 + j\omega} \\ \therefore V_{out}(j\omega) &= V_0 T \,\text{sinc}^2(\omega T/2) \frac{j\omega}{1 + j\omega} \\ \text{or, } |V_{out}(j\omega)| &= V_0 T \,\text{sinc}^2(\omega T/2) \frac{\omega}{\sqrt{1 + \omega^2}} \\ \text{and, } \angle V_{out}(j\omega) &= \pi/2 - \tan^{-1} \omega \end{aligned}$$

6.9 Unit summary

- Cos and sin functions at a frequency and its harmonics are orthogonal. The set of functions can be used to describe any periodic function.
- A periodic function can be broken up into its Fourier series. The Fourier series comprises of sines and cosines at the fundamental frequency and all its harmonics. The Fourier series can also be written as a sum of complex exponentials at the fundamental frequency and its harmonics.

- The class of functions that can be analyzed as a Fourier series is very broad. All practical functions are covered. Functions that are (a) not absolutely integrable within the period, (b) have an infinite number of maxima and minima within the period, (c) have an infinite number of discontinuities within the period cannot be analyzed as a Fourier series.

- There is an error between the sum of terms in the Fourier series and the periodic function at the points of discontinuity. This is known as Gibb's phenomenon and amounts to approximately 9% in amplitude around the points of discontinuity. The width of Gibb's error reduces as the number of Fourier terms increases.

- The Fourier series for even functions comprises only of cosine terms.

- The Fourier series for odd functions comprises only of sine terms.

- For functions with half-wave symmetry, $f(x + T/2) = -f(x)$. Such functions do not have any even harmonics.

- The exponential Fourier series is related to the trigonometric Fourier series. The cosine coefficient, a_n, is $2\,\mathrm{Re}[C_n]$. The sine coefficient, b_n, is $-2\,\mathrm{Im}[C_n]$. Further, the exponential Fourier series coefficients have conjugate symmetry, i.e., $C_{-n} = C_n^*$.

- The sinc function is ordinarily $\sin(x)/x$. At $x = 0$, $\mathrm{sinc}(0) = 1$.

- For a 0 to 1 square wave, the Fourier series coefficients, C_n, are given by $\eta\,\mathrm{sinc}(n\pi\eta)$, where η is the duty cycle of the square wave.

- A periodic function comprises of a DC, a fundamental frequency, and its harmonics. The Fourier coefficients of the function can be described in a plot with discrete points at the different frequencies.

- For circuit analysis, a periodic input can be analyzed as its Fourier series. The transfer function of the circuit multiplies the individual Fourier coefficients, and the Fourier series at the output is obtained.

- Parseval's relation: The mean-square of a periodic signal is the sum of the squares of all its exponential Fourier series coefficients.

- As the period of a periodic signal increases, the fundamental frequency decreases. In the limit, all frequencies appear in the spectrum of a non-periodic signal. The Fourier series sum is converted into a Fourier transform integral in this limit.

- Fourier transforms are closely related to the Laplace transform. For signals that are undefined for $t < 0$, the Laplace and Fourier transforms are identical. The Laplace transform method is used to analyze circuits where both the homogeneous and steady-state responses are required. For circuits in steady state, (no zero-input response), the Fourier series / Fourier transform method is required.

6.10 Exercises

Multiple choice type questions

6.1 A function $\sin(5\omega_0 t + \pi/3)$ has a fundamental frequency of
(a) ω_0 rad/s (b) $\omega_0/(2\pi)$ Hz (c) $5\omega_0$ Hz (d) $5\omega_0/(2\pi)$ Hz

6.2 The Fourier series of a square wave with 50% duty cycle will have no
(a) sine components (b) cosine components (c) even harmonics (d) odd harmonics

6.3 A 0 to 5 V saw-tooth waveform is expressed as a Fourier series. In every period, the waveform ramps from 0 to 5 V over the entire period, and then abruptly jumps back to 0 at the beginning of the next period. The Fourier series sum will exhibit an error at the discontinuities of
(a) 45 mV (b) 450 mV (c) 2.5 V (d) 5 V

6.4 The voltage across a 50 Ω resistor is $3\cos(\omega_0 t) + \sin(3\omega_0 t) - 0.1\cos(5\omega_0 t)$. The power dissipated in the resistor is:
(a) 10.01 W (b) 5.005 W (c) 200.2 mW (d) 100.1 mW

Short answer type questions

6.5 The exponential Fourier series coefficients for a periodic even function are ______, while the exponential Fourier series coefficients for a periodic odd function are ______.

6.6 The Fourier transform of a voltage, $v(t)$, is $V(j\omega)$. What are the units of $V(j\omega)$? What are the units of the Fourier series coefficients (if $v(t)$ is periodic)?

6.7 In the waveform sketches of Fig. 6.13, deduce the different kinds of symmetry, if any, in each waveform. Shift the $t = 0$ point to any convenient location to achieve symmetry.

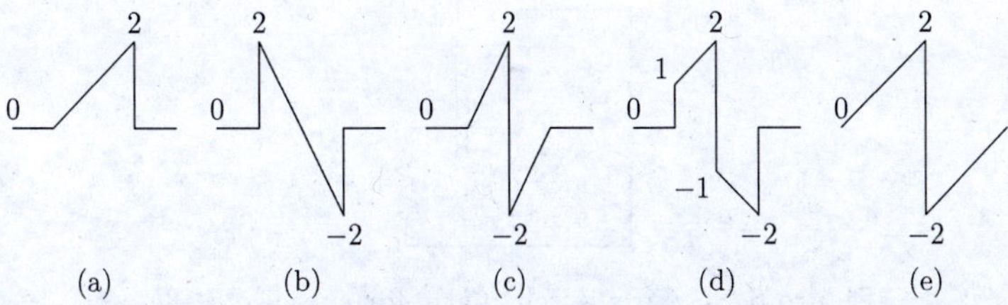

Figure 6.13: Waveforms for exercises 6.7 and 6.8.

Numericals and long answer type questions

6.8 Evaluate the Fourier series coefficients for the signals shown in Fig. 6.13. The waveforms are only shown for one period.

6.9 A non-linear circuit block behaves as a system with an input-output relationship given as $y = ax^3 + bx^2 + cx + d$, where y is the output and x is the input. If the input is $V_0 \cos(\omega t)$, express the output as a Fourier series. If the input is $V_0 \cos(\omega_1 t) + V_0 \cos(\omega_2 t)$, what will be the output? Express the output as a sum of cosines (or complex exponentials).

6.10 A non-linear circuit block behaves as a system with an input-output relationship given as $y = ax^3 + bx^2 + cx + d$, where y is the output and x is the input. Prove that if we increase the input amplitude by 1 dB, the 3rd harmonic content in the output will increase by 3 dB, and the 2nd harmonic content in the output will increase by 2 dB.

6.11 Develop the Fourier series for a half-wave rectified sine wave. Shift the time axis to a point such that the waveform is even.

6.12 A 50% duty cycle square wave at 1 Hz from -1 V to +1 V is applied across an RLC series circuit, with 0.1 Ω, inductance of 0.16 H and capacitance of 0.16 F. Evaluate the voltage waveform across the resistor.

6.13 An RF power amplifier circuit comprises of a (controlled) current source, and an RLC circuit as shown in Fig. 6.14. In practice, the (controlled) current source is designed with a transistor. In one period, $i(t)$ is defined as follows.

$$i(t) = \begin{cases} I_a(\cos(\omega t) - \cos\phi) & \text{for } |t| < \phi/\omega, \\ 0 & \text{for } |t| \geq \phi/\omega \end{cases}$$

First evaluate the Fourier series corresponding to $i(t)$ for arbitrary values of ϕ and I_a. Find the power delivered to R_L at the fundamental frequency, if $R_L = 50\ \Omega$, $L = 1$ mH, $C = 1\ \mu$F, ϕ is $\pi/3$, I_a is 2 mA, V_D is 5 V, and ω_0 is 10^9 rad/s. Find the total power dissipated by the current source.

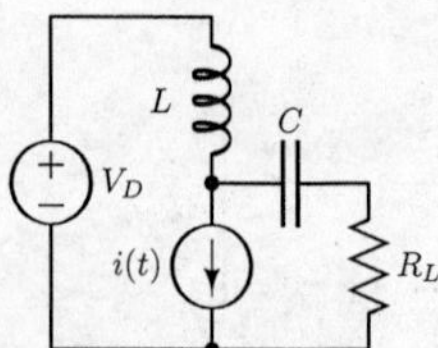

Figure 6.14: Circuit schematic for exercise 6.13.

6.14 The current through a diode, $i_D = I_0(e^{v_D/v_T} - 1)$, where I_0 and v_T are constants, and v_D is the voltage across the diode. Expand the exponential as a series and curtail i_D to its five leading terms. If $v_D(t)$ is $1\cos(\omega_0 t)$, find the Fourier series of $i_D(t)$. I_0 is 1 pA, v_T is 25 mV. Plot the estimated $i_D(t)$.

Know more

Historical profiles

Joseph Fourier (1768 to 1830) was a French mathematician and engineer. The Fourier series and transform, named after him, have profound applications in Electrical Engineering, in the analysis of periodic signals and heat transfer. It is also well known that Fourier discovered the greenhouse effect, a phenomenon that can now alter the course of human history.

Fourier was a contemporary of Laplace, although the course of his early life was very different. The son of a tailor, Fourier was an orphan at the young age of nine. He was brought up and educated through the Benedectine Order of the Convent of St. Mark. He took a prominent part in the French revolution and was with Napoleon as a scientific advisor. Napoleon appointed him as the governor of Isère, a department of France in the Alps region. He worked on heat propagation, even while involved in administrative activities. Fourier's name is inscribed on the northeast side of the Eiffel tower.

Understand in depth

Fourier series and Fourier transforms are covered in detail in most standard text books on signals and systems. The student is referred to:

1. Signals & Systems, by Alan V. Oppenheim, Alan S. Willsky, and S. Hamid Nawab, Prentice Hall, chapters 3, 4.

2. Signals and Systems, by Simon Haykin and Barry Van Veen, John Wiley and Sons, chapter 3.

Unit 7

Two-port networks

Unit specifics

In this unit we have discussed the following:

- Two port networks and terminal pairs
- Various two port parameters: Z, Y, H, G, T
- Relationship between two port variables, Z, Y, H, G, T
- Interconnection of two port networks

Rationale

Two-port networks are the starting point of general multi-port networks. The unit will help students formalize their understanding of multi-port networks. In the unit, we first discuss the different sets of two-port parameters. This is followed by developing relationships between the parameters, and an understanding of when to use which parameter set. The input and output impedance of a two-port network, and the transfer function of a circuit with a two-port network are derived. Specific circuit configurations for two-port networks are detailed.

The two-port network is a very convenient abstraction regime for devices that may be introduced in subsequent courses.

Pre-requisites

- Units 2, 4, 5 of this text. A clear understanding of network theorems, circuits in sinusoidal steady state, and Laplace transforms is assumed in this unit.

Unit outcomes

The list of outcomes of this unit are as follows.

U7-O1: Be able to analyze and derive two-port parameters for any given two-port network.

U7-O2: Be able to use the given two-port parameters to deduce the function of a circuit.

U7-O3: Be able to simplify larger circuits into smaller two-port circuit blocks.

Unit-1 outcomes	**Expected mapping with course outcomes** *(1: Weak, 2: medium, and 3: strong correlation)*					
	CO-1	**CO-2**	**CO-3**	**CO-4**	**CO-5**	**CO-6**
U7-O1	3	-	-	-	-	-
U7-O2	-	2	-	-	2	-
U7-O3	-	3	-	-	-	-

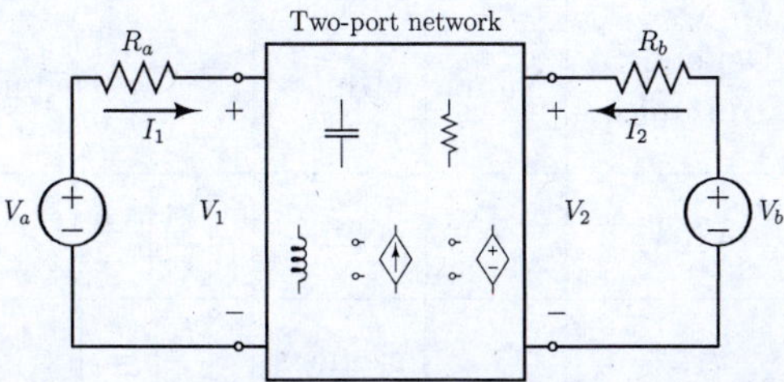

Figure 7.1: A two-port network in a circuit. The two-port network is an arbitrary connection of components. The voltage and current conventions are as shown.

7.1 Preliminaries

Two-port networks have two ports, $\mathcal{P}_1$ and $\mathcal{P}_2$. A port, as defined in definition 2.4, is nothing but a pair of terminals. For the discussion in this unit, a requirement is that the **current going out through a wire of the port comes back through the other wire of the same port.** This can easily be imposed. For example, in Fig. 7.1, V_a and R_a are in series at $\mathcal{P}_1$. The current going into the + terminal of $\mathcal{P}_1$ is guaranteed by KCL to be the current going out of the − terminal of $\mathcal{P}_1$. The same constraint has been imposed at $\mathcal{P}_2$.

In this discussion, we will be performing a detailed analysis of linear 2-port networks. It is assumed that a linear 2-port will only contain linear circuit elements, namely resistors, capacitors, inductors, mutual inductances, dependent voltage sources, and dependent current sources. While working with inductors, capacitors, and mutual inductances, we will work with either Laplace transforms or phasors. We will assume there are no independent voltage and current sources inside the two-port networks.

Fig. 7.1 shows a general two-port network in a circuit. For $\mathcal{P}_1$, the voltage and the current are V_1 and I_1, whereas, for $\mathcal{P}_2$, the voltage and the current are V_2 and I_2. The convention for defining the direction of the current is as shown in Fig. 7.1, i.e., **going into the two-port network.**

7.1.1 General idea behind two-port analyses

The circuit in Fig. 7.1 can be simplified, eliminating the resistors R_a and R_b from the picture, and directly applying voltage or current sources to the two ports. For example, we can apply I_1 and I_2 as two current sources to the two ports in the directions indicative of I_1 and I_2. Now we can think

Independent variables	Dependent variables	Name	Units	Definition
I_1, I_2	V_1, V_2	**Z**	ohms	$\begin{bmatrix} V_1 \\ V_2 \end{bmatrix} = \begin{bmatrix} z_{11} & z_{12} \\ z_{21} & z_{22} \end{bmatrix} \begin{bmatrix} I_1 \\ I_2 \end{bmatrix}$
V_1, V_2	I_1, I_2	**Y**	siemens	$\begin{bmatrix} I_1 \\ I_2 \end{bmatrix} = \begin{bmatrix} y_{11} & y_{12} \\ y_{21} & y_{22} \end{bmatrix} \begin{bmatrix} V_1 \\ V_2 \end{bmatrix}$
I_1, V_2	V_1, I_2	**H**	mixed	$\begin{bmatrix} V_1 \\ I_2 \end{bmatrix} = \begin{bmatrix} h_{11} & h_{12} \\ h_{21} & h_{22} \end{bmatrix} \begin{bmatrix} I_1 \\ V_2 \end{bmatrix}$
V_1, I_2	I_1, V_2	**G**	mixed	$\begin{bmatrix} I_1 \\ V_2 \end{bmatrix} = \begin{bmatrix} g_{11} & g_{12} \\ g_{21} & g_{22} \end{bmatrix} \begin{bmatrix} V_1 \\ I_2 \end{bmatrix}$
V_2, I_2	V_1, I_1	**T**	mixed	$\begin{bmatrix} V_1 \\ I_1 \end{bmatrix} = \begin{bmatrix} t_{11} & t_{12} \\ t_{21} & t_{22} \end{bmatrix} \begin{bmatrix} V_2 \\ -I_2 \end{bmatrix}$
V_1, I_1	V_2, I_2	$\mathbf{T}^{-1}$	mixed	$\begin{bmatrix} V_2 \\ -I_2 \end{bmatrix} = \begin{bmatrix} t'_{11} & t'_{12} \\ t'_{21} & t'_{22} \end{bmatrix} \begin{bmatrix} V_1 \\ I_1 \end{bmatrix}$

Table 7.1: All possible combinations of independent and dependent variables for two-port networks, with their names and definitions.

of the response to these two currents as the voltages V_1 and V_2. Applying superposition, because there are no independent sources inside the two-port network, the voltages V_1 and V_2 can be written as:

$$\begin{aligned} V_1 &= a_1 I_1 + a_2 I_2 \\ \text{and, } V_2 &= a_3 I_1 + a_4 I_2, \end{aligned}$$

where a_1, a_2, a_3, a_4 are arbitrary constants. These constants have units of resistance in this case.

Instead of applying two current sources at the two ports, we could have applied two voltage sources and have measured the currents. We could have applied a voltage source at $\mathcal{P}_1$, a current source at $\mathcal{P}_2$ and measured the current at $\mathcal{P}_1$ and voltage at $\mathcal{P}_2$. We could have mixed things up in any other combination as well and written up the equations. We have listed all possible combinations of independent variables (inputs) and dependent variables (outputs) in Table 7.1.

Once we have decided the dependent and independent variables in the two ports, we can always express the relationships as a set of two linear equations. We can readily summarize the two linear equations in the form

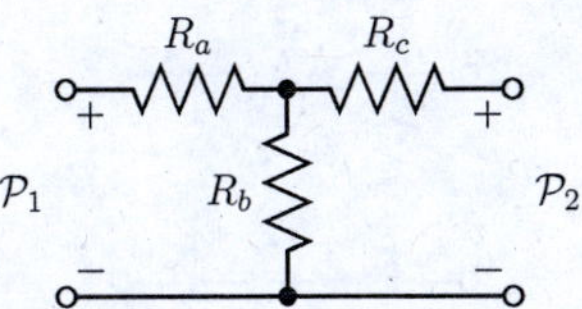

Figure 7.2: A T-network. It is convenient to use the **Z**-parameters to abstract a T-network.

of a matrix equation. We express the definitions of the two-port parameters with the help of these matrix equations.

7.2 Impedance (Z) parameters

We can apply current sources I_1 and I_2 at the two ports and measure the voltages V_1 and V_2. The relationships will give us the impedance parameters, or **Z**-parameters. For any given two-port network, the **Z**-parameters are defined as follows.

$$\begin{aligned} V_1 &= z_{11}I_1 + z_{12}I_2 \\ V_2 &= z_{21}I_1 + z_{22}I_2 \\ \text{or,} \begin{bmatrix} V_1 \\ V_2 \end{bmatrix} &= \begin{bmatrix} z_{11} & z_{12} \\ z_{21} & z_{22} \end{bmatrix} \begin{bmatrix} I_1 \\ I_2 \end{bmatrix} \end{aligned} \tag{7.1}$$

(7.1) can be reversed to work out the individual **Z**-parameters.

$$\begin{aligned} z_{11} &= \left.\frac{V_1}{I_1}\right|_{I_2=0} , \quad z_{12} = \left.\frac{V_1}{I_2}\right|_{I_1=0} \\ z_{21} &= \left.\frac{V_2}{I_1}\right|_{I_2=0} \quad \text{and} \quad z_{22} = \left.\frac{V_2}{I_2}\right|_{I_1=0} \end{aligned} \tag{7.2}$$

The condition $I_2 = 0$ implies an experiment performed when $\mathcal{P}_2$ is an open circuit. Similarly, the condition $I_1 = 0$ implies an experiment performed when $\mathcal{P}_1$ is an open circuit.

Example 7.1. Evaluate the **Z**-parameters for the two-port of Fig. 7.2. The definitions of the voltages and currents at the two ports should follow easily from the information shown in the schematic. The currents are always defined in a direction going into the + terminal of the corresponding port.

There are two experiments to be performed.

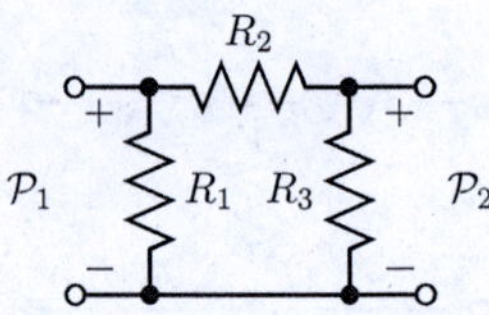

Figure 7.3: A Π-network. It is convenient to abstract a Π-network with the help of the **Y**-parameters.

1. We will open-circuit $\mathcal{P}_2$ and push a current I_1 into $\mathcal{P}_1$. In this case, the voltage $V_1 = I_1(R_a + R_b)$ and the voltage $V_2 = I_1R_b$. Then we have obtained z_{11} and z_{21} as follows.

$$z_{11} = \left.\frac{V_1}{I_1}\right|_{I_2=0} = R_a + R_b \text{ , and, } z_{21} = \left.\frac{V_2}{I_1}\right|_{I_2=0} = R_b$$

2. Next, we will open-circuit $\mathcal{P}_1$ and push a current I_2 into $\mathcal{P}_2$. The configuration will imply $V_2 = I_2(R_b + R_c)$ and $V_1 = I_2R_b$. We have obtained z_{22} and z_{12} as follows.

$$z_{22} = \left.\frac{V_2}{I_2}\right|_{I_1=0} = R_b + R_c \text{ , and, } z_{12} = \left.\frac{V_1}{I_2}\right|_{I_1=0} = R_b$$

The **Z**-parameters work out as:

$$\mathbf{Z} = \begin{bmatrix} R_a + R_b & R_b \\ R_b & R_b + R_c \end{bmatrix}$$

The circuit in the previous example is known as a T-network. In the following example, we will analyze two-port parameters for a Π-network.

Example 7.2. Evaluate the **Z**-parameters for the two-port network of Fig. 7.3.

1. First, let us open-circuit $\mathcal{P}_2$ and push a current I_1 into $\mathcal{P}_1$. This configuration results in $V_1 = I_1(R_1 \parallel (R_2 + R_3))$. The fraction of I_1 that chooses to go through $R_2 + R_3$ is $R_1/(R_1 + R_2 + R_3)$. $\therefore$ $V_2 = I_1R_1R_3/(R_1 + R_2 + R_3)$.
2. Next, we will open-circuit $\mathcal{P}_1$ and push a current I_2 into $\mathcal{P}_2$. The configuration results in $V_2 = I_2(R_3 \parallel (R_2 + R_1))$. The fraction of

I_2 that chooses to go through $R_1 + R_2$ is $R_3/(R_1 + R_2 + R_3)$. $\therefore$ $V_1 = I_2 R_1 R_3/(R_1 + R_2 + R_3)$.

Overall the **Z**-parameters work out as follows.

$$\mathbf{Z} = \begin{bmatrix} \frac{R_1R_2+R_1R_3}{R_1+R_2+R_3} & \frac{R_1R_3}{R_1+R_2+R_3} \\ \frac{R_1R_3}{R_1+R_2+R_3} & \frac{R_1R_3+R_2R_3}{R_1+R_2+R_3} \end{bmatrix}$$

Corollary 7.1. For a passive (reciprocal) two-port network, $z_{12} = z_{21}$.

Proof. The proof follows directly from the reciprocity theorem, Theorem 2.7. ■

7.3 Conductance (Y) parameters

When the independent variables are V_1 and V_2, and the dependent variables are I_1 and I_2, the two-port parameters obtained are the conductance or **Y** parameters.

$$\begin{aligned} I_1 &= y_{11}V_1 + y_{12}V_2 \\ I_2 &= y_{21}V_1 + y_{22}V_2 \\ \text{or, } \begin{bmatrix} I_1 \\ I_2 \end{bmatrix} &= \begin{bmatrix} y_{11} & y_{12} \\ y_{21} & y_{22} \end{bmatrix} \begin{bmatrix} V_1 \\ V_2 \end{bmatrix} \end{aligned} \tag{7.3}$$

(7.3) can be reversed to work out the individual **Y**-parameters.

$$\begin{aligned} y_{11} &= \left.\frac{I_1}{V_1}\right|_{V_2=0}, \quad y_{12} = \left.\frac{I_1}{V_2}\right|_{V_1=0} \\ y_{21} &= \left.\frac{I_2}{V_1}\right|_{V_2=0} \quad \text{and} \quad y_{22} = \left.\frac{I_2}{V_2}\right|_{V_1=0} \end{aligned} \tag{7.4}$$

The condition $V_2 = 0$ implies an experiment performed when $\mathcal{P}_2$ is a short circuit. Similarly, the condition $V_1 = 0$ implies an experiment performed when $\mathcal{P}_1$ is a short circuit. Also, note that by the definitions of **Y** and **Z** parameters as given in (7.3) and (7.1), $\mathbf{Y} = \mathbf{Z}^{-1}$.

As illustrations, we will work out the two-port parameters for T-networks and Π-networks in the following examples.

Example 7.3. Let us now evaluate the **Y**-parameters of the Π-network of Fig. 7.3. We have already worked out the **Z**-parameters of the circuit in example 7.2. As such, we can just invert the **Z** matrix to obtain **Y**. However, for instruction, we will work out the **Y**-parameters directly, step by step.

1. First, we will short-circuit $\mathcal{P}_2$ (i.e., guarantee $V_2 = 0$) and apply a voltage V_1 across $\mathcal{P}_1$. The current I_1 will be the sum of the current going through R_1 and the current going through R_2. (R_3 is short-circuited, and no current flows through R_3. As such, R_3 can be removed from the circuit while performing this experiment.) $\therefore I_1 = V_1/R_1 + V_1/R_2$. Further, the current going through the short circuit will be the total current flowing through R_2. Keep the direction of current in mind; the direction of I_2 needs to be **going into** $\mathcal{P}_2$. $\therefore I_2 = -V_1/R_2$. The results for I_1 and I_2 give $y_{11} = 1/R_1 + 1/R_2$ and $y_{21} = -1/R_2$.
2. Next, we will short-circuit $\mathcal{P}_1$ and apply a voltage V_2 across $\mathcal{P}_2$. The current I_2 will be the sum of the current going through R_3 and the current going through R_2. (R_1 is short-circuited, and no current flows through R_1. R_1 can be removed during this experiment.) $\therefore I_2 = V_2/R_3 + V_2/R_2$. Further, the current through the short circuit will be the total current flowing through R_2 in the reverse direction. $\therefore I_1 = -V_2/R_2$. The results for I_1 and I_2 give $y_{22} = 1/R_3 + 1/R_2$ and $y_{12} = -1/R_2$.

Overall, the **Y** parameter matrix is as follows.

$$\mathbf{Y} = \begin{bmatrix} 1/R_1 + 1/R_2 & -1/R_2 \\ -1/R_2 & 1/R_2 + 1/R_3 \end{bmatrix} = \begin{bmatrix} G_1 + G_2 & -G_2 \\ -G_2 & G_2 + G_3 \end{bmatrix}$$

G_1, G_2, G_3 are $1/R_1$, $1/R_2$, and $1/R_3$, respectively.

We can quickly check if **Y** and **Z** from example 7.2 are inverses of each other. Using Octave to check this will be satisfying. If we have values of the resistors, the check is trivial. Octave can also use its "symbolic" toolbox for algebraic manipulations without knowing the values of the resistors.

```
octave:1> pkg load symbolic
octave:2> syms r1 r2 r3
Symbolic pkg v2.8.0: Python communication link active, SymPy v1.5.1.
octave:3> Y = [1/r1+1/r2, -1/r2; -1/r2, 1/r2+1/r3];
octave:4> Z = [r1*r2+r1*r3, r1*r3; r1*r3, r1*r3+r2*r3]/(r1+r2+r3);
octave:5> simplify(Z*Y)
ans = (sym 2x2 matrix)
```

$$\begin{bmatrix} 1 & 0 \\ 0 & 1 \end{bmatrix}$$

Example 7.4. Let us now evaluate the **Y**-parameters of the T-network in Fig. 7.2. We have already worked out the **Z**-parameters of the circuit in example 7.1. As such, $\mathbf{Y} = \mathbf{Z}^{-1}$. Nevertheless, let us work out the **Y**-parameters directly from the circuit.

1. First, we will short-circuit $\mathcal{P}_2$ while applying a voltage V_1 at $\mathcal{P}_1$. The current at $\mathcal{P}_1$ is $V_1/(R_a+(R_b \parallel R_c))$, or $V_1(R_b+R_c)/(R_aR_b+R_aR_c+R_bR_c)$. This current breaks up (current division) between R_b and R_c. The current in R_c is the current through $\mathcal{P}_2$, backward. The current in $\mathcal{P}_2$ is, therefore, $-V_1R_b/(R_aR_b+R_aR_c+R_bR_c)$.
2. Next, we will short-circuit $\mathcal{P}_1$ while applying a voltage V_2 at $\mathcal{P}_2$. The current at $\mathcal{P}_2$ is $V_2(R_a+R_b)/(R_aR_b+R_aR_c+R_bR_c)$. The current at $\mathcal{P}_1$ is $-V_2R_b/(R_aR_b+R_aR_c+R_bR_c)$.

Overall the admittance parameters have been worked out as follows.

$$\begin{aligned} \mathbf{Y} &= \frac{1}{R_aR_b+R_bR_c+R_cR_a}\begin{bmatrix} R_b+R_c & -R_b \\ -R_b & R_a+R_b \end{bmatrix} \\ &= \frac{1}{G_a+G_b+G_c}\begin{bmatrix} G_cG_a+G_aG_b & -G_cG_a \\ -G_cG_a & G_bG_c+G_cG_a \end{bmatrix} \end{aligned}$$

G_a, G_b, G_c are $1/R_a$, $1/R_b$, and $1/R_c$, respectively.

Corollary 7.2. For a passive (reciprocal) two-port network, $y_{12} = y_{21}$.

Proof. The proof follows directly from the reciprocity theorem, Theorem 2.6. ■

7.4 Star-Delta and Delta-Star conversion

The circuit configuration in example 7.1 is known as a T-network (because it looks like the alphabet T), or a Y-network (because if you tilt the arms, it looks like the alphabet Y), or sometimes a star network (a 3-arm star). On the other hand, the circuit configuration in example 7.2 is known as a

Π-network (because it looks like the Greek alphabet Π), or a delta-network (because if you tilt the arms and draw it upside-down, it looks like the Greek alphabet Δ).

If two two-port networks have the same two-port parameter values, it does not matter what they contain; they will behave identically looking from the outside. A delta network can be replaced with a star network and vice-versa if the two-port parameter values of both are the same.

The two networks of example 7.1 and example 7.2 are equivalent if their **Z**-parameters are equal. That is,

$$\begin{bmatrix} R_a + R_b & R_b \\ R_b & R_b + R_c \end{bmatrix} = \frac{1}{R_1 + R_2 + R_3} \begin{bmatrix} R_1R_2 + R_1R_3 & R_1R_3 \\ R_1R_3 & R_1R_3 + R_2R_3 \end{bmatrix} \tag{7.5}$$

(7.5) indicates that:

1. Let us observe z_{12} and z_{21}. $R_b = R_1R_3/(R_1 + R_2 + R_3)$. The value of R_b will equalize both z_{12} and z_{21}.

2. Let us observe z_{11}. $R_a + R_b = (R_1R_2 + R_1R_3)/(R_1 + R_2 + R_3)$. Of this, from the earlier observation, R_b is $R_1R_3/(R_1 + R_2 + R_3)$. $\therefore R_a = R_1R_2/(R_1 + R_2 + R_3)$.

3. Lastly, let us observe z_{22}. $R_b + R_c = (R_1R_3 + R_2R_3)/(R_1 + R_2 + R_3)$. Of this, from the earlier observation, R_b is $R_1R_3/(R_1 + R_2 + R_3)$. $\therefore$ $R_c = R_2R_3/(R_1 + R_2 + R_3)$.

If we have a known delta network, we can replace it with a star network with the given conversion relationships.

The two networks are also equivalent if their **Y**-parameters are equal. The **Y**-parameters obtained in example 7.3 and example 7.4 are equated.

$$\begin{bmatrix} G_1 + G_2 & -G_2 \\ -G_2 & G_2 + G_3 \end{bmatrix} = \frac{1}{G_a + G_b + G_c} \begin{bmatrix} G_cG_a + G_aG_b & -G_cG_a \\ -G_cG_a & G_bG_c + G_cG_a \end{bmatrix} \tag{7.6}$$

1. Comparing y_{12} and y_{21}, $G_2 = G_cG_a/(G_a + G_b + G_c)$, or $R_2 = (R_bR_c + R_cR_a + R_aR_b)/R_b$.

2. Next, we can equate the y_{11} terms. We can use the result obtained for G_2 in the previous step. Therefore, $G_1 = G_aG_b/(G_a + G_b + G_c)$, or $R_1 = (R_bR_c + R_cR_a + R_aR_b)/R_a$.

3. We need to equate the y_{22} terms. Let us substitute the result obtained for G_2. We obtain $G_3 = G_bG_c/(G_a + G_b + G_c)$. In terms of resistance, $R_3 = (R_bR_c + R_cR_a + R_aR_b)/R_c$.

Star / T / Y	Delta / Δ / Π
R_z R_y R_x	R_1 R_2 R_3
R_x	$\frac{R_2 R_3}{R_1+R_2+R_3}$
R_y	$\frac{R_3 R_1}{R_1+R_2+R_3}$
R_z	$\frac{R_1 R_2}{R_1+R_2+R_3}$
$\frac{G_y G_z}{G_x+G_y+G_z}$	G_1
$\frac{G_z G_x}{G_x+G_y+G_z}$	G_2
$\frac{G_x G_y}{G_x+G_y+G_z}$	G_3

Table 7.2: Conversion between star (T or Y) and delta (Δ or Π)

If we have a known star network, we can replace it with a delta network with the given conversion relationships.

Conversion between star-delta configurations is useful while simplifying a network. We have shown the transformations between star and delta in a memorizable form in Table 7.2. Typically, the star to delta transformation (second in the table) is more useful, because it reduces a node in the circuit.

Example 7.5. In the circuit shown below, find the current i_x.

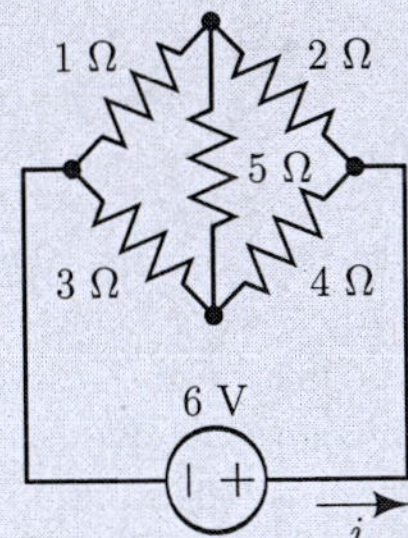

The 1 Ω, 5 Ω, 2 Ω resistors form a star. The node joining the three resistors can be reduced if we convert the set of resistors into a delta. The resistor opposite to 5 Ω reduces to $(5 \times 1 + 5 \times 2 + 2 \times 1)/5$ or 3.4 Ω. The resistor opposite to 2 Ω works out to $(5 \times 1 + 5 \times 2 + 2 \times 1)/2$ or 8.5 Ω. The resistor opposite to 1 Ω works out to 17 Ω. The circuit is redrawn after the transformation in the following schematics.

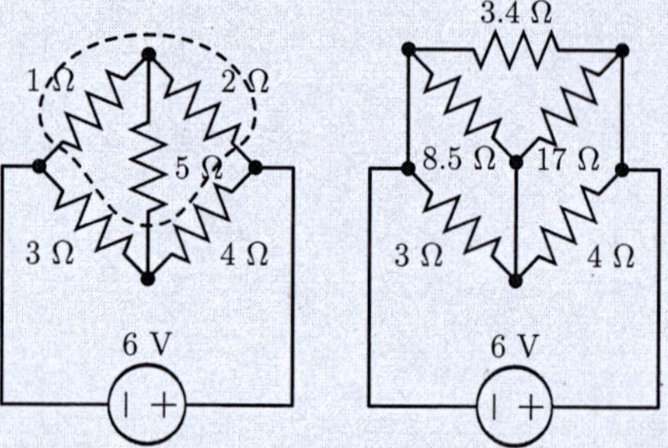

The parallel branches can now be combined, and the series effective resistors can be further combined. The entire network has a resistance of $[\{(3 \parallel 8.5) + (17 \parallel 4)\} \parallel 3.4]$ Ω, or 2.095 Ω. i_x is 2.865 A.

Star-Delta and Delta-Star conversion are often used to simplify circuits through repeated application.

Example 7.6. In the circuit shown below, all resistors are of value 1 Ω. Find the resistance between the two terminals.

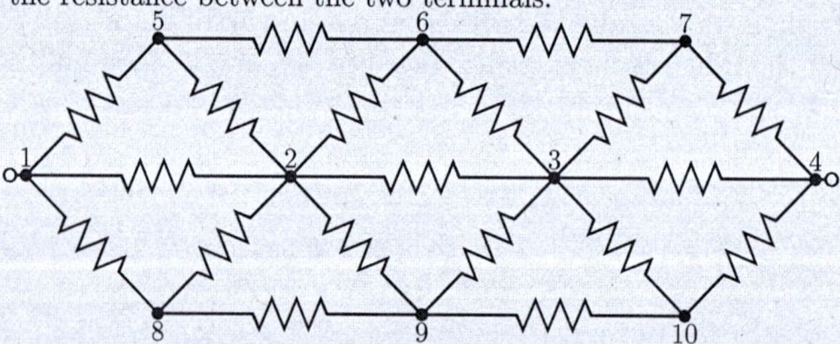

There are many ways to solve the problem. We can take advantage of the symmetry in the network. All resistors are of equal value and it will be sufficient if we obtain the resistance of the network till the axis of symmetry, at the middle. Although significantly simpler, in the following discussion we will not use this path, but illustrate the solution with repeated applications of delta-star transformations.

First, star-delta transformations are applied at nodes 5, 7, 8, 10 and these nodes are eliminated. The 1 Ω, 1 Ω, 1 Ω star connection transforms into a 3 Ω, 3 Ω, 3 Ω delta connection. Resistors in parallel are combined. The transformed circuit is given below. Unnamed resistors continue to be 1 Ω.

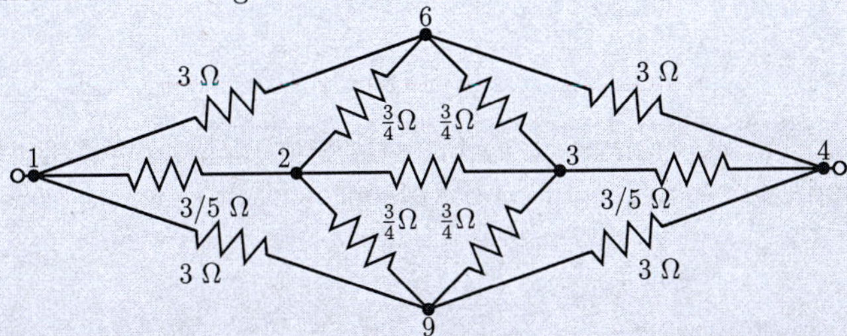

There are no three-connection star nodes any more and star-delta transformations will no longer be useful. However, a delta-star transformation will be useful to simplify connections at nodes 6 and 9. We will convert the delta-connected resistors between nodes 2, 3, 6 into a star, with a new central node labeled as 11. We will convert the delta-connected resistors between 2, 3, 9 into a star, with a new central node labeled as 12. To apply both transformations, first we split the 1 Ω resistor between 2 and 3 into two 2 Ω resistors in shunt. The transformed circuit is given below.

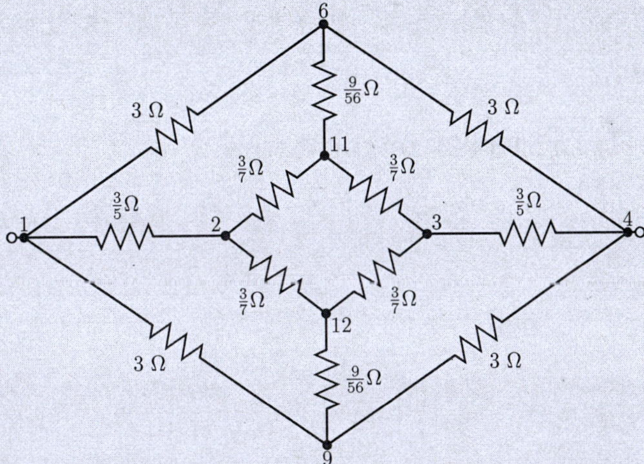

Stars formed at nodes 6 and 9 can now be eliminated to further simplify. However, we will not do this. We can now use the symmetry in the network and infer that no current will flow through the $\frac{9}{56}$ Ω resistors. As such, these two resistors can be treated as open circuits.

We now evaluate the equivalent resistance using series-parallel combina-

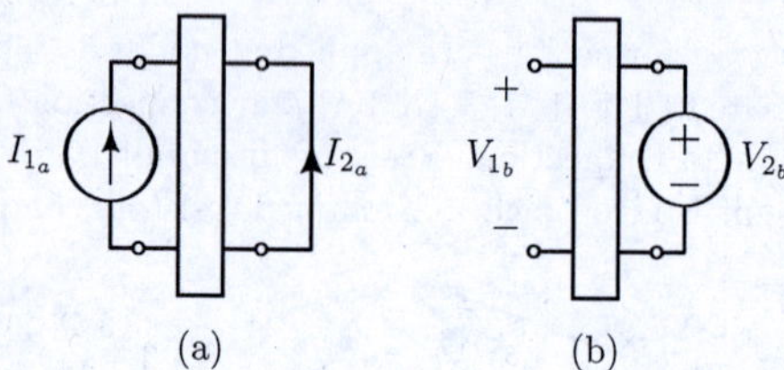

Figure 7.4: Two experiments to evaluate the **H**-parameters. (a) Short circuit $\mathcal{P}_2$ and apply a current I_{1_a} at $\mathcal{P}_1$, and (b), open-circuit $\mathcal{P}_1$ and apply a voltage V_{2_b} at $\mathcal{P}_2$.

tions. The resistance between 1 and 4 is given by:

$$6 \parallel 6 \parallel \{\frac{3}{5} + (\frac{6}{7} \parallel \frac{6}{7}) + \frac{3}{5}\} = \frac{19}{18}\ \Omega$$

If exploiting symmetry is not straightforward, or in the absence of symmetry, the reader can eliminate nodes 6 and 9 through star-delta transformations, followed by elimination of nodes 2 and 3 with further star-delta transformations. Finally, either node 11 or node 12 can be eliminated with a star-delta transformation. This will completely simplify the network.

7.5 Hybrid (H) parameters

I_1 and V_2 are chosen as the independent variables in the hybrid **H**-parameters representation of a two-port network.

$$\begin{aligned} V_1 &= h_{11}I_1 + h_{12}V_2 \\ I_2 &= h_{21}I_1 + h_{22}V_2 \\ \text{or,} \begin{bmatrix} V_1 \\ I_2 \end{bmatrix} &= \begin{bmatrix} h_{11} & h_{12} \\ h_{21} & h_{22} \end{bmatrix} \begin{bmatrix} I_1 \\ V_2 \end{bmatrix} \end{aligned} \tag{7.7}$$

(7.7) can be reversed to work out the individual **H**-parameters.

$$\begin{aligned} h_{11} &= \left.\frac{V_1}{I_1}\right|_{V_2=0} , \quad h_{12} = \left.\frac{V_1}{V_2}\right|_{I_1=0} \\ h_{21} &= \left.\frac{I_2}{I_1}\right|_{V_2=0} \quad \text{and} \quad h_{22} = \left.\frac{I_2}{V_2}\right|_{I_1=0} \end{aligned} \tag{7.8}$$

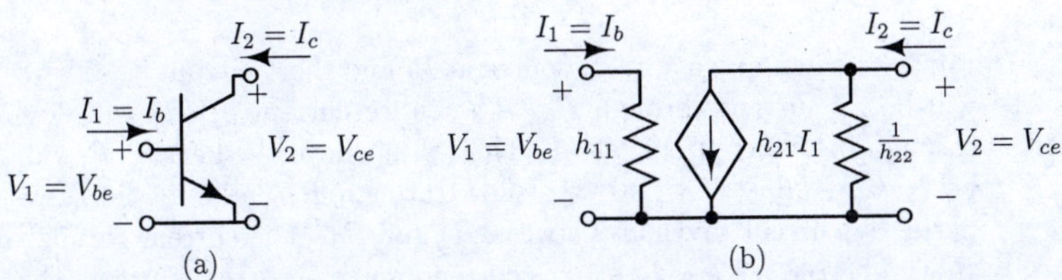

Figure 7.5: The model for a BJT is often expressed with **H**-parameters. The figure shows (a) the BJT configured as a two-port network, and (b) the hybrid model for the BJT.

The condition $V_2 = 0$ implies an experiment performed when $\mathcal{P}_2$ is a short circuit. Similarly, the condition $I_1 = 0$ implies an experiment performed when $\mathcal{P}_1$ is an open circuit. The hybrid parameters are so-called because their dimensions are not consistent. h_{11} has units of resistance, h_{22} has units of conductance, while h_{12} and h_{21} are dimensionless. To work out the **H**-parameters, we need to open-circuit $\mathcal{P}_1$ but short-circuit $\mathcal{P}_2$.

The two experiments required to evaluate the **H**-parameters of a two-port network are given in Fig. 7.4.

It is convenient to remember the **H**-parameters with the help of the BJT (bipolar junction transistor). The base-emitter junction of the BJT is $\mathcal{P}_1$, while the collector-emitter output is $\mathcal{P}_2$. Fig. 7.5(a) shows the general configuration of the BJT with its two ports, while Fig. 7.5(b) shows the model of the BJT. The current in the collector, I_c, is ideally βI_b; in this case I_2 is $h_{21}I_1 + h_{22}V_2$. The h_{22} term is a correction term to account for a non-ideal BJT.

Example 7.7. Evaluate the **H**-parameters for the two-port network in the schematic below.

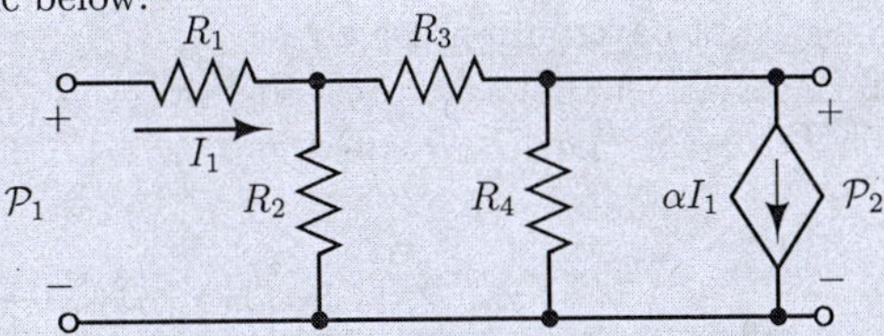

To work out the hybrid parameters, we will perform two experiments; (1) with a current source at $\mathcal{P}_1$ and a short-circuit at $\mathcal{P}_2$; (2) with a voltage source at $\mathcal{P}_2$ and an open-circuit at $\mathcal{P}_1$.

1. First, we will apply a current source at $\mathcal{P}_1$ and short-circuit $\mathcal{P}_2$. There will be no current through R_4. The entire current of αI_1 will flow through the short-circuit at $\mathcal{P}_2$. The voltage developed across $\mathcal{P}_1$ will be $I_1(R_1 + (R_2 \parallel R_3))$. I_1 will flow through R_1 and split into two parts, by current division, between R_2 and R_3. The current through R_3 will be $I_1R_2/(R_2+R_3)$. This current, and the current through the controlled current source, will add and flow through the short circuit at $\mathcal{P}_2$. So $I_2 = \alpha I_1 - I_1R_2/(R_2 + R_3)$.
2. Next, we will open circuit $\mathcal{P}_1$ and apply a voltage source at $\mathcal{P}_2$. When we open circuit $\mathcal{P}_1$, I_1 is zero, and the current through the controlled current source is zero. R_4 and $R_2 + R_3$ appear in parallel. $I_2 = V_2/R_4 + V_2/(R_2 + R_3)$. The voltage across R_2 is the same as the voltage across $\mathcal{P}_1$. The voltage across R_2 is $V_2R_2/(R_2 + R_3)$.

The overall hybrid parameters are as follows.

$$\mathbf{H} = \begin{bmatrix} R_1 + (R_2 \parallel R_3) & R_2/(R_2+R_3) \\ \alpha - R_2/(R_2+R_3) & 1/R_4 + 1/(R_2+R_3) \end{bmatrix}$$

It is essential to double-check the dimensions. h_{11} has dimensions of resistance, h_{22} has dimensions of conductance, h_{12} and h_{21} are dimensionless.

Corollary 7.3. For a passive (reciprocal) two-port network, $h_{12} = -h_{21}$.

Proof. Let us prove this using Tellegen's theorem.

1. To evaluate h_{21}, we short-circuit $\mathcal{P}_2$, and apply a current I_{1_a} at $\mathcal{P}_1$, as shown in Fig. 7.4. The current I_{2_a} is $h_{21}I_{1_a}$.

2. To evaluate h_{12}, we open-circuit $\mathcal{P}_1$, and apply a voltage V_{2_b} at $\mathcal{P}_2$, as shown in Fig. 7.4. The voltage at $\mathcal{P}_1$, V_{1_b}, is $h_{12}V_{2_b}$.

Let the two-port network have only passive resistors (phasors can be used in case of a general impedance). Let these resistors be $\{R_k\}$. Let the currents through them be $\{I_{k_a}\}$ for the first experiment and $\{I_{k_b}\}$ for the second experiment.

	$\mathcal{P}_1$	Inside two-port	$\mathcal{P}_2$
Experiment a, voltages	V_{1_a}	$I_{k_a}R_k$	0
Experiment a, currents	$-I_{1_a}$	I_{k_a}	$-I_{2_a}$
Experiment b, voltages	V_{1_b}	$I_{k_b}R_k$	V_{2_b}
Experiment b, currents	0	I_{k_b}	$-I_{2_b}$

All the voltages and currents measured in these two experiments are listed. Using Tellegen's theorem, we can cross-multiply, i.e., the sum of the voltages from the first experiment times the currents from the second experiment will add up to zero, and vice-versa.

$$\begin{aligned}
V_{1_a}\cdot 0+\sum I_{k_a}R_kI_{k_b}+0\cdot(-I_{2_b})=0,\quad \text{or,}\ \sum I_{k_a}R_kI_{k_b} &= 0.\\
\text{Also,}\ (-I_{1_a})\cdot V_{1_b}+\sum I_{k_a}I_{k_b}R_k+(-I_{2_a})V_{2_b} &= 0\\
\therefore I_{1_a}V_{1_b} &= -I_{2_a}V_{2_b}\\
\text{or,}\ V_{1_b}/V_{2_b} &= -I_{2_a}/I_{1_a}.\\
\therefore h_{12} &= -h_{21}
\end{aligned}$$

■

The **H**-parameters are related to the other parameters. For example, with the **Z**-parameters, conversions from **Z** to **H** are derived below. Similar derivations are possible between any two sets of parameters.

$$\begin{aligned}
V_1 &= z_{11}I_1+z_{12}I_2\\
V_2 &= z_{21}I_1+z_{22}I_2\\
\text{or,}\ I_2 &= -\frac{z_{21}}{z_{22}}I_1+\frac{1}{z_{22}}V_2\\
\text{and,}\ V_1 &= (z_{11}-\frac{z_{12}z_{21}}{z_{22}})I_1+\frac{z_{12}}{z_{22}}V_2\\
\therefore \mathbf{H} &= \begin{bmatrix} z_{11}-\frac{z_{12}z_{21}}{z_{22}} & \frac{z_{12}}{z_{22}}\\ -\frac{z_{21}}{z_{22}} & \frac{1}{z_{22}}\end{bmatrix}
\end{aligned}$$

7.6 Inverse hybrid (G) parameters

V_1 and I_2 are the independent variables in the inverse hybrid or **G**-parameter representation of a two-port network.

$$\begin{aligned}
I_1 &= g_{11}V_1+g_{12}I_2\\
V_2 &= g_{21}V_1+g_{22}I_2\\
\text{or,}\ \begin{bmatrix} I_1\\ V_2\end{bmatrix} &= \begin{bmatrix} g_{11} & g_{12}\\ g_{21} & g_{22}\end{bmatrix}\begin{bmatrix} V_1\\ I_2\end{bmatrix} \qquad (7.9)
\end{aligned}$$

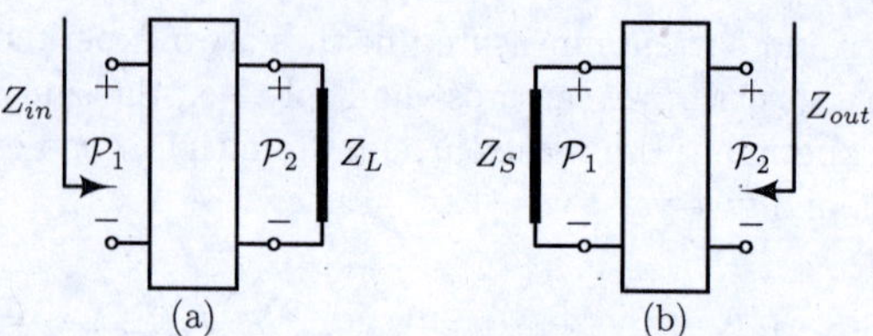

Figure 7.6: Circuit structures for evaluating driving point impedance. (a) When we terminate $\mathcal{P}_2$ with Z_L, the driving point impedance looking in from $\mathcal{P}_1$ is the input impedance, Z_{in}. (b) When we terminate $\mathcal{P}_1$ with Z_S, the driving point impedance looking in from $\mathcal{P}_2$ is the output impedance, Z_{out}.

(7.9) can be reversed to work out the individual **G**-parameters.

$$g_{11} = \left.\frac{I_1}{V_1}\right|_{I_2=0}, \quad g_{12} = \left.\frac{I_1}{I_2}\right|_{V_1=0}$$
$$g_{21} = \left.\frac{V_2}{V_1}\right|_{I_2=0} \quad \text{and} \quad g_{22} = \left.\frac{V_2}{I_2}\right|_{V_1=0} \tag{7.10}$$

The condition $I_2 = 0$ implies an experiment performed when $\mathcal{P}_2$ is an open circuit. Similarly, the condition $V_1 = 0$ implies an experiment performed when $\mathcal{P}_1$ is a short circuit. If we interchange $\mathcal{P}_1$ and $\mathcal{P}_2$, the **H**-parameters for the new circuit will equal the **G**-parameters for the original circuit because of the relationship between (7.8) and (7.10). Also, without reversing the circuit, the **H** and **G** parameters are inverses of each other.

Corollary 7.4. For a passive (reciprocal) network, $g_{12} = -g_{21}$.

Proof. The **G**-matrix for a network is the reverse (i.e., ports one and two interchanged) of the **H**-matrix of the reversed network. Also $h_{12} = -h_{21}$. $\therefore g_{12} = -g_{21}$. ■

7.7 Input and output impedance

Two-port parameters are convenient when we need to compute driving point impedance functions, looking in, either from $\mathcal{P}_1$ or from $\mathcal{P}_2$. Working out the driving point impedance in the configurations shown in Fig. 7.6 is particularly useful.

Definition 7.1 (Input impedance): For a two-port network, we define the input impedance as the driving point impedance looking in from $\mathcal{P}_1$, when we terminate $\mathcal{P}_2$ with an impedance Z_L.

Definition 7.2 (Output impedance): For a two-port network, the output impedance is the driving point impedance looking in from $\mathcal{P}_2$, when we terminate $\mathcal{P}_1$ with an impedance Z_S.

We can use any two-port parameters to work out the input impedance. As an example, let us use the **Z**-parameters. If **Z** for the two-port is known, we can write the equations for the two sides.

$$\begin{aligned} V_1 &= z_{11}I_1 + z_{12}I_2 \\ V_2 &= z_{21}I_1 + z_{22}I_2 = -Z_L I_2 \\ \therefore I_2 &= -z_{21}I_1/(z_{22} + Z_L) \\ \therefore V_1 &= z_{11}I_1 - \frac{z_{12}z_{21}}{z_{22} + Z_L} \cdot I_1 \\ \therefore Z_{in} = V_1/I_1 &= z_{11} - \frac{z_{12}z_{21}}{z_{22} + Z_L} \end{aligned} \tag{7.11}$$

We can repeat an identical derivation looking in from $\mathcal{P}_2$, with $\mathcal{P}_1$ terminated by Z_S. In (7.11), we have to interchange ports 1 and 2, and replace Z_L with Z_S.

$$Z_{out} = z_{22} - \frac{z_{21}z_{12}}{z_{11} + Z_S} \tag{7.12}$$

The fascinating fact is that we could use any set of parameters, and the expressions for the input and output impedance would have the same form, as long as the dimensions are consistent. For example, if we use the **Y**-parameters,

$$1/Z_{in} = y_{11} - \frac{y_{12}y_{21}}{y_{22} + G_L}$$

Notice that to make the dimensions consistent, we have obtained $1/Z_{in}$, not Z_{in}. Also, notice that we have adjusted the denominator term as $y_{22} + G_L$, where G_L is $1/Z_L$, to make the dimensions consistent.

Exercise 7.1. *Prove the following results.*

1. $1/Z_{in} = y_{11} - \frac{y_{12}y_{21}}{y_{22}+G_L}$
2. $1/Z_{out} = y_{22} - \frac{y_{12}y_{21}}{y_{11}+G_S}$
3. $Z_{in} = h_{11} - \frac{h_{12}h_{21}}{h_{22}+G_L}$
4. $1/Z_{out} = h_{22} - \frac{h_{12}h_{21}}{h_{11}+Z_S}$
5. $1/Z_{in} = g_{11} - \frac{g_{12}g_{21}}{g_{22}+Z_L}$
6. $Z_{out} = g_{22} - \frac{g_{12}g_{21}}{g_{11}+G_S}$

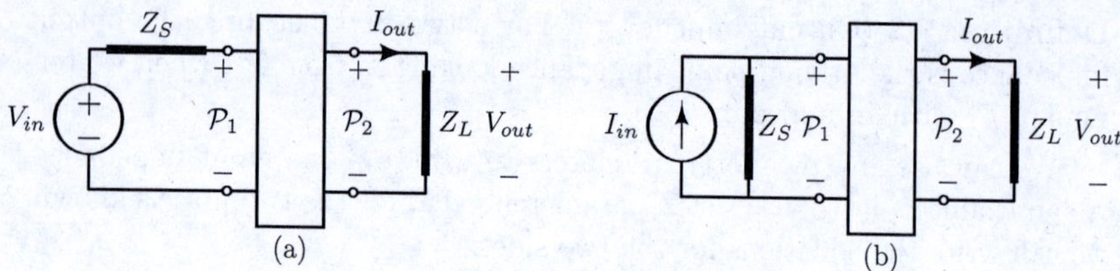

Figure 7.7: A two-port network used as a system. We have shown the input source as a voltage source in (a), as a current source in (b). The output quantity is either V_{out} or I_{out}.

7.8 Transfer function

The two-port network is often used as a system, with input at $\mathcal{P}_1$ and the output taken at $\mathcal{P}_2$. The network configurations are as shown in Fig. 7.7.

Definition 7.3 (System transfer function): The system transfer function is V_{out}/V_{in}, V_{out}/I_{in}, I_{out}/V_{in}, or I_{out}/I_{in}.

V_{out} and I_{out} are interchangeable, with a ratio of Z_L between them. Further, $V_{out} = V_2$. $I_{out} = V_2/Z_L = -I_2$. Also, V_{in} and I_{in} are interchangeable, as a source transformation can convert one network topology to the other. As such, V_{in} is related to I_{in} by a factor of Z_S.

Without any loss of generality, let us assume that we know the **Y**-parameters for the two-port network, and the network configuration is that of Fig. 7.7(b). On the load side, $I_2 = -V_2 G_L$.

$$
\begin{aligned}
I_1 &= y_{11}V_1 + y_{12}V_{out} \\
I_2 = -G_L V_{out} &= y_{21}V_1 + y_{22}V_{out} \\
\therefore y_{21}V_1 &= -V_{out}(G_L + y_{22}) \\
\therefore V_{out} &= -V_1 y_{21}/(y_{22} + G_L) \\
\therefore I_1 &= V_1(y_{11} - \frac{y_{12}y_{21}}{y_{22} + G_L}) \\
I_{in} &= V_1(y_{11} - \frac{y_{12}y_{21}}{y_{22} + G_L}) + V_1 G_S \\
\therefore V_1 &= \frac{I_{in}}{(y_{11} + G_S) - \frac{y_{12}y_{21}}{y_{22}+G_L}} \\
\therefore V_{out} &= \frac{y_{21}}{y_{12}y_{21} - (y_{11} + G_S)(y_{22} + G_L)} I_{in} \qquad (7.13)
\end{aligned}
$$

Interestingly, the form of the system transfer function is the same, regardless of the two-port parameters. Only minor changes need to be made, based on dimensions, to (7.13), to obtain the various forms. The specific system transfer function obtained is based on $1/p_{12}$. If the 1-2 parameter is current by voltage (admittance), the system transfer function will be I_{out}/V_{in}. If the 1-2 parameter is current by current (inverse hybrid parameters), the system transfer function will be I_{out}/I_{in}. Further, if the output quantity is I_{out}, we will have an extra minus sign, because $I_{out} = -I_2$, whereas, $V_{out} = V_2$.

Exercise 7.2. *Prove the following relationships.*

1. $I_{out}/V_{in} = -\frac{z_{21}}{z_{12}z_{21}-(z_{11}+Z_S)(z_{22}+Z_L)}$
2. $V_{out}/V_{in} = \frac{h_{21}}{h_{12}h_{21}-(h_{11}+Z_S)(h_{22}+G_L)}$
3. $I_{out}/I_{in} = -\frac{g_{21}}{g_{12}g_{21}-(g_{11}+G_S)(g_{22}+Z_L)}$

There are two important observations here.

Observation 7.1. If p_{21} is very large, then the magnitude of the system transfer function is approximately $1/p_{12}$. We can use any parameter, impedance, admittance, hybrid, or inverse hybrid. The observation is the same.

Observation 7.2. To obtain a system that has a very high gain, we will have to maximize p_{21} and make $p_{12} = 0$.

Observations 7.1 and 7.2 form the basis of analog electronics, amplifier design, and feedback circuits.

Example 7.8. Design a 6 dB attenuator. The attenuator is a two-port resistive network with 50 Ω input impedance, 50 Ω output impedance. The correct usage of the attenuator is by applying a 50 Ω load resistance, when the input source is a 50 Ω source. Without the attenuator, a 50 Ω source will drive a 50 Ω load with a voltage gain of 1/2. The attenuator will further attenuate the voltage gain by its rated attenuation.

In our case, the voltage gain from the input voltage source to the output voltage is 1/4. A sketch of the system is drawn below.

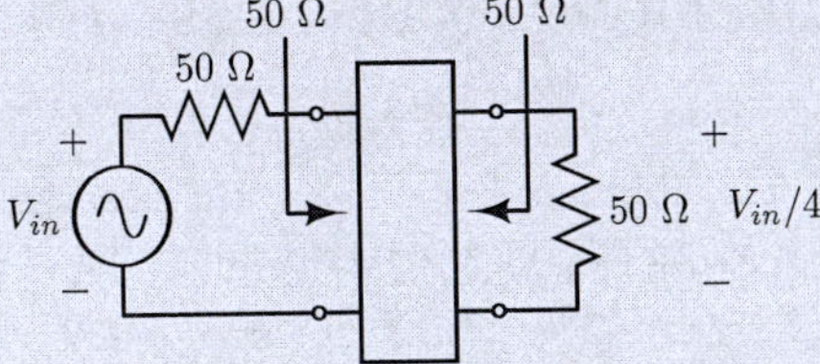

Let us choose to use the **Z**-parameters to design our attenuator. Since the network is resistive (passive), $z_{12} = z_{21}$. From the input and output impedance relationships we have:

$$\begin{aligned}
z_{11} - \frac{z_{12}^2}{z_{22}+50} &= 50 \\
z_{22} - \frac{z_{12}^2}{z_{11}+50} &= 50 \\
\text{or, } z_{11}(z_{22}+50) &= z_{22}(z_{11}+50) \\
\therefore\ z_{11} &= z_{22} \\
\text{and, } z_{11} - z_{12}^2/(z_{11}+50) &= 50 \\
\therefore\ z_{11}^2 - 50^2 &= z_{12}^2
\end{aligned}$$

Next, we will use the gain relationship. In our case, V_{out} is $V_{in}/4$. I_{out} is $V_{in}/200$.

$$\begin{aligned}
\therefore -\frac{z_{12}}{z_{12}^2 - (z_{11}+50)^2} &= \frac{1}{200} \\
\text{or, } 200z_{12} &= (z_{11}+50)^2 - z_{12}^2
\end{aligned}$$

Solving the two relationships we obtain z_{11} as 250/3 and z_{12} as 200/3. Finally we have:

$$\mathbf{Z} = \frac{1}{3}\begin{bmatrix} 250 & 200 \\ 200 & 250 \end{bmatrix} \Omega$$

A **Z** matrix of the above form can easily be designed with a T-network. Refer to Example 7.1. The final two-port network is shown below.

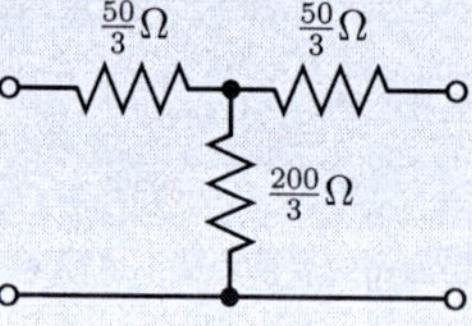

7.9 Transmission parameters

We have defined the transmission parameters of a two-port network in the fifth and sixth rows of Table 7.1. For **T**, V_2 and I_2 are the independent variables, while for $\mathbf{T}^{-1}$, V_1 and I_1 are the independent variables. The **T** matrix is also known as the $ABCD$-matrix. Notice how $-I_2$ has been used instead of I_2. We will be discussing only the **T** matrix. An identical discussion for

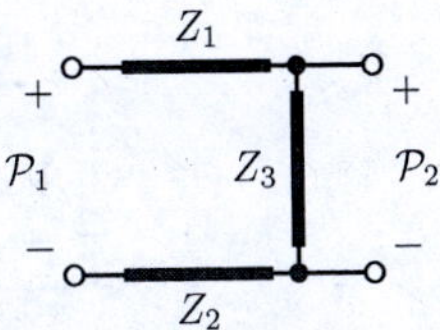

Figure 7.8: Two-port network for example-7.9.

$\mathbf{T}^{-1}$ may follow and will be omitted.

$$\begin{aligned} V_1 &= t_{11}V_2 - t_{12}I_2 = AV_2 - BI_2 \\ I_1 &= t_{21}V_2 - t_{22}I_2 = CV_2 - DI_2 \end{aligned} \tag{7.14}$$

(7.14) can be reversed to extract the individual transmission parameters.

$$\begin{aligned} t_{11} = A = \left.\frac{V_1}{V_2}\right|_{I_2=0} \quad &, \quad t_{12} = B = -\left.\frac{V_1}{I_2}\right|_{V_2=0} \\ t_{21} = C = \left.\frac{I_1}{V_2}\right|_{I_2=0} \quad &\text{and} \quad t_{22} = D = -\left.\frac{I_1}{I_2}\right|_{V_2=0} \end{aligned} \tag{7.15}$$

We need to perform two experiments to evaluate the transmission parameters; (1) with $\mathcal{P}_2$ as an open circuit ($I_2 = 0$) and (2) with $\mathcal{P}_2$ as a short circuit ($V_2 = 0$).

1. We will open circuit $\mathcal{P}_2$ and force $I_2 = 0$. Now, unfortunately, we cannot apply a voltage at V_2. Instead, we will apply a voltage at $\mathcal{P}_1$, V_1, and measure the open-circuit voltage, V_2. V_1/V_2 will give t_{11}. Next, we will push a current I_1 into $\mathcal{P}_1$ and measure the open-circuit voltage V_2. I_1/V_2 will give t_{21}.

2. Next, we will short circuit $\mathcal{P}_2$ and force $V_2 = 0$. Again we will apply a voltage source at $\mathcal{P}_1$, V_1, and measure the short-circuit current I_2. $V_1/(-I_2)$ will give t_{12}. Further, we will apply a current source I_1 into $\mathcal{P}_1$ and measure the short-circuit current, I_2. $I_1/(-I_2)$ will give t_{22}.

A few notes about the technique. Once again, we emphasize that for the transmission parameters, $-I_2$ is used instead of I_2. $-I_2$ is used to accommodate cascading multiple networks, as explained in Corollaries 7.6 and 7.7. Further, notice that the two experiments involve open and short circuits only at port-2, while the stimulus is applied only at port-1.

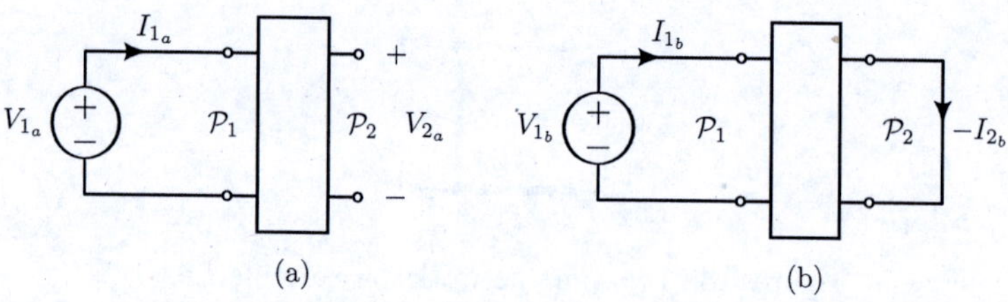

Figure 7.9: We need to perform two experiments to evaluate the transmission parameters. (a) open circuit at $\mathcal{P}_2$ and (b) short circuit at $\mathcal{P}_2$.

Example 7.9. Evaluate the **T** parameters for the two-port network of Fig. 7.8.

1. First, we open $\mathcal{P}_2$ to guarantee $I_2 = 0$. Now we apply V_1 at $\mathcal{P}_1$. $V_2 = V_1 Z_3/(Z_1 + Z_2 + Z_3)$. $\therefore V_1/V_2 = 1 + (Z_1 + Z_2)/Z_3$.
2. We continue with the open circuit at $\mathcal{P}_2$, but now we apply I_1 at $\mathcal{P}_1$. $V_2 = I_1 Z_3$. $\therefore I_1/V_2 = 1/Z_3$.
3. Next, we short $\mathcal{P}_2$ to guarantee $V_2 = 0$. Now we apply V_1 at $\mathcal{P}_1$. $-I_2 = V_1/(Z_1 + Z_2)$. $\therefore V_1/(-I_2) = Z_1 + Z_2$.
4. We continue with the short circuit at $\mathcal{P}_2$. Now we apply I_1 at $\mathcal{P}_1$. $-I_2 = I_1$. $\therefore I_1/(-I_2) = 1$.

Finally, we obtain the following **T** matrix for the two-port network.

$$\mathbf{T} = \begin{bmatrix} 1 + \frac{Z_1+Z_2}{Z_3} & Z_1 + Z_2 \\ 1/Z_3 & 1 \end{bmatrix}$$

Corollary 7.5. For a reciprocal (passive) two-port network, $|\mathbf{T}| = t_{11}t_{22} - t_{12}t_{21} = 1$.

Proof. We will apply Tellegen's theorem while performing two experiments, as shown in Fig. 7.9.

1. $\mathcal{P}_2$ is open-circuited, and at $\mathcal{P}_1$, we apply a voltage V_{1_a}. The current at $\mathcal{P}_1$ is I_{1_a}, and the voltage at $\mathcal{P}_2$ is V_{2_a}. The current at $\mathcal{P}_2$ is 0.

2. We short-circuit $\mathcal{P}_2$ and apply a voltage V_{1_b} at $\mathcal{P}_1$. The current at $\mathcal{P}_1$ is I_{1_b}, and the current at $\mathcal{P}_2$ is I_{2_b}. The voltage at $\mathcal{P}_2$ is 0.

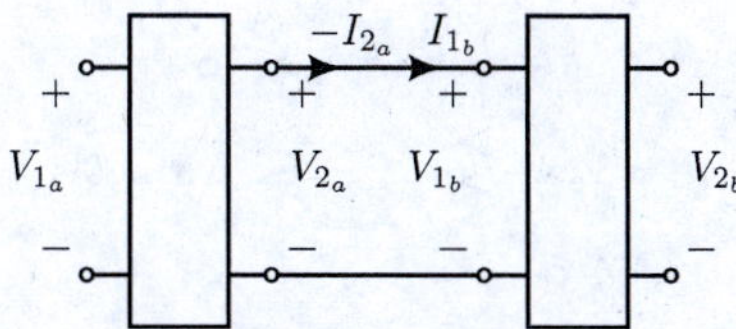

Figure 7.10: Two networks cascaded. Clearly $V_{2_a} = V_{1_b}$ and $-I_{2_a} = I_{1_b}$.

Let us assume that the two-port network only has passive resistors (phasors can be used in case of a general impedance.) Let these resistors be $\{R_k\}$. Let the currents through these resistors be $\{I_{k_a}\}$ for the first experiment and $\{I_{k_b}\}$ for the second experiment.

	$\mathcal{P}_1$	Inside two-port	$\mathcal{P}_2$
Experiment a, voltages	V_{1_a}	$I_{k_a}R_k$	V_{2_a}
Experiment a, currents	$-I_{1_a}$	I_{k_a}	0
Experiment b, voltages	V_{1_b}	$I_{k_b}R_k$	0
Experiment b, currents	$-I_{1_b}$	I_{k_b}	$-I_{2_b}$

By Tellegen's theorem, the currents and voltages of experiments a and b can be cross-multiplied and added to get 0.

$$\begin{aligned}
\therefore -V_{1_a}I_{1_b} + \sum I_{k_a}R_kI_{k_b} - V_{2_a}I_{2_b} &= 0 = -V_{1_b}I_{1_a} + \sum I_{k_b}R_kI_{k_a} + 0 \\
\therefore -V_{1_a}I_{1_b} + V_{1_b}I_{1_a} - V_{2_a}I_{2_b} &= 0 \\
\therefore V_{1_a}I_{1_b} - V_{1_b}I_{1_a} &= -V_{2_a}I_{2_b} \\
\text{or, } (V_{1_a}/V_{2_a}) \cdot (I_{1_b}/(-I_{2_b})) &- (V_{1_b}/(-I_{2_b})) \cdot (I_{1_a}/V_{2_a}) = 1 \\
\text{or, } t_{11}t_{22} - t_{21}t_{12} &= 1
\end{aligned}$$

■

Corollary 7.6. Two two-port networks, **a** and **b**, are cascaded. We have connected Port-2 of network **a** to port-1 of network **b**. The transmission parameters from port-1 of network **a** to port-2 of network **b** are the product of the transmission matrices **a** and **b**. That is, $\mathbf{T} = \mathbf{T_a} \cdot \mathbf{T_b}$.

Proof. We have shown the schematic of the cascaded two ports in Fig. 7.10. Clearly $V_{2_a} = V_{1_b}$ and $-I_{2_a} = I_{1_b}$. If the transmission parameters for network **a** and network **b** are $\mathbf{T_a}$ and $\mathbf{T_b}$, then:

$$\begin{bmatrix} V_{1_a} \\ I_{1_a} \end{bmatrix} = \mathbf{T_a} \begin{bmatrix} V_{2_a} \\ -I_{2_a} \end{bmatrix} = \mathbf{T_a} \begin{bmatrix} V_{1_b} \\ I_{1_b} \end{bmatrix} = \mathbf{T_a T_b} \begin{bmatrix} V_{2_b} \\ I_{2_b} \end{bmatrix}$$

$\therefore$ the overall transmission parameters, $\mathbf{T}$ is $\mathbf{T_a T_b}$. ■

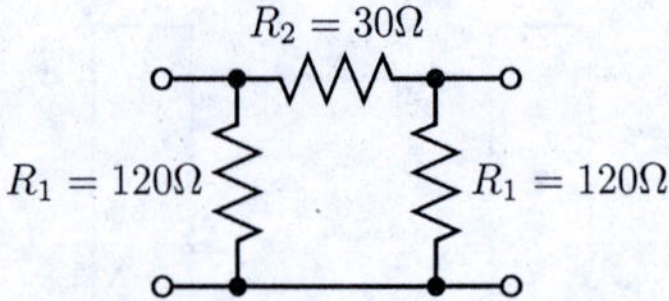

Figure 7.11: Two-port network for example-7.10.

Corollary 7.7. If many two-port networks with transmission parameters $\mathbf{T_a}$, $\mathbf{T_b}$, $\ldots \mathbf{T_N}$ are cascaded, the overall transmission parameters are the product, $\mathbf{T} = \mathbf{T_a} \cdot \mathbf{T_b} \ldots \mathbf{T_N}$.

Proof. We can prove the corollary by repeated application of corollary 7.6. ∎

We can also obtain the system transfer function using the **T**-parameters. Let us assume a network configuration as in Fig. 7.7(a). $-I_2 = V_2 G_L$. Also, $V_{in} = V_1 + I_1 Z_S$.

$$\begin{aligned} V_1 = t_{11}V_2 - t_{12}I_2 &= (t_{11} + t_{12}G_L)V_2 \\ I_1 = t_{21}V_2 - t_{22}I_2 &= (t_{21} + t_{22}G_L)V_2 \end{aligned}$$

$$\begin{aligned} \therefore V_{in} &= V_2(t_{11} + t_{12}G_L + Z_S(t_{21} + t_{22}G_L)) \\ \therefore \frac{V_2}{V_{in}} &= \frac{1}{Z_S(t_{21} + t_{22}G_L) + (t_{11} + t_{12}G_L)} \end{aligned}$$

Example 7.10. The network of Fig. 7.11 is cascaded four times as N_1, N_2, N_3, N_4. Port-1 of N_1 is excited by 1 V in series with 40 Ω, while Port-2 of N_4 is terminated with 40 Ω. Find the voltage across the load.

The **T**-matrix for the two-port network is:

$$T = \begin{bmatrix} 1 + R_2/R_1 & R_2 \\ \frac{R_2 + 2R_1}{R_1^2} & 1 + R_2/R_1 \end{bmatrix}$$

So the T-matrix for the cascaded network is $\mathbf{T'} = \mathbf{T}^4$.

$$V_{out}/V_{in} = \frac{40}{(40t_{21} + t_{22}) \cdot 40 + (40t_{11} + t_{12})}$$

V_{out}/V_{in} works out to $1/32$.

$$
\begin{array}{ccccccccc}
\mathbf{Z} & = & \mathbf{Y}^{-1} & = & \dfrac{\begin{bmatrix} |H| & h_{12} \\ -h_{21} & 1 \end{bmatrix}}{h_{22}} & = & \dfrac{\begin{bmatrix} 1 & -g_{12} \\ g_{21} & |G| \end{bmatrix}}{g_{11}} & = & \dfrac{\begin{bmatrix} t_{11} & |T| \\ 1 & t_{22} \end{bmatrix}}{t_{21}} \\
\mathbf{Z}^{-1} & = & \mathbf{Y} & = & \dfrac{\begin{bmatrix} 1 & -h_{12} \\ h_{21} & |H| \end{bmatrix}}{h_{11}} & = & \dfrac{\begin{bmatrix} |G| & g_{12} \\ -g_{21} & 1 \end{bmatrix}}{g_{22}} & = & \dfrac{\begin{bmatrix} t_{22} & -|T| \\ -1 & t_{11} \end{bmatrix}}{t_{12}} \\
\dfrac{\begin{bmatrix} |Z| & z_{12} \\ -z_{21} & 1 \end{bmatrix}}{z_{22}} & = & \dfrac{\begin{bmatrix} 1 & -y_{12} \\ y_{21} & |Y| \end{bmatrix}}{y_{11}} & = & \mathbf{H} & = & \mathbf{G}^{-1} & = & \dfrac{\begin{bmatrix} t_{12} & |T| \\ -1 & t_{21} \end{bmatrix}}{t_{22}} \\
\dfrac{\begin{bmatrix} 1 & -z_{12} \\ z_{21} & |Z| \end{bmatrix}}{z_{11}} & = & \dfrac{\begin{bmatrix} |Y| & y_{12} \\ -y_{21} & 1 \end{bmatrix}}{y_{22}} & = & \mathbf{H}^{-1} & = & \mathbf{G} & = & \dfrac{\begin{bmatrix} t_{21} & -|T| \\ 1 & t_{12} \end{bmatrix}}{t_{11}} \\
\dfrac{\begin{bmatrix} z_{11} & |Z| \\ 1 & z_{22} \end{bmatrix}}{z_{21}} & = & -\dfrac{\begin{bmatrix} y_{22} & 1 \\ |Y| & y_{11} \end{bmatrix}}{y_{21}} & = & -\dfrac{\begin{bmatrix} |H| & h_{11} \\ h_{22} & 1 \end{bmatrix}}{h_{21}} & = & \dfrac{\begin{bmatrix} 1 & g_{22} \\ g_{11} & |G| \end{bmatrix}}{g_{21}} & = & \mathbf{T}
\end{array}
$$

Table 7.3: Conversion between different two-port parameters.

7.10 Relationship between variables

The two-port parameter sets are inter-related. If we know any one parameter-set, we can compute all others. The **H**-parameters were shown to be a function of the **Z**-parameters. All other combinations are listed in Tab. 7.3.

7.11 Circuit configurations

The **Z**, **Y**, **H**, and **G** parameters are beneficial in a few specific circuits. However, for all of the following circuits, one needs to be careful. The definition of a port requires the currents going into and coming out of a port to be equal. The equality can be easily satisfied if we attach a port to a single external element or a single loop. However, in the following circuits, this is not satisfied so simply, as there are connections between ports of multiple two-port networks. The following discussion is applicable only in situations where the currents entering and leaving each port are equal.

7.11.1 Series-series

In the series-series topology in Fig. 7.12, I_1 and I_2 are the same for both the two-port networks. Further, $V_1 = V_{1_a} + V_{1_b}$ and $V_2 = V_{2_a} + V_{2_b}$. We can use the **Z**-parameters for the two networks to set up equations as follows.

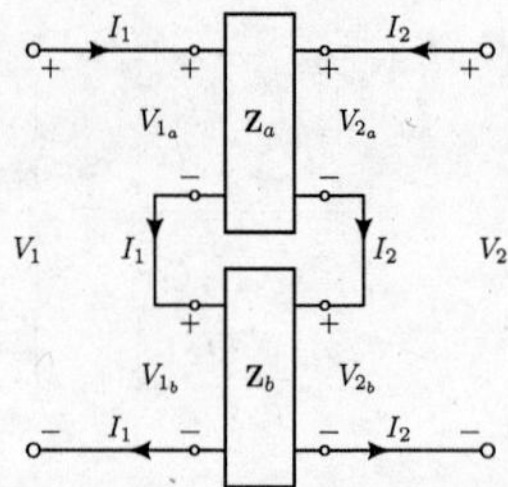

Figure 7.12: Series-series configuration. The impedance parameters of the overall two-port network are $\mathbf{Z}$, the impedance parameters of the two internal networks are $\mathbf{Z}_a$ and $\mathbf{Z}_b$.

$$\begin{aligned} V_{1_a} &= z_{11_a}I_1 + z_{12_a}I_2 \text{ and } V_{2_a} = z_{21_a}I_1 + z_{22_a}I_2 \\ V_{1_b} &= z_{11_b}I_1 + z_{12_b}I_2 \text{ and } V_{2_b} = z_{21_b}I_1 + z_{22_b}I_2 \end{aligned}$$

Adding the (a) equations with the corresponding (b) equations, we get:

$$\begin{aligned} V_1 = V_{1_a} + V_{1_b} &= (z_{11_a} + z_{11_b})I_1 + (z_{12_a} + z_{12_b})I_2 \\ V_2 = V_{2_a} + V_{2_b} &= (z_{21_a} + z_{21_b})I_1 + (z_{22_a} + z_{22_b})I_2 \end{aligned}$$

In other words, $\mathbf{Z}$ for the entire two-port is $\mathbf{Z} = \mathbf{Z}_a + \mathbf{Z}_b$.

7.11.2 Series-shunt

In the series-shunt topology of Fig. 7.13, I_1 and V_2 are the same for both the two-port networks. Further, $V_1 = V_{1_a} + V_{1_b}$ and $I_2 = I_{2_a} + I_{2_b}$. The **H**-parameters for the two networks are convenient because we can use I_1 and V_2 as the common independent variables.

$$\begin{aligned} V_{1_a} &= h_{11_a}I_1 + h_{12_a}V_2 \text{ and } I_{2_a} = h_{21_a}I_1 + h_{22_a}V_2 \\ V_{1_b} &= h_{11_b}I_1 + h_{12_b}V_2 \text{ and } I_{2_b} = h_{21_b}I_1 + h_{22_b}V_2 \ . \end{aligned}$$

Adding the (a) equations with the corresponding (b) equations we obtain:

$$\begin{aligned} V_1 = V_{1_a} + V_{1_b} &= (h_{11_a} + h_{11_b})I_1 + (h_{12_a} + h_{12_b})V_2 \\ I_2 = I_{2_a} + I_{2_b} &= (h_{21_a} + h_{21_b})I_1 + (h_{22_a} + h_{22_b})V_2 \end{aligned}$$

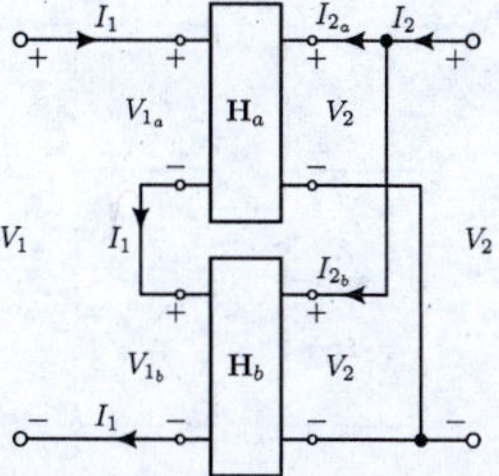

Figure 7.13: Series-shunt configuration. The hybrid parameters of the overall two-port network are $\mathbf{H}$, of the two internal networks are $\mathbf{H}_a$ and $\mathbf{H}_b$.

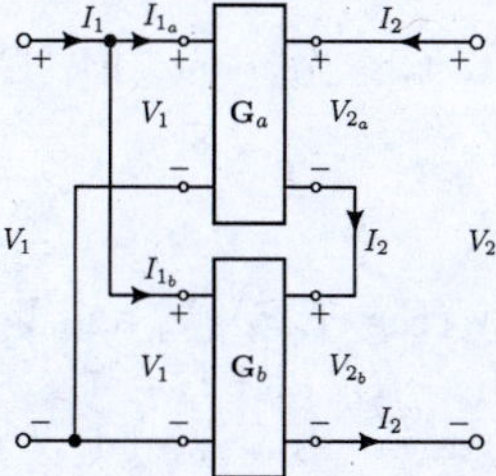

Figure 7.14: Shunt-series configuration. The inverse hybrid parameters of the overall two-port network are $\mathbf{G}$, of the internal networks are $\mathbf{G}_a$ and $\mathbf{G}_b$.

In summary, the $\mathbf{H}$-parameters for the entire two-port is $\mathbf{H} = \mathbf{H}_a + \mathbf{H}_b$.

7.11.3 Shunt-series

We have shown the shunt-series topology in Fig. 7.14. V_1 and I_2 are the same for both the two-port networks. Also, $I_1 = I_{1_a} + I_{1_b}$ and $V_2 = V_{2_a} + V_{2_b}$. We choose V_1, I_2 as the independent variables and analyze using $\mathbf{G}$-parameters.

$$\begin{aligned} I_{1_a} &= g_{11_a}V_1 + g_{12_a}I_2 \text{ and } V_{2_a} = g_{21_a}V_1 + g_{22_a}I_2 \\ I_{1_b} &= g_{11_b}V_1 + g_{12_b}I_2 \text{ and } V_{2_b} = g_{21_b}V_1 + g_{22_b}I_2 \end{aligned}$$

Adding the (a) equations with the corresponding (b) equations:

$$\begin{aligned} I_1 = I_{1_a} + I_{1_b} &= (g_{11_a} + g_{11_b})V_1 + (g_{12_a} + g_{12_b})I_2 \\ V_2 = V_{2_a} + V_{2_b} &= (g_{21_a} + g_{21_b})V_1 + (g_{22_a} + g_{22_b})I_2 \end{aligned}$$

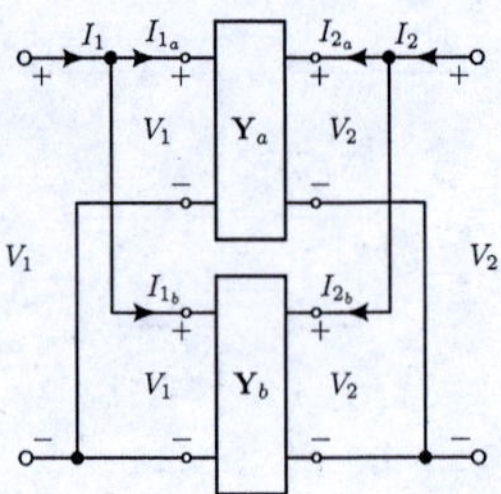

Figure 7.15: Shunt-shunt configuration. The admittance parameters of the two-port are $\mathbf{Y}$. The $\mathbf{Y}$-parameters of the internal networks are $\mathbf{Y}_a$ and $\mathbf{Y}_b$.

To summarize, for the entire two-port network, $\mathbf{G} = \mathbf{G}_a + \mathbf{G}_b$.

7.11.4 Shunt-shunt

In the shunt-shunt topology (Fig. 7.15), V_1 and V_2 are the same for both the two-ports. Further, $I_1 = I_{1_a} + I_{1_b}$ and $I_2 = I_{2_a} + I_{2_b}$. We should choose V_1 and V_2 as the independent variables and analyze using the $\mathbf{Y}$-parameters.

$$\begin{aligned} I_{1_a} &= y_{11_a} V_1 + y_{12_a} V_2 \text{ and } I_{2_a} = & y_{21_a} V_1 + y_{22_a} V_2 \\ I_{1_b} &= y_{11_b} V_1 + y_{12_b} V_2 \text{ and } I_{2_b} = & y_{21_b} V_1 + y_{22_b} V_2 \end{aligned}$$

Adding the (a) equations with the corresponding (b) equations:

$$\begin{aligned} I_1 = I_{1_a} + I_{1_b} &= (y_{11_a} + y_{11_b}) V_1 + (y_{12_a} + y_{12_b}) V_2 \\ I_2 = I_{2_a} + I_{2_b} &= (y_{21_a} + y_{21_b}) V_1 + (y_{22_a} + y_{22_b}) V_2 \end{aligned}$$

The $\mathbf{Y}$-parameter set for the entire two-port network is $\mathbf{Y} = \mathbf{Y}_a + \mathbf{Y}_b$.

7.11.5 Overall remarks

Placing two-port networks in series or parallel may not preserve the port properties. One has to be careful to determine if the current going into a port comes out of opposite terminal of the same port. Typical structures that work well are given in Fig. 7.16.

Recall observations 7.1 and 7.2. Feedback circuits are designed in either series-series, or series-shunt, or shunt-series, or shunt-shunt configurations.

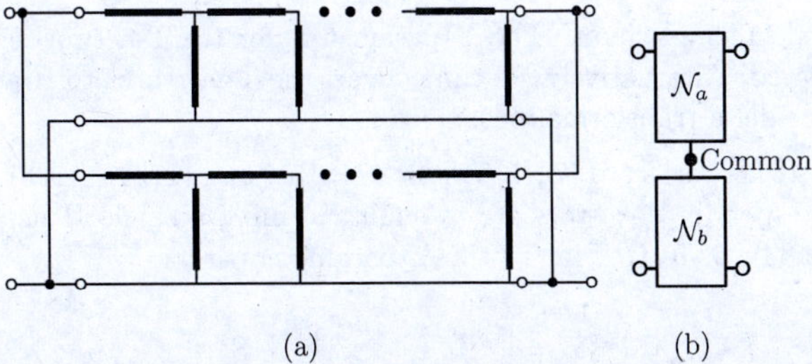

Figure 7.16: Common interconnections that preserve port properties. (a) We can interconnect two ladder networks in parallel. (b) Two networks are connected in series. Notice how $\mathcal{N}_b$ is connected upside-down. Both $\mathcal{N}_a$ and $\mathcal{N}_b$ have a common terminal for both ports. For instance, imagine ladder networks inside $\mathcal{N}_a$ and $\mathcal{N}_b$, with $\mathcal{N}_b$ upside down.

As such, the relevant two-port parameters simply add up to give us the overall two-port parameters. In these circuits, one of the two sub-circuits is often designed with very large p_{21} and zero p_{12}. One may infer that this sub-circuit can not be passive and must have controlled voltage or current sources. Then the overall system transfer function ends up being approximately $1/p_{12}$ of the second sub-circuit.

Example 7.11. Find any two-port parameters of the networks shown below.

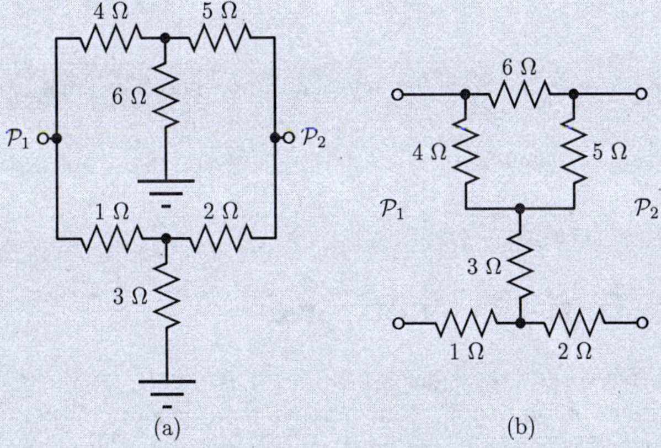

(a) The ground/earth connections indicate the common reference terminal, and the negative terminal for both $\mathcal{P}_1$ and $\mathcal{P}_2$. There are two T-networks connected in shunt-shunt. The **Y**-parameters for the T-networks need to be evaluated. Alternatively, we can convert the T-networks to Π-networks using star-delta transformations.

The two Π networks are (74/5)-(74/6)-(74/4) Ω and (11/2)-(11/3)-(11) Ω. The corresponding resistors are in shunt, forming a single Π network of (4.01)-(2.83)-(6.90) Ω. Finally, the **Y**-parameters are:

$$\mathbf{Y} = \begin{bmatrix} 0.603 & -0.354 \\ -0.354 & 0.499 \end{bmatrix} \text{ S}$$

(b) Two two-port networks are in series, just like in Fig. 7.16(b). The common node for both two-ports is the node between 3 Ω, 4 Ω and 5 Ω resistors. We will choose the **Z**-parameters to evaluate.

The upper network is a Π, which can be converted into a T-network with a delta-star transformation. A T-network is preferable because we will work with the impedance parameters. The T-network is (8/5)-(4/3)-(2) Ω and has a **Z**-matrix:

$$\mathbf{Z}_a = \begin{bmatrix} 2.933 & 1.333 \\ 1.333 & 3.333 \end{bmatrix} \Omega$$

The lower network is a T-network with **Z**-matrix:

$$\mathbf{Z}_b = \begin{bmatrix} 4 & 3 \\ 3 & 5 \end{bmatrix} \Omega$$

The overall impedance parameters are:

$$\mathbf{Z} = \mathbf{Z}_a + \mathbf{Z}_b = \begin{bmatrix} 6.933 & 4.333 \\ 4.333 & 8.333 \end{bmatrix} \Omega$$

The two examples illustrate that one may or may not choose to use the two-port interconnection properties. As long as we choose convenient two-port parameters, the computations will be straightforward.

7.12 Multi-port networks

Multi-port networks are a generalization of two-port networks. One can easily conceive of three-port, four-port, and in general, N-port networks. The impedance parameters for an N-port network are an $N \times N$ matrix. z_{ii} is the impedance looking into the ith port, with all other ports left as open circuits. z_{ij} with $i \neq j$ is the ratio of the voltage across the ith port and a current input into the jth port, where all ports except j are open circuits. Likewise, we can evaluate the admittance parameters. The hybrid parameters are ill-defined, with many possibilities. If hybrid parameters need to be used, they have to be defined while being used.

The reciprocity rules hold for multi-port networks as well. For a reciprocal (passive) network, $z_{ij} = z_{ji}$, $y_{ij} = y_{ji}$. If the hybrid parameters are defined, and h_{ij}, h_{ji} are dimensionless, then it can be shown that $h_{ij} = -h_{ji}$. However, if they have dimensions of impedance or admittance, it can be shown that $h_{ij} = h_{ji}$.

7.13 Unit summary

- There are six possible combinations of parameters for two-port networks. We have summarized these in table 7.1. The independent variables are applied to the two-port, the dependent variables are measured.
- The **Z**-parameters can be computed by open-circuiting the two ports, one at a time, and applying a current at the other port. The **Z**-parameters have dimensions of impedance.
- We can quickly obtain the **Z**-parameters for the T-network.
- The **Y**-parameters can be computed by short-circuiting the two ports, one at a time, and applying a voltage at the other port. The **Y**-parameters have dimensions of admittance.
- The Π network is easily expressed with the help of **Y**-parameters.
- For passive (reciprocal) two-port networks, $z_{12} = z_{21}$ and $y_{12} = y_{21}$.
- Star-delta or delta-star conversion can be memorized with the help of a single mnemonic. The impedances have to be more in the delta (because it appears as if the impedances are in parallel) and less in the star (because they appear to be in series). The single formula is $R_x = R_2R_3/(R_1 + R_2 + R_3)$. R_x must be the impedance in the star configuration because R_x has roughly become smaller than R_1, R_2, R_3. The reverse expression must therefore be, $G_1 = G_yG_z/(G_x + G_y + G_z)$.
- For the hybrid parameters, first, we short-circuit port-2 and apply a current at port-1. Next, we open-circuit port-1 and apply a voltage at port-2.
- For reciprocal (passive) two-port networks, $h_{21} = -h_{12}$.

- The inverse-hybrid parameters are the opposite of **H**-parameters. Here too, $g_{21} = -g_{12}$ for reciprocal networks.

- The input impedance/admittance with any two-port parameter-set (**Z**, **Y**, **H**, **G**) is $p_{11} - p_{12}p_{21}/(p_{22} + P_L)$. Here P_L has to be impedance or admittance depending on the dimensions of p_{22}. Likewise, the result will have the dimensions of p_{11}. The output impedance/admittance is similar, with 1 and 2 interchanged.

- We can express the transfer function with any two-port parameter-set (**Z**, **Y**, **H**, **G**). The quantities of the input and output depend on the parameters; however, these are easily convertible. The transfer function is always $p_{21}/(p_{12}p_{21} - (p_{11} + P_S)(p_{22} + P_L))$. P_S is either Z_S or Y_S, depending on the dimensions of p_{11}. P_L is either Z_L or Y_L, depending on the dimensions of p_{22}. The transfer function will have the dimensions of $1/p_{12}$. For example, if p_{12} is voltage by current, the transfer function will be the current at port-2 by the voltage at port-1.

- **Z**, **Y**, **H**, and **G** parameters are beneficial for series-series, shunt-shunt, series-shunt, and shunt-series network configurations, respectively. The combined two-port parameters are the sum of the individual two-port parameters.

- The transmission (**T**) parameters have V_2 and $-I_2$ as the independent variables. The emphasis is on the negative sign of $-I_2$.

- For a reciprocal or passive network, the determinant of **T** is 1.

- The transmission parameters are beneficial when two networks are cascaded. The combined transmission parameters are the product of the individual **T** matrices.

7.14 Exercises

Multiple choice type questions

7.1 A passive two-port network has the parameters given. What is x?

$$\mathbf{H} = \begin{bmatrix} 1\ \text{k}\Omega & x \\ -0.4 & 1\ \text{mS} \end{bmatrix}$$

(a) -2.5 (b) -0.4 (c) 0.4 (d) 2.5

7.2 A passive two-port network has the parameters given. What is t_{12}?

$$\mathbf{T} = \begin{bmatrix} 10 & t_{12} \\ 3\text{ S} & 4 \end{bmatrix}$$

(a) -3 Ω (b) 3 Ω (c) 13 Ω (d) 13.33 Ω

7.3 A Π-network with three resistors 3Ω-6Ω-9Ω can be transformed to a Y-network with resistors:
(a) 1Ω-2Ω-3Ω (b) 3Ω-1.5Ω-1Ω (c) 1.83Ω-3.67Ω-5.5Ω (d) 9Ω-18Ω-27Ω

7.4 A Y-network with three resistors 3Ω-6Ω-9Ω can be transformed to a Π-network with resistors:
(a) 1Ω-2Ω-3Ω (b) 3Ω-1.5Ω-1Ω (c) 1.83Ω-3.67Ω-5.5Ω (d) 11Ω-16.5Ω-33Ω

7.5 Two two-port networks are interconnected in shunt-shunt feedback. Which two-port parameters will be the most convenient for analysis?
(a) **Z**-parameters (b) **Y**-parameters (c) **H**-parameters (d) **G**-parameters

7.6 Two two-port networks are interconnected in series-series feedback. Which two-port parameters will be the most convenient for analysis?
(a) **Z**-parameters (b) **Y**-parameters (c) **H**-parameters (d) **G**-parameters

7.7 Two two-port networks are cascaded. Which two-port parameters will be the most convenient for analysis?
(a) **Z**-parameters (b) **Y**-parameters (c) **T**-parameters (d) **H**-parameters

Short answer type questions

7.8 To work out the **H**-parameters, we will first ______-circuit port-2 of the circuit. h_{11} will be the input ______, while h_{21} will be the ______ in port-2 by the ______in port-1. Next, we will ______ circuit port-1 of the circuit. h_{22} will be the ______ looking into port-2, while h_{12} will be the ______ at port-1 by the ______ at port-2.

7.9 To work out the **Y**-parameters, we will first ______-circuit port-2 of the circuit. y_{11} will be the input ______, while y_{21} will be the ______ in port-2 by the ______ in port-1. Next, we will ______-circuit port-1 of the circuit. y_{22} will be the ______ looking into port-2, while y_{12} will be the ______ at port-1 by the ______ at port-2.

7.10 To work out the **T**-parameters, we will first ______-circuit port-2 of the circuit. If we apply a 1 V at port-______, the ______ at port-______gives t_{11} and the ______ at port-______gives t_{21}. Now we will ______-circuit port-2 of the circuit. If we apply 1 A at port-______, t_{12} is negative of the ______ at port-______, while t_{22} is negative of the ______ at port-______.

7.11 A system is to be designed with a specific voltage gain of A. Show that it is possible to design this as a series-shunt configuration of two circuits. The first circuit (with hybrid parameters as $\mathbf{H}_a$) can have $h_{21_a} \to \infty$ and h_{12_a} of 0, while the second circuit (with hybrid parameters $\mathbf{H}_b$) needs to have h_{12_b} of $1/A$. The other parameters of the two two-port networks are unimportant. What are the additional constraints that should be introduced to make this design a good voltage-controlled voltage source?

7.12 Evaluate the $\mathbf{Z}$ parameters for a mutual inductance, where the terminal pairs of the two coils are the two ports.

7.13 Use Tellegen's theorem to prove that the impedance matrix of a passive two-port network, $\mathbf{Z}$, will be such that $|\mathbf{Z}| \geq 0$ for all ω. In mathematical terms, prove that $\mathbf{Z}$ is positive semi-definite.

Numericals and long answer type questions

7.14 Work out the two-port parameters ($\mathbf{Z}$, $\mathbf{Y}$, $\mathbf{H}$, $\mathbf{G}$, $\mathbf{T}$) for the circuits shown in Fig. 7.17. Use $R_1 = 1\ \Omega$, $R_2 = 2\ \Omega$, $R_3 = 3\ \Omega$, $R_4 = 4\ \Omega$, $R_5 = 5\ \Omega$, and $R_6 = 6\ \Omega$ for all your calculations.

7.15 An attenuator circuit is to be designed such that (a) its input impedance is $50\ \Omega$ when its output port is terminated by $50\ \Omega$, (b) its output impedance is $50\ \Omega$ when $50\ \Omega$ terminates the input port, and (c) its voltage gain is $1/20$ when the source and load impedances are both $50\ \Omega$. Design the two-port network as a Π-network of three resistors.

7.16 Consider the two-port network in Fig. 7.18. First, evaluate any convenient two-port parameters for the network. Next, assume an impedance $Z_L = \sqrt{L/C}$ terminates $\mathcal{P}_2$. Find the input impedance of the circuit looking in from $\mathcal{P}_1$. What will be the input impedance if multiple such circuits are cascaded, with the final output terminated by $Z_L = \sqrt{L/C}$?

7.17 For the two-port network of Fig. 7.18, assume $Z_0 = \sqrt{L/C}$. A voltage source V_{in}, with a source resistance of Z_0, excites the network at $\mathcal{P}_1$. The load impedance at $\mathcal{P}_2$ is also Z_0. Evaluate the voltage gain as a function of frequency. Examine both the magnitude and the phase of the voltage gain. What are the magnitude and phase of the voltage gain at DC? *(Hint: At DC, use the assumption that the inductor is a short circuit and the capacitor is an open circuit.)* What are the magnitude and phase of the voltage gain at an infinitely large frequency? *(Hint: At an infinitely large frequency, use the assumption that the inductor is an open circuit and the capacitor is a short circuit.)* What are the magnitude and the phase of the voltage gain at the frequency $\omega_0 = 1/\sqrt{LC}$?

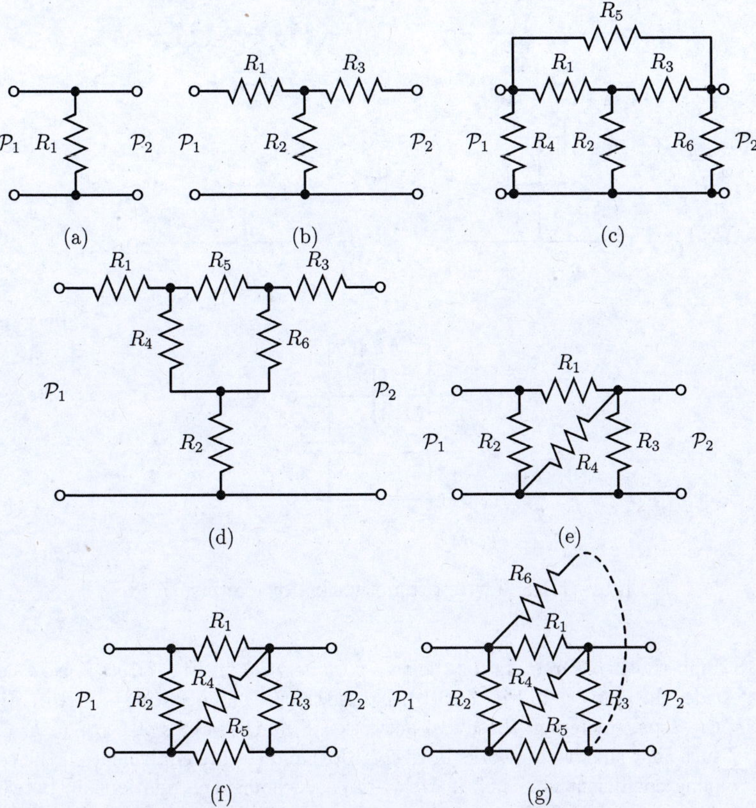

Figure 7.17: Two-port networks for exercise 7.14.

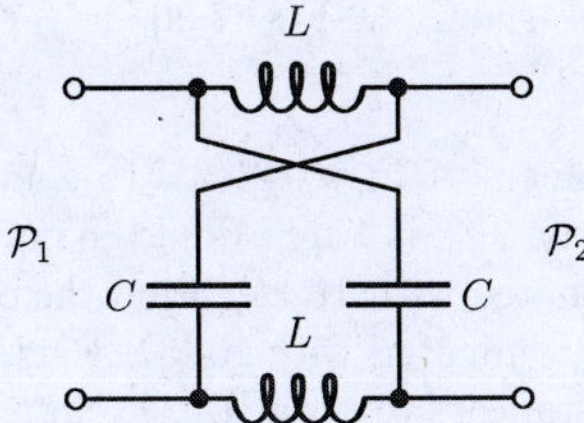

Figure 7.18: Lattice network for exercise 7.16, 7.17

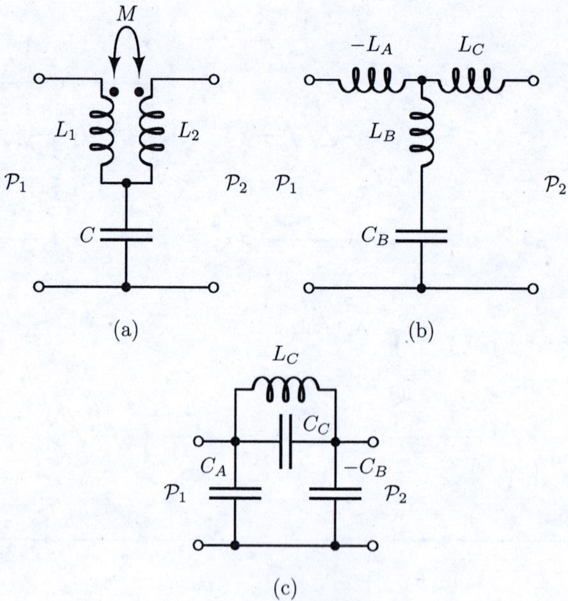

Figure 7.19: Two port networks for exercise 7.18.

7.18 Find the **Z**-parameters for the two-port network in Fig. 7.19(a). Now consider the network in Fig. 7.19(b). Notice the use of a negative inductor. Find the **Z**-parameters of these two networks. Find the values of L_A, L_B, L_C, C, such that the two networks in Fig. 7.19(a) and (b) are equivalent. If the coupling coefficient is 1, i.e., if $M = \sqrt{L_1 L_2}$, what is the relationship between L_A, L_B, and L_C? Find the **Z**-parameters for the network in Fig. 7.19(c). If we assume C_A, C_B and C_C are such that $-C_A C_B - C_B C_C + C_C C_A = 0$, can the network in Fig. 7.19(c) be re-cast as the networks in Fig. 7.19(a) and (b)? What should be the relationships between the components?

7.19 For the twin-T filter shown in Fig. 7.20, use **Y**-parameters to work out $V_L(s)/V_S(s)$.

7.20 In the circuit schematics of Fig. 7.21(a), (b), and (c), R_A is 1 kΩ, R_B is 5 kΩ, R_{AB} is 20 kΩ, and g_m is 1 mS. Find the input impedance looking into $\mathcal{P}_1$ when $\mathcal{P}_2$ is terminated with 10 kΩ. Find the output impedance, looking into $\mathcal{P}_2$, when $\mathcal{P}_1$ is terminated with 10 kΩ. Find the voltage gain when the source and load resistances are both 10 kΩ. Note: In Fig. 7.21(c), v_{12} is the voltage across $\mathcal{P}_1$ minus the voltage across $\mathcal{P}_2$.

7.21 For the two-port networks A and B, the following two-port parameters are

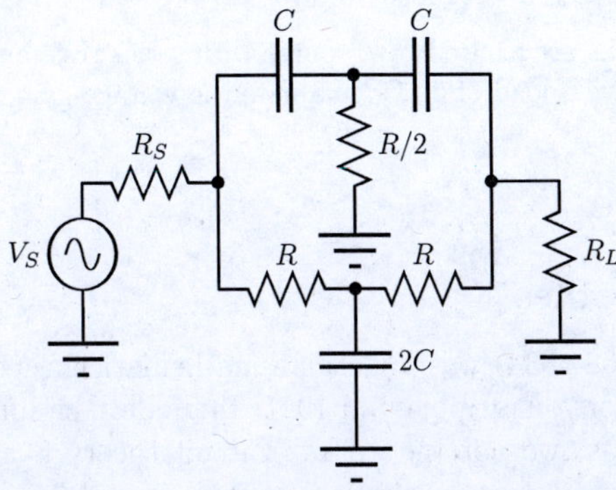

Figure 7.20: Twin-T filter for exercise 7.19.

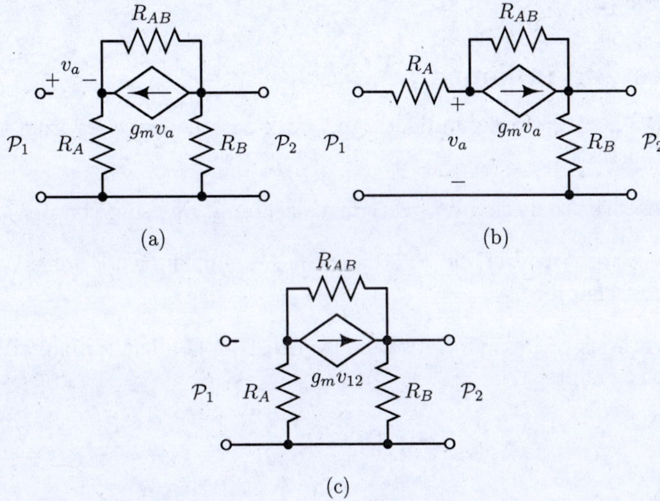

Figure 7.21: Schematics for exercise 7.20.

given.

$$\mathbf{Z}_A = \begin{bmatrix} 50 & 0 \\ 10000 & 50 \end{bmatrix} \Omega, \ \mathbf{Y}_B = \begin{bmatrix} 0.1 & 100 \\ 0.2 & 0.1 \end{bmatrix} \text{S}$$

Both networks are excited by a source with series resistance 50 Ω and both networks drive a load of 50 Ω. Evaluate the voltage gain in each case.

Know more

Historical profiles

Franz Breisig (1868-1934) was a German mathematician who came up with the abstract idea of "quadripoles" in 1921. In modern circuit theory, quadripoles are known as two-port networks. Circuit theory is a development of abstractions; we start from a physical carbon or metal device known as a resistor and abstract it into a lumped element which follows $v = iR$. Similar abstractions for other components followed. The abstraction of a two-terminal pair network into a lumped matrix of values followed Breisig's work in 1921. [23]

An understanding of two-port network theory led to the development of synthesis techniques for filters, namely by Otto Brune [24] and by Sidney Darlington [25].

Understand in depth

Study two-port networks in depth and work out the exercises from the following sources.

1. Network Analysis, by M.E. Van Valkenburg, Pearson, chapter 11.
2. Network Analysis and Synthesis, by Franklin F. Kuo, John Wiley & Sons, chapter 9.
3. Synthesis of Passive Networks, by Ernst A. Guillemin, John Wiley & Sons, chapter 6.

Unit 8

Basic filters

Unit specifics

In this unit we have discussed the following:

- Basic low pass filters
- Basic high pass filters
- Band pass and band stop filters
- Identification of filters

Rationale

Filters are two-port networks that are designed that allow certain frequencies or block certain frequencies. Earlier we have introduced the concept of a system for a network. Filters are useful system blocks that shape the frequency response of signals that pass through it. Communication circuits and power circuits often require filtering to achieve desired operational characteristics.

The study and design of filters forms one of the objectives of advanced circuit theory. As such, classical filter design is a course by itself. In this unit, we will give a flavor of a few basic filters and filter design. Further, we will learn to recognize basic filter structures by inspection.

Filters shape the frequency domain character of the input. It is imperative for us to work in the frequency domain for the design and analyses of filters.

Pre-requisites

- Units 2, 4, 5 of this text. A clear understanding of network theorems, circuits in sinusoidal steady state, and Laplace transforms is assumed in this unit.

Unit outcomes

The list of outcomes of this unit are as follows.

U8-O1: Be able to visually identify low-pass, high-pass, band-pass, and band-stop two-port networks.

U8-O2: Be able to design simple low-pass, high-pass, band-pass and band-stop circuits.

Unit-1 outcomes	**Expected mapping with course outcomes** *(1: Weak, 2: medium, and 3: strong correlation)*					
	CO-1	**CO-2**	**CO-3**	**CO-4**	**CO-5**	**CO-6**
U8-O1	-	3	2	-	3	-
U8-O2	-	3	2	-	3	-

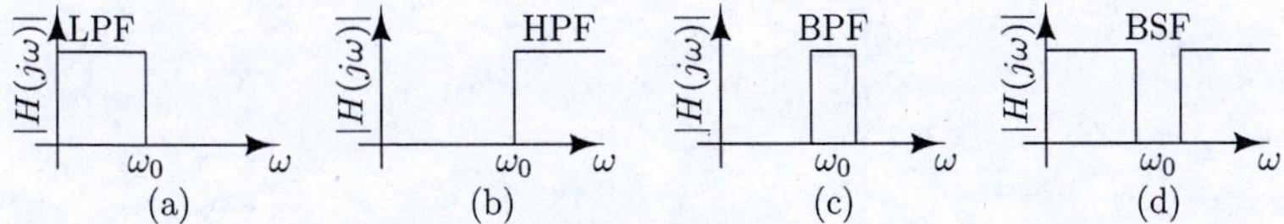

Figure 8.1: Ideal filter characteristics. (a) Low-pass, (b) high-pass, (c) band-pass, and (d) band-stop filter characteristics.

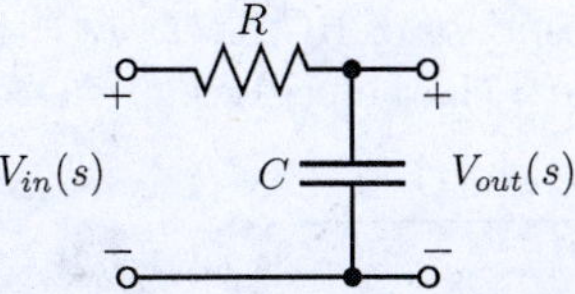

Figure 8.2: Basic RC filter

Filters are two-port networks with specific frequency domain properties. The two-port network is used as a system, with a transfer function of V_{out}/V_{in}. The system transfer function, as a function of frequency, is of a specific character for different kinds of filters. The ideal frequency-domain character of four different filters are shown in Fig. 8.1.

The impulse responses corresponding to $h(t)$ for the four transfer functions, $H(j\omega)$, shown in Fig. 8.1, will all require a sinc function. As such, none of the impulse responses are causal; in short, none of these ideal filters are practical.

8.1 Low-pass filters

The most basic low-pass filter is the R-C filter, given in Fig. 8.2. The transfer function of the filter is given by $H(s) = V_{out}(s)/V_{in}(s) = 1/(1+sCR)$. In terms of frequency, we have:

$$\begin{aligned} H(j\omega) &= \frac{1}{1+j\omega RC} \\ \text{or, } |H(j\omega)|^2 &= \frac{1}{1+\omega^2R^2C^2} \\ \text{If } \omega_0 = \frac{1}{RC}, \text{ then } |H(j\omega)|^2 &= \frac{1}{1+(\omega/\omega_0)^2} \end{aligned} \tag{8.1}$$

The frequency response of the filter, given in (8.1), may be plotted. For frequencies much lesser than ω_0, $|H(j\omega)|$ is nearly 1. For frequencies much

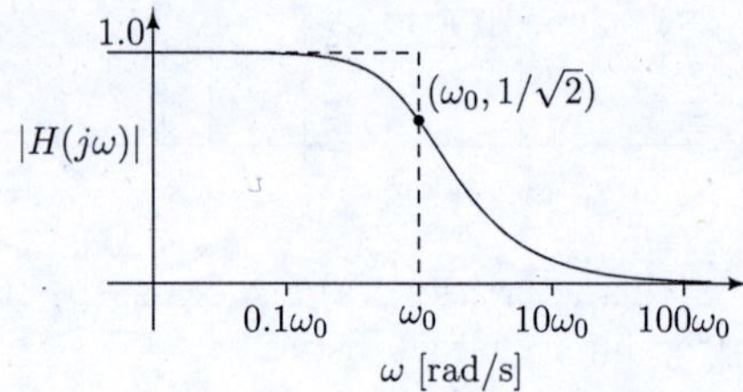

Figure 8.3: Frequency response of RC filter. $\omega_0 = 1/(RC)$. The dashed line represents the desired ideal characteristics.

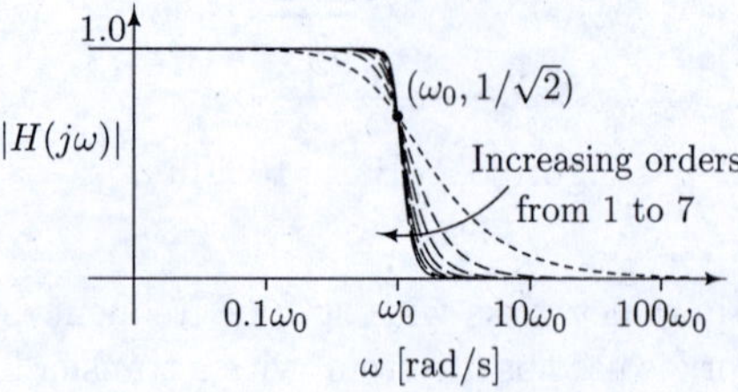

Figure 8.4: The response of the Butterworth filter for different orders, from $n = 1$ to $n = 7$.

higher than ω_0, $|H(j\omega)|$ is nearly 0. The transfer characteristics is plotted in Fig. 8.3. Since only low frequencies pass through from the input to the output, the circuit is a low-pass filter. ω_0 is the cut-off frequency.

It is clear from the filter characteristics in Fig. 8.3 that the frequency response leaves a lot more to be desired. The RC circuit in Fig. 8.2 is a first order circuit, as there is only one capacitor. Increasing the order of the filter can give us better performance.

8.1.1 High order Butterworth filters

An example of a high order filter is the Butterworth filter. The response of a Butterworth low-pass filter is nominally given below:

$$|H(j\omega)|^2 = \frac{1}{1 + (\omega/\omega_0)^{2n}}, \tag{8.2}$$

where n is the order of the filter, and ω_0 is the cut-off frequency. (8.2) is plotted in Fig. 8.4 for different values of n. As we increase the order of the filter, we obtain a response closer and closer to the desired ideal response.

The Butterworth filter is designed with both inductors and capacitors, as a loss-less ladder two-port network as shown in Fig. 8.5(a). Typically, the

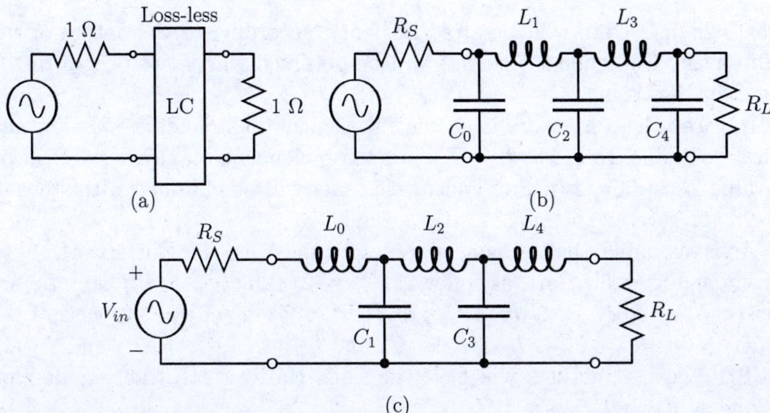

Figure 8.5: (a) A loss-less two-port ladder network is configured as a filter. Source and load impedances of 1 Ω are chosen. Two possible LC-ladder networks as filters are shown in (b) and (c).

source resistance and the load resistance at the two ports are initially kept as 1 Ω resistors. Examples of 5th order LC ladder networks, configured as filters, are shown in Fig. 8.5(b) and (c).

The structure of the ladder network is such that low frequencies pass. At low frequencies, all the inductors may be visualized as short circuits and the capacitors as open circuits. The output voltage will be simply 1/2 of the input because of resistive division. At high frequencies, the inductors are open circuits and the capacitors are short circuits. As such, the output is disconnected from the input. This gives the low-pass property to the network.

For a Butterworth filter, the values of the different components in the filter are given by a single concise expression.

$$L_k \text{ or } C_k = 2\cos\left(\frac{k}{n} + \frac{1}{2n} - \frac{1}{2}\right)\pi \tag{8.3}$$

An explanation of (8.3) follows. In an nth order filter, there will be a total of n inductors and capacitors. If n is even, there will be $n/2$ inductors, $n/2$ capacitors. If n is odd, the number of capacitors will be one more than the number of inductors, as in Fig. 8.5(b), or the number of inductors will be one more than the number of capacitors, as in Fig. 8.5(c).

Let us first examine the case where n is odd. The central element will have a value of $2\cos 0$ or 2 F. The two elements on each side will have values of $2\cos(\pm\pi/n)$. The next two elements on each side will have values of

$2\cos(\pm 2\pi/n)$. Units will be farad or henry, according to the nature of the component. Subsequent elements will be obtained with steps of π/n in the angle for the cosine.

In case n is even, there is no central element. The middle two elements will be of value $2\cos(\pm\pi/2n)$. The next two elements on either side will be of value $2\cos(\pm 3\pi/2n)$. Subsequent elements will be obtained with steps of $2\pi/2n$ (or π/n) in the angle for the cosine.

As an example, the component values for the 5th order Butterworth filter shown in Fig. 8.5(b) are as follows. C_2 is $2\cos 0 = 2$ F. L_1 and L_3 are $2\cos(\pi/5)$ or 1.618 H. C_0 and C_4 are $2\cos(2\pi/5)$ or 0.618 F. In case of the structure in Fig. 8.5(c), L_2 is 2 H, C_1 and C_3 are 1.618 F, L_0 and L_4 are 0.618 H. Both structures will have the same Butterworth transfer function as long as R_S and R_L are 1 Ω.

As a second example, consider a 6th order Butterworth filter. This will have one more inductor, L_5, beyond the components shown in Fig. 8.5(b). Now the two central components are C_2 and L_3. The values of these two elements will be $2\cos(\pi/12)$. C_2 is 1.9319 F and L_3 is 1.9319 H. Next the values of L_1 and C_4 will be $2\cos(3\pi/12)$. L_1 is 1.4142 H and C_4 is 1.4142 F. Next the values of C_0 and L_5 will be $2\cos(5\pi/12)$. The value of C_0 is 0.5176 F and L_5 is 0.5176 H.

A filter so-designed, is known as a **prototype filter**, and has a cut-off frequency (ω_0) of **1 rad/s**. The transfer function of the filter (V_{out}/V_{in}) has the characteristic given in (8.2), **but with an extra factor of 1/2**.

8.1.2 Low pass filter transformation

Typically a filter design process starts with a low-pass filter for a matched load and source resistance of 1 Ω and a cut-off frequency of 1 rad/s. For example, the Butterworth filter of section 8.1.1 was a filter with a cut-off frequency of 1 rad/s and source/load resistances of 1 Ω. The filter designer then uses this low-pass **prototype** filter and transforms it to suit the requirements at hand. This transformation is typically a scaling of impedances and a warping of the entire frequency axis.

We can transform the low-pass prototype filter for source/load of 1 Ω and 1 rad/s cut-off to any arbitrary matched source/load of Z_0 and any arbitrary cut-off frequency ω_0 in the following manner.

1. All impedances in the circuit can be scaled uniformly by a constant. If we scale the load/source resistances from 1 Ω to Z_0, all impedances in the circuit need to be scaled by Z_0. An inductor will therefore be scaled by Z_0; a capacitor will therefore be scaled by $1/Z_0$.

2. If we want to change the cut-off frequency from 1 rad/s to ω_0, the impedances/admittances of the circuit at ω_0 should be equal to the impedances/admittances of the prototype at 1 rad/s. As such, resistors should remain the same; inductors and capacitors should scale by $1/\omega_0$.

Overall, we need to scale all resistors by Z_0, inductors by Z_0/ω_0, and capacitors by $1/(Z_0\omega_0)$.

Example 8.1. Design a 5th order Butterworth filter for $R_S = R_L = 50\ \Omega$ and a cut-off frequency of 1 MHz.

First we will set up the prototype low-pass filter for 1 rad/s cut-off and 1 Ω load/source resistance, according to Fig. 8.5(c). This has components of 0.618 F, 1.618 H, 2 F, 1.618 H and 0.618 F.

Now we can scale the prototype filter to an impedance level of 50 Ω and a cut-off frequency of 1 MHz as follows. All resistors are to be scaled by 50. All inductors are to be scaled by $50/(2\pi \times 10^6)$, while all capacitors are to be scaled by $1/(100\pi \times 10^6)$. The new Butterworth filter is shown below.

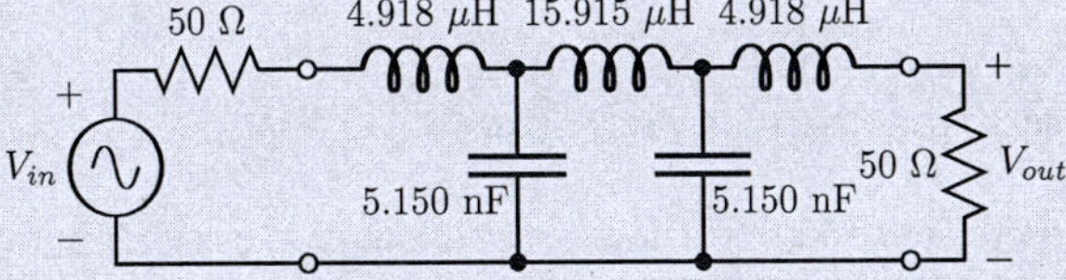

We have simulated the circuit and have plotted the magnitude of the frequency response ($|V_{out}/V_{in}|$) in Fig. 8.6. The graph is in decibels, and the x-axis is in the log scale. The −3 dB cut-off frequency is the frequency scaling constant of the circuit. Of note, in this case, the −3 dB cut-off is at −9 dB because the circuit offers a 6 dB attenuation at DC (extra factor of 1/2 as mentioned earlier).

8.2 High-pass filters

The most common high-pass filter is the C-R structure. The input voltage is applied across the series combination of a resistor and a capacitor as shown in Fig. 8.2. The output voltage is taken across the resistor, and not across the capacitor. The transfer function obtained is $sRC/(1 + sRC)$. A frequency response plot will indicate the high pass nature of the transfer function.

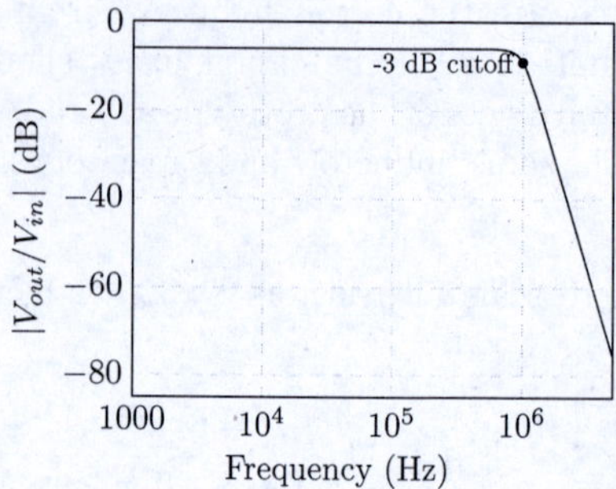

Figure 8.6: Simulated magnitude of the frequency response of the scaled Butterworth low-pass filter of Example 8.1.

The other common technique to generate a high pass filter is with the help of a frequency transformation on the prototype low-pass filter.

8.2.1 Low-pass to high-pass transformation

We can transform any low-pass prototype filter to a high-pass filter with impedance scaling in the following steps.

1. As earlier discussed in the low-pass to low-pass case, all impedances can be scaled uniformly by Z_0. All inductors scale by Z_0. All capacitors scale by $1/Z_0$.

2. Let us label the frequency axis of the prototype filter as Ω and the frequency axis of the target high pass filter as ω. 1 rad/s in the prototype axis needs to map to ω_0. 0 in the prototype axis must map to $\pm\infty$, while $\pm\infty$ in the prototype frequency axis must map to 0. We can achieve this mapping by plugging in $\omega = -\omega_0/\Omega$.[1]

3. We scale the impedances of the resistors to Z_0.

4. An inductor L has an impedance $j\Omega L Z_0$ in the prototype filter after scaling. The impedance changes to $-j\omega_0 L Z_0/\omega$ in the final filter and corresponds to the impedance of a capacitor of value $1/(\omega_0 L Z_0)$.

5. After scaling, a capacitor C has an impedance $-jZ_0/(\Omega C)$ in the prototype filter. The impedance changes to $jZ_0\omega/(\omega_0 C)$ in the final filter and corresponds to the impedance of an inductor of value $Z_0/(\omega_0 C)$.

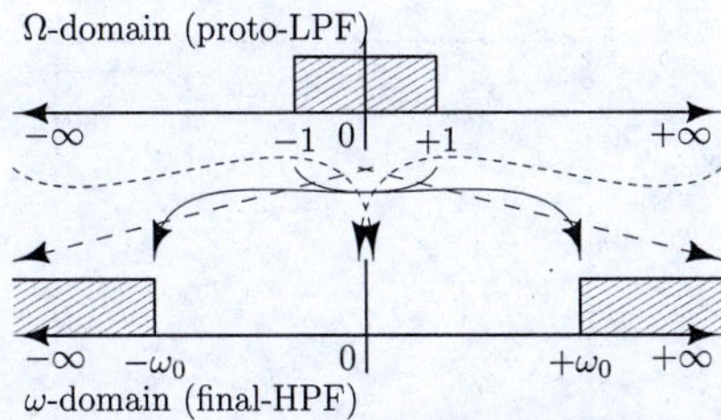

Figure 8.7: Illustration of the frequency mapping plan for a low-pass to high-pass transformation.

The sketch in Fig. 8.7 illustrates the $1 \mapsto 1$ frequency mapping plan. In summary, the following recipe needs to be followed:

1. Replace every 1 Ω resistor with Z_0.
2. Replace every inductor of value L with a capacitor of value $1/(\omega_0 L Z_0)$.
3. Replace every capacitor of value C with an inductor of value $Z_0/(\omega_0 C)$.

Example 8.2. Design a 5th order Butterworth high pass filter with a cut-off of 1 MHz for load and source impedances of 50 Ω.

First we will set up the prototype low-pass filter for 1 rad/s cut-off and 1 Ω load/source resistance according to Fig. 8.5(c). This has components of 0.618 F, 1.618 H, 2 F, 1.618 H and 0.618 F.

We can transform the prototype 5th order Butterworth filter to a high-pass filter of cut-off 1 MHz for an impedance level of 50 Ω. The circuit below shows the transformed Butterworth 1 MHz high-pass filter for 50 Ω.

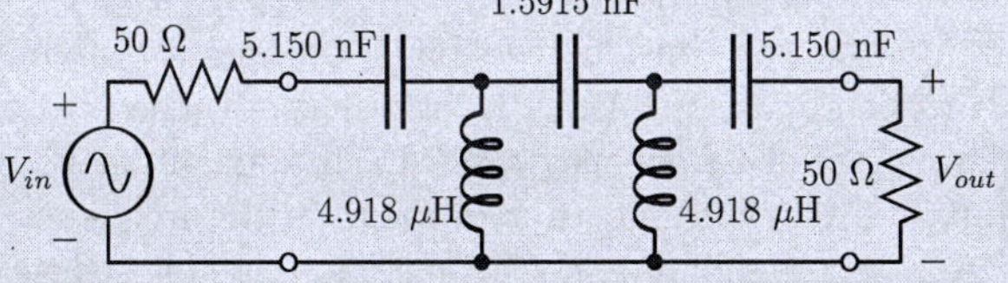

We simulated the circuit and have plotted the transfer function $|V_{out}/V_{in}|$ in Fig. 8.8. The x-axis is in log-scale, while the y-axis is in dB.

[1]Both + and − signs work for this; however, we will use the − sign.

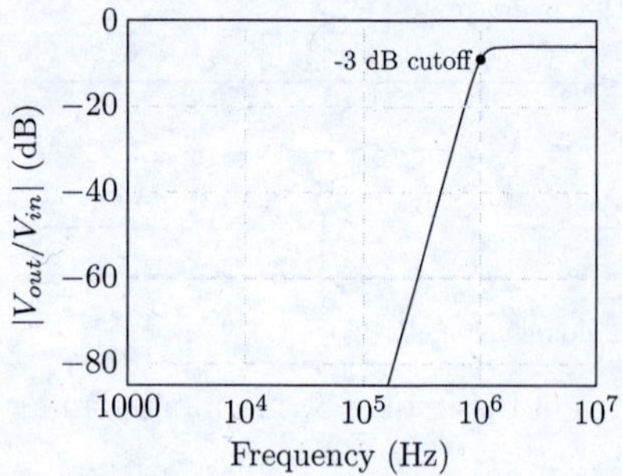

Figure 8.8: Simulated transfer function of the circuit in Example 8.2.

8.3 Band-pass filters

The most common band-pass filter is the series RLC circuit. An input voltage is applied across the series RLC circuit, the output voltage is drawn from across the resistor, as shown earlier, in Fig. 5.6.

At low frequencies the capacitor is an open circuit and the output is disconnected from the input. At very high frequencies the inductor is an open circuit and the output is disconnected from the input. Only at the resonance frequency, the series combination of L and C forms a short circuit, and the output voltage is equal to the input voltage. This gives the band-pass nature of the circuit.

Most generally, band-pass circuits can be constructed from the low-pass prototype filter with the help of a low-pass to band-pass frequency transformation.

8.3.1 Low-pass to band-pass transformation

A band-pass filter can be constructed from a low-pass prototype filter using the low-pass to band-pass frequency transformation. The transformation is a $1 \mapsto 2$ mapping of the frequency axis. Each point in the frequency axis of the prototype filter will map to two points in the frequency axis of the final filter. A $1 \mapsto 2$ mapping is not a function, and the reverse is easier to understand. The frequency axis of the final filter is therefore mapped to the frequency axis of the prototype using a $2 \mapsto 1$ mapping. Two different points on the frequency axis of the final filter will map to the same point on the frequency axis of the prototype. We have illustrated the mapping in Fig. 8.9.

Two points, $\pm\omega_0$, need to map to 0. If there is a polynomial mapping

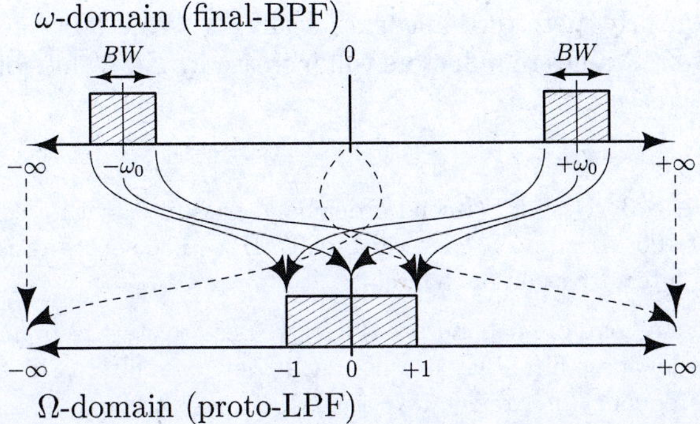

Figure 8.9: $2 \mapsto 1$ map of a low-pass to band-pass transformation. Two points on the frequency axis of the final filter map to one point on the frequency axis of the prototype.

function, both $\pm\omega_0$ will be the roots of this function. Also, ± 0, $\pm\infty$ need to map to $\pm\infty$. Therefore, a function of the following form will suffice.

$$\Omega = K(\omega - \omega_0)(\omega + \omega_0)/\omega$$

We can refine the function to obtain:

$$\Omega = \frac{1}{BW}\frac{\omega^2 - \omega_0^2}{\omega}$$

We can substitute $j\Omega$ as S and $j\omega$ as s to convert to the Laplace domain. Here S is the Laplace domain variable for the prototype filter, while s is the Laplace domain variable for the final filter.

$$S = \frac{1}{BW}\frac{s^2 + \omega_0^2}{s} = \frac{1}{BW}(s + \frac{\omega_0^2}{s})$$

Therefore, we must substitute an impedance SL in the prototype filter with a series combination of an impedance sL/BW and an impedance $\omega_0^2 L/(sBW)$. Similarly, we must replace an admittance SC in the prototype filter with the parallel combination of admittances sC/BW and $\omega_0^2 C/(sBW)$. In summary, we can follow the recipe below:

1. Replace every inductor L with a series combination of an inductor of value L/BW and a capacitor of value $BW/(\omega_0^2 L)$.

2. Replace every capacitor C with the parallel combination of a capacitor of value C/BW and an inductor of value $BW/(\omega_0^2 C)$.

3. Subsequently, perform impedance scaling to scale every resistor from $1\ \Omega$ to Z_0. Values of inductors will increase by Z_0. Values of capacitors will decrease by Z_0.

Example 8.3. Design a second order band-pass filter with a center frequency of 100 MHz and a bandwidth of 1 MHz. The source resistance is $1\ \text{k}\Omega$, the filter drives an open circuit.

To design a second order band-pass filter, we need to start from an order-1 prototype low-pass filter. We can use an RC low pass filter (Fig. 8.2). For a cut-off frequency of 1 rad/s, if R is $1\ \text{k}\Omega$, C needs to be 1 mF.

Now we will transform this prototype RC low-pass filter into a band-pass filter with center frequency of 100 MHz and bandwidth of 1 MHz. ω_0 is $100 \times 10^6 \times 2\pi$, and BW is $10^6 \times 2\pi$.

According to the recipe above, we will replace the capacitor with an LC-shunt combinations. The 1 mF capacitor will transform into a capacitor of $1\ \text{mF}/(10^6 \times 2\pi)$, or 159 pF, in shunt with an inductor of value $10^6 \times 2\pi/(4\pi^2 \times 10^{16} \times 10^{-3})$ or 15.9 nH. The circuit below shows our final design.

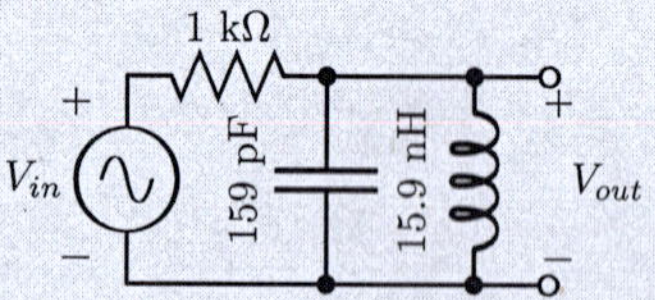

One can use quality-factor based calculations to quickly estimate the bandwidth of this band-pass filter and verify the results. Since the tank is a parallel RLC tank (think of when the input is zero, or a short circuit), Q is $R/(\omega L)$. In this case, Q will be 100, and the bandwidth will be $0.01\times$ the center frequency.

We can also transform the low pass Butterworth filter to band pass. The order of the filter will necessarily double.

Example 8.4. Design a 6th order band-pass filter for WiFi, with a center frequency of 2.4 GHz and bandwidth of 100 MHz. The source and load resistances are both $50\ \Omega$.

For a 6th order band-pass filter, we will start from a prototype 3rd order low-pass filter, with a structure as in Fig. 8.5(b). This has components of $2\cos\pi/3$ F (1 F), 2 H, and $2\cos-\pi/3$ F (1 F).

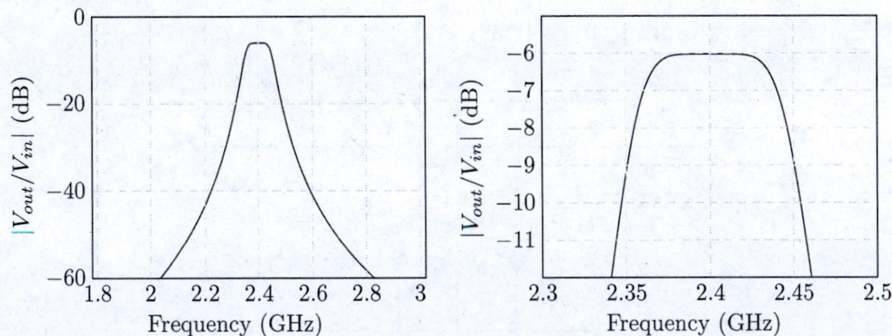

Figure 8.10: Simulated transfer function of the Butterworth band-pass filter of Example 8.4. The zoomed-in view on the right shows the flat nature of the pass-band.

Now we will transform this prototype RC low-pass filter into a band-pass filter, with center frequency of 2.4 GHz, bandwidth of 100 MHz, and an impedance level of 50 Ω. ω_0 is $2.4 \times 10^9 \times 2\pi$ rad/s; BW is $100 \times 10^6 \times 2\pi$ rad/s.

The 2 H inductor will be replaced by an inductor of $2 \times 50/(100 \times 10^6 \times 2\pi)$ H (159 nH) in series with a capacitor of 27.6 fF. Each 1 F capacitor will be replaced by a capacitor of 31.8 pF in shunt with 138 pH. The complete circuit is shown below.

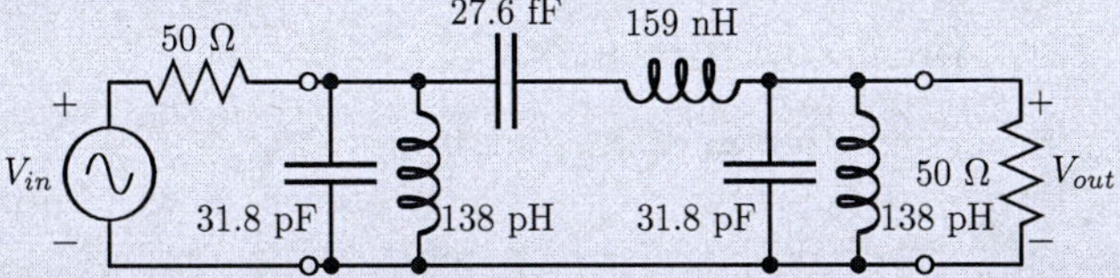

We simulated the circuit and have plotted $|V_{out}/V_{in}|$ in Fig. 8.10. We have plotted with a linear x-axis and a dB-scale y-axis to demonstrate the band-pass nature of the circuit. A zoomed-in view of the pass-band response is also shown in the figure to highlight the bandwidth and cut-offs of the filter.

8.4 Band-stop filters

The most common band-stop filter is a circuit with shunt-LC resonance, shown in Fig. 5.9. At low frequencies the inductor is a short circuit and the

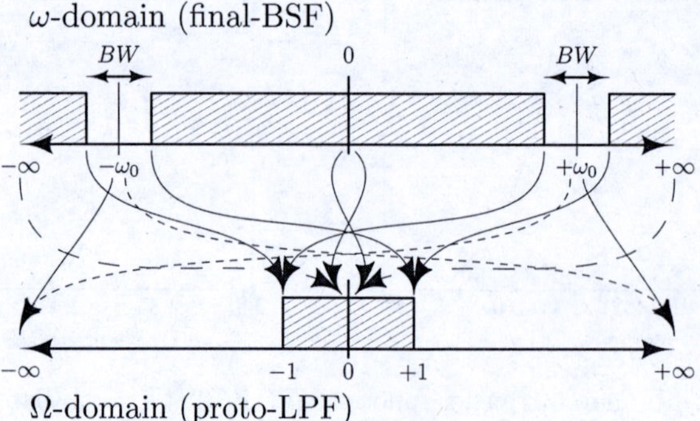

Figure 8.11: Frequency mapping scheme for a low-pass to band-stop transformation.

input voltage appears across the load. At very high frequencies the capacitor is a short circuit and the input voltage appears across the load. At the resonance frequency, the shunt-LC combination behaves as an open circuit. This frequency is blocked from passing through the filter.

Most generally, band-stop filters can be constructed from a low-pass prototype filter with the help of a low-pass to band-stop transformation.

8.4.1 Low-pass to band-stop

The low-pass to band-stop transformation is very similar to the low-pass to band-pass transformation of section 8.3.1. We have shown the frequency mapping in Fig. 8.11. With the same notation (Ω, S) as in section 8.3.1, the $2 \mapsto 1$ mapping function from the ω band-stop filter domain to the Ω prototype filter domain is given by:

$$\begin{aligned}\Omega &= \frac{BW \cdot \omega}{\omega^2 - \omega_0^2} \\ \text{or, } S &= BW \cdot \frac{s}{s^2 + \omega_0^2} = \frac{1}{\frac{s}{BW} + \frac{\omega_0^2}{s \cdot BW}}\end{aligned}$$

The statement implies that we must replace an impedance of SL in the prototype filter with an admittance s/BW in shunt with an admittance $\omega_0^2/(sBW)$. Further, we must replace an admittance of SC in the prototype filter with an impedance s/BW in series with an impedance $\omega_0^2/(sBW)$. The following summarized recipe needs to be followed.

1. Replace every inductor of value L with a parallel combination of a capacitor of value $1/(L \cdot BW)$ and an inductor of value $BW \cdot L/\omega_0^2$.

2. Replace every capacitor of value C with the series combination of an inductor of value $1/(C \cdot BW)$ and a capacitor of value $BW \cdot C/\omega_0^2$.

3. Proceed with impedance scaling. Scale all resistors by Z_0, all inductors by Z_0, all capacitors by $1/Z_0$.

Example 8.5. Design a 2nd order band-stop filter to filter out the third harmonic in a power line, with a stop band bandwidth of 1 Hz at 150 Hz, and a load of 0.1 Ω.

Let us first set up a first order low-pass prototype filter to drive a resistor of 0.1 Ω. The load in an RC filter is not a resistor; as such we can choose to use an LR configuration instead. The inductor will be an open circuit at high frequencies, a short circuit at low frequencies. A cut-off frequency of 1 rad/s can be designed if we set R as 0.1 Ω and L as 0.1 H.

Next let us transform this prototype low-pass filter to a band-stop filter centered at 150 Hz and with a bandwidth of 1 Hz. ω_0 is $2\pi \times 150$ rad/s; BW is 2π. The inductor needs to be replaced with a parallel combination of an inductor of value $2\pi \times 0.1/(150 \times 2\pi)^2$ or 707 nH, and a capacitor of value $1/(0.1 \times 2\pi)$ or 1.59 F.

The complete circuit and its simulated characteristics are shown below.

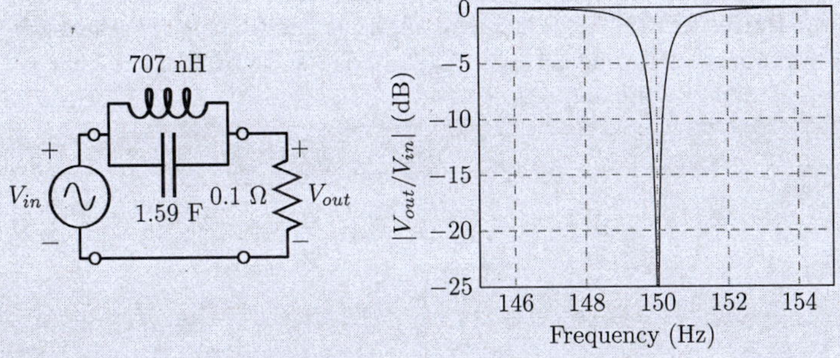

We can also transform high order filters to band-stop. This will give us a flatter pass-band and sharper characteristics in the transition to the stop-band.

Example 8.6. Design an 8th order band-stop filter with a bandwidth of 5 MHz and a center frequency of 900 MHz, for an impedance level of 50 Ω.

We will need to start from a prototype 4th order Butterworth filter. If we go with the structure in Fig. 8.5(c), the components in the prototype filter are $2\cos 3\pi/8$ or 0.765 H in series, $2\cos \pi/8$ or 1.848 F in shunt, $2\cos -\pi/8$ or 1.848 H in series, and $2\cos -3\pi/8$ or 0.765 F in shunt.

For our transformation, ω_0 is $900\times10^6\times2\pi$ rad/s, while BW is $5\times10^6\times2\pi$.

As given in the recipe, parallel LC combinations replace all inductors, and series LC combinations replace all capacitors. The impedance level is to be scaled to 50 Ω. The complete circuit schematic for the band-stop filter is shown below.

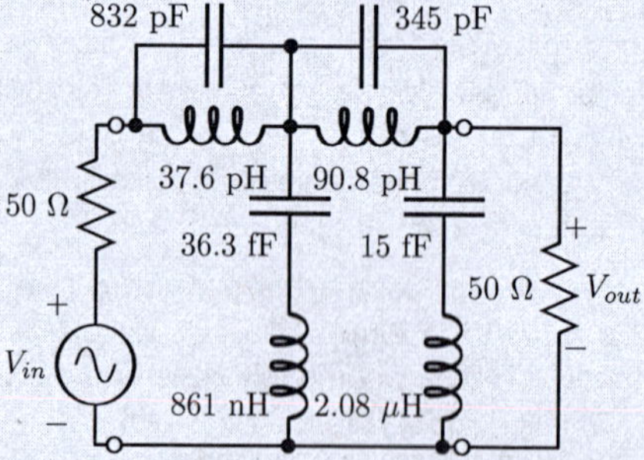

The simulated frequency response of the circuit is given below. In the frequency response, note the flat nature of the pass-band. This is a specialty of the Butterworth filter. The pass-band is at −6 dB; this is because when the signal passes through, V_{out} is $V_{in}/2$. The band-width of the filter is to be measured at −9 dB.

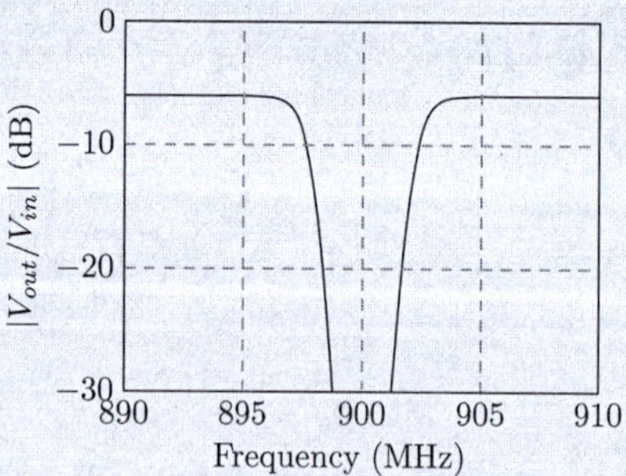

Band-stop filters are extremely useful to block interfering signals in communication systems, and to suppress harmonics in power systems.

8.5 Unit summary

- There are four major classes of filters: low-pass, high-pass, band-pass and band-stop.
- Ideal filters cannot be designed.
- The performance of a filter improves as we increase its order.
- A series R-C circuit, with output taken across the capacitor serves as a first-order low-pass filter. A series R-L circuit may also serve as a first-order low-pass filter, with the output taken across the resistor.
- A series R-C circuit, with output taken across the capacitor, serves as a high-pass filter. A series R-L circuit, with output taken across the inductor, serves as a high-pass filter.
- Band-pass and band-stop filters can be designed using a series LC or a shunt LC combination.
- A prototype low-pass high-order Butterworth filter, for a 1 rad/s cut-off and a 1 Ω source and load impedance, can be configured as a lossless LC ladder network with n components. The values of the components are given as $2\cos(k/n + 1/(2n) - 1/2)\pi$. The Butterworth filter is maximally flat in the pass-band.
- A circuit may be impedance-scaled by multiplying all impedances by the same factor. The transfer function will remain unchanged in this case. The transfer function may be frequency-scaled by dividing all inductance and capacitance values by a factor. In the case of frequency scaling, the resistors must not be scaled.
- A low-pass transfer function may be converted to a high-pass function by replacing all inductors with capacitors and vice-versa.
- A low-pass transfer function may be converted to a band-pass function by replacing all inductors with series-LC tanks and all capacitors with shunt-LC tanks.
- A low-pass transfer function may be converted to a band-stop function by replacing all inductors with shunt-LC tanks and all capacitors with series-LC tanks.

8.6 Exercises

Multiple choice type questions

8.1 An RC filter is set up as a low-pass filter. The capacitor is 10 pF. For a cut-off frequency of 1 MHz, the resistor needs to be:
(a) 2.53 kΩ (b) 10 kΩ (c) 15.9 kΩ (d) 100 kΩ

8.2 A series RL circuit is set up as a low-pass filter. The inductor is 1 μH and the resistor is 1 kΩ. What is the cut-off frequency?
(a) 159 Hz (b) 1 kHz (c) 159 MHz (d) 1 GHz

Short answer type questions

8.3 In case of a low-pass to high-pass transformation, every inductor in the original circuit is replaced with a ______ and every capacitor in the original circuit is replaced with a ______.

8.4 In case of a low-pass to band-pass transformation, every inductor in the original circuit is replaced with the ______ combination of an inductor and a capacitor. Every capacitor of the original circuit is replaced with the ______ combination of an inductor and a capacitor.

8.5 In case of a low-pass to band-stop transformation, every inductor in the original circuit is replaced with the ______ combination of an inductor and a capacitor. Every capacitor of the original circuit is replaced with the ______ combination of an inductor and a capacitor.

Numericals

8.6 Design a 2nd order band-pass filter with center frequency of 1 MHz and bandwidth of 1 kHz, for a load resistance of 50 Ω.

8.7 Design a 2nd order band-pass filter, with both source and load resistances of 50 Ω, using a parallel LC resonator, for a center frequency of 1 MHz and bandwidth of 1 kHz.

8.8 Design a 2nd order band-stop filter, with both source and load resistances of 50 Ω, using a series LC resonator, for a center frequency of 1 MHz and bandwidth of 1 kHz.

Know more

Historical profiles

Stephen Butterworth (1885-1958) was a British physicist, most famous for the Butterworth filter. His career was mainly as a government scientist at the Admiralty's Research Laboratory. Most of his work involved submarine cables, underwater explosions, torpedoes, and the like. As such, his work was invariably classified and not published.

Pafnuty Lvovich Chebyshev (1821-1894) is known as the founding father of mathematics in Russia. His career was mainly in Moscow and St. Petersburg, as an academic at Moscow University and then at St. Petersburg University, sometimes under extreme financial hardships. Chebyshev worked in the areas of probability, statistics, number theory, and mechanics. Chebyshev's polynomials and Chebyshev's inequality (weak law of large numbers) are essential concepts named after him. In filter design, Chebyshev polynomials are used to generate the Chebyshev filter.

Wilhelm Cauer (1900-1945) was a brilliant German mathematician who migrated to the USA in 1920 under severe financial duress. Cauer filters (elliptic filters) are known to be the most optimal filters.

Postage stamp in honor of Chebyshev
(image taken from the public domain)

Butterworth filters, Chebyshev filters and elliptic (Cauer) filters are the most commonly used filters in various applications.

Understand in depth

Filter design is a subject by itself. It is straightforward to list a large set of texts dedicated to filter design. In the very short list of two books that follow, the book by Ernst A. Guillemin is the most authoritative text on circuit synthesis. The book by Van Valkenburg brings classical and modern filter design together.

- Synthesis of Passive Networks, by Ernst A. Guillemin, John Wiley & Sons Inc., 1964.
- Design of Analog Filters, by Rolf Schaumann and Mac E. Van Valkenburg, Oxford University Press, 2001.

Appendix A

Octave usage in this text

This appendix lists all the Octave/MATLAB functions used in this text, with a brief overview. Most of these functions can operate over an entire array or vector. For example, `cos(X)` will return the cosine of every element in `X` in an array of the same shape. The descriptions here may be insufficient for a working knowledge. Please use the `help` in Octave to learn more about the functions. The built-in documentation additionally gives usage examples. Detailed online documentation is available at [2].

Octave is available for free, with source code, for all operating systems. Instructions to download and install Octave are available at:
`https://www.gnu.org/software/octave/download`
On Ubuntu/Debian, installing Octave is as simple as issuing the command `sudo apt install octave` from the terminal.

`abs` : The `abs` function computes the magnitude.

`angle` : The function returns the phase of a complex number in radians.

`assume` : This function is used to place constraints on a symbolic variable.

`clf` : `clf` clears the current figure.

`columns` : The function returns the number of columns of its input matrix.

`conj` : The `conj` function computes the complex conjugate.

`conv` : Two (row or column) vectors can be convolved using the `conv` function. The operation is handy for polynomial multiplication as the polynomial coefficients convolve.

`cos` : `cos` computes the cosine. The angle is in radians.

`cosh` : The function computes the hyperbolic cosine of its input. The cosine and hyperbolic cosine of an imaginary quantity are given as follows:

$$\cos(jx) = \cosh x \text{ and, } \cosh(jx) = \cos x$$

`deal` : `deal` copies the input parameters into the corresponding outputs.

`diag` : `diag` returns a diagonal matrix with the diagonal elements given by the vector input.

`diff` : The function performs differencing. Consecutive elements of a vector are differenced.

`double` : The function converts the input to a double-precision numerical value.

`end` : `end` could refer to the end of a `for` loop, the end of a `while` loop, or the last element of a vector.

`exp` : The function computes the exponential. If the input is an array, element-by-element exponentials are computed. The function does not give the matrix exponential.

`eye` : `eye` generates an identity matrix with the given number of rows and columns.

`fclose` : The `fclose` function closes a file.

`feof` : `feof` is a flag that rises if an input file reaches its end.

`figure` : The function opens a new figure window.

`find` : The function `finds` elements in an array that match a specific condition.

`floor` : The function computes the largest integer not greater than the input.

`fopen` : The function is used to open a file handle. Read the manual for more details. It is similar to implementations in other languages like C.

`for` : The `for` statement is the beginning of a `for` loop.

`fscanf` : The function is used to analyze a line in a file. It is similar to implementations in other languages like C.

`function` : The command is used to start the body of a function.

`function_handle` : The function creates a function based on a symbolic expression.

`grid` : The grid in a plot can be turned on and off using the function.

`hold` : To view multiple curves in the same plot, the `hold on` command may be issued.

`imag` : The function computes the imaginary part of its input.

`ilaplace` : The inverse Laplace transform using the symbolic toolbox.

`inf` : The phrase refers to infinity.

`int` : The function performs symbolic integration. It is a part of the symbolic package in Octave. Both definite and indefinite integrals can be handled.

`inv` : The function computes the inverse of a square matrix.

`ismember` : Returns a logical matrix with the same shape as the first input. The elements are true if the element in the first input is found in the second; false if not found.

`length` : The function returns the length of a vector.

`log` : The `log` routine gives the natural logarithm of its input.

`log10` : The `log10` routine gives the logarithm to base 10.

`max` : The function reports the maximum of a vector.

`min` : The function reports the minimum of a vector.

`ones` : The function generates a matrix full of 1s. The input gives the size of the matrix to the function.

`pi` : The phrase refers to the value of π.

`pkg load symbolic` : The command is required before using any function from the symbolic toolbox. MATLAB does not recognize the command.

`plot` : This command is used to generate a plot.

`poly` : This function is used to generate a polynomial with the given roots.

`polyval` : The function is used to compute the value of a polynomial at a given point.

`real` : The function computes the real part of its input.

`residue` : The function can be used to break a polynomial by a polynomial into partial fractions. It can also be used to reconstruct a polynomial by polynomial function from its partial fractions.

`roots` : The function computes the roots of a polynomial.

`round` : The function is used to round the input to its nearest integer.

`rows` : The function returns the number of rows of its input matrix.

`semilogx` : The function generates a plot with an x-axis in log-scale and a y-axis in linear scale.

`simplify` : The `simplify` command from the symbolic toolbox is used to simplify a symbolic expression.

`sin` : The function computes the sine of its input. The input is in radians.

`sinh` : The function computes the hyperbolic sine of its input. The sine and hyperbolic sine of an imaginary quantity are given as follows:

$$\sin(jx) = j\sinh(x) \text{ and, } \sinh(jx) = j\sin x$$

`sqrt` : The function computes the square root of its input.

`stem` : The function plots a line plot, with a baseline of zero.

`sum` : The function generates the sum of a vector.

`syms` : This function from the symbolic toolbox declares symbolic variables.

`transpose` : The function generates the transpose of a matrix. A row vector is transposed into a column vector.

`while` : A `while` loop begins with this declaration.

`ylim` : The function is used to adjust the limits of the y-axis in a plot.

`zeros` : The function generates a matrix full of 0. The input gives the size of the matrix.

Bibliography

[1] D. J. Griffiths, *Introduction to Electrodynamics*, 3rd ed. Upper Saddle River, NJ 07458, USA: Prentice Hall, 1999.

[2] J. W. Eaton, D. Bateman, S. Hauberg, and R. Wehbring, *GNU Octave version 5.2.0 manual: a high-level interactive language for numerical computations*, 2020. [Online]. Available: https://www.gnu.org/software/octave/doc/v5.2.0/

[3] G. Monteath, "Application of the compensation theorem to certain radiation and propagation problems," *Proceedings of the IEE-Part IV: Institution Monographs*, vol. 98, no. 1, pp. 23–30, 1951.

[4] F. Judd and P. Chirlian, "The application of the compensation theorem in the proof of Thevenin's and Norton's theorems," *IEEE Transactions on Education*, vol. 13, no. 2, pp. 87–88, 1970.

[5] H. Vogt, M. Hendrix, P. Nenzi, and D. Warning, *Ngspice User's Manual*, 2022. [Online]. Available: http://ngspice.sourceforge.net/docs/ngspice-manual.pdf

[6] *IEEE Standard Definitions for Selected Quantities, Units, and Related Terms, with Special Attention to the International System of Units (SI)*, Institute of Electrical and Electronics Engineers (IEEE) Std. 270, Rev. R2012, Sep. 2006.

[7] W. Jackson, "Life and work of Oliver Heaviside (May 18, 1850-February 3, 1925)," *Nature*, vol. 165, pp. 1476–4687, 1950.

[8] I. Yavetz, *From Obscurity to Enigma: The Work of Oliver Heaviside, 1872–1889.* Birkhauser, 1995.

[9] A. E. Kennelly, "Impedance," *American Institute of Electrical Engineers Transactions*, vol. 10, pp. 175–232, 1893.

[10] A. E. Kennelly, "Impedance," *Proceedings of the IEEE*, vol. 72, no. 4, pp. 463–492, 1984.

[11] C. P. Steinmetz, "Complex quantities and their use in electrical engineering," in *Proceedings of the International Electrical Congress*, 1893, pp. 33–74.

[12] A. Sen, "Sir J.C. Bose and radio science," in *1997 IEEE MTT-S International Microwave Symposium Digest*, vol. 2, 1997, pp. 557–560 vol.2.

[13] P. Bondyopadhyay, "Sir J.C. Bose diode detector received Marconi's first transatlantic wireless signal of december 1901 (the "Italian Navy Coherer" scandal revisited)," *Proceedings of the IEEE*, vol. 86, no. 1, pp. 259–285, 1998.

[14] K. Ghosh, *Unsung Genius: A Life of Jagadish Chandra Bose.* 7/16 Ansari Road, New Delhi: Aleph Book Company, 2022.

[15] C. L. Fortescue, "Method of symmetrical co-ordinates applied to the solution of polyphase networks," *Transactions of the American Institute of Electrical Engineers*, vol. 37, pp. 1027–1140, Jul. 1918.

[16] (2022, Jul.) Pradhan mantri sahaj bijli har ghar yojana. Dashboard. Government of India, Ministry of Power. [Online]. Available: https://saubhagya.gov.in/

[17] E. Kreyszig, *Advanced Engineering Mathematics*, 10th ed. John Wiley & Sons Inc., 2011, sec. 16.3.

[18] H. Carslaw and J. Jaeger, *Operational Methods in Applied Mathematics.* London, U.K.: Oxford University Press, 1941.

[19] F. F. Kuo, *Network Analysis and Synthesis*, 2nd ed. John Wiley & Sons Inc., 1966, pp. 153–155.

[20] A. Zygmund, *Trigonometric series*, 3rd ed., ser. Cambridge Mathematical Library. Cambridge: Cambridge University Press, 2003.

[21] J. W. Gibbs, "Fourier's series," *Nature*, vol. 59, no. 1539, pp. 606–606, Apr. 1899.

[22] E. W. Weisstein. Sinc function. Wolfram Alpha. From MathWorld – A Wolfram web resource. [Online]. Available: mathworld.wolfram.com/SincFunction.html

[23] V. Belevitch, "Summary of the history of circuit theory," *Proceedings of the IRE*, vol. 50, no. 5, pp. 848–855, May 1962.

[24] O. Brune, "Synthesis of a finite two-terminal network whose driving-point impedance is a prescribed function of frequency," Ph.D. dissertation, Massachussets Institute of Technology (MIT), Boston, MA, USA, 1931.

[25] S. Darlington, "Synthesis of reactance 4-poles which produce prescribed insertion loss characteristics including special applications to filter design," Ph.D. dissertation, Columbia University, New York, NY, USA, 1939.

Index

ABCD parameters, 261
active power, 132–134
admittance, 118, 119
admittance parameters, 245
apparent power, 133, 138
available power, 60

balanced load, 141
band pass response, 190
band stop response, 192
bandwidth, 190, 192, 193
Butterworth filter, 282–285, 287, 291, 294

capacitor, 5, 118, 119
CCCS, 7, 13
CCVS, 7, 13
characteristic equation, 85, 87, 88, 90, 91, 93, 97
compensation theorem, 51, 129
complex power, 133, 134
continuous spectra, 229, 230
controlled sources, 7, 13
convolution, 180, 181
convolution integral, 113, 180, 181
copper, 139
coupling coefficient, 5, 124
critically damped response, 88
critically damped system, 91
current gain, 259
current source, 7

damping factor, 91
datum, 8
delta function, 112
delta-connected load, 140
differential equation
    first order, 71
    linear, 72
Dirac delta function, 112
Dirichlet conditions, 206
discrete spectra, 219, 220
    square wave, 220
dot convention, 122
driving point impedance, 45, 59, 80, 118, 123
dual network, 27, 93
duality, 47

exponential Fourier series, 216
    coefficients, 216
    full-wave rectifier, 225, 226
    square wave, 217, 218

fall time, 77
feedback circuits, 269
filter
    band pass, 190, 288–291
    band stop, 192, 292–294
    Butterworth, 282–285, 287, 291, 294
    characteristics, 281
    first order, 282
    high pass, 285–287
    low pass, 281–285
    prototype, 284
final value theorem, 166
forced response, 73, 86

Fortescue's theorem, 143
Fourier transform, 229–232
  properties, 232
free response, 72, 73, 85, 89
frequency map, 286, 288, 289, 292–294
frequency scaling, 285
frequency transformation, 284
  low-pass to band-pass, 288–291
  low-pass to band-stop, 292–294
  low-pass to high-pass, 286
full-wave rectifier, 224
fundamental period, 204

G-parameters, 255
Gibbs' phenomenon, 211

H-parameters, 252
Heaviside expansion theorem, 170
homogeneity, 56, 131
homogeneous equation, 85
homogeneous solution, 72, 85
hybrid parameters, 252

impedance, 118, 119
  parallel combination, 118
  series combination, 118
impedance parameters, 243
impedance scaling, 284, 290, 293
impulse function, 112, 161
impulse response, 112, 183, 281
incidence matrix, 24
  reduced, 24
inductor, 4, 117, 119
initial conditions, 73, 78, 82, 93, 96
initial value theorem, 165
input impedance, 256
instantaneous power, 137
inverse Fourier transform, 231
inverse hybrid parameters, 255
inverse Laplace transform, 166–169

KCL, 3
KVL, 3

Laplace transform, 157, 173
  capacitor, 174
  circuit analysis, 174–179
  convergence, 158
  inductor, 175
  linearity, 158
  properties, 159–162
  solving differential equations, 173
line current, 141
line voltage, 141
linear circuit, 114
linear system, 42, 44, 111
linearity, 44, 111
LTI system, 44

maximum power transfer, 57, 59, 129, 135
mesh method, 126, 127
multi-port network, 271
mutual inductance, 5, 122, 124

natural frequency, 91
neutral, 139
node-voltage method, 8, 128
non-planar network, 27
Norton's theorem, 46, 129

Ohm's law, 4
order, 80, 83
orthogonality, 203
oscilloscope probes, 222
output impedance, 256, 257
over-damped response, 87
over-damped system, 91

parallel resonance, 188, 191, 193
parallel RLC, 92
Parseval's equality, 227, 233
partial fraction expansion, 168, 183

- repeated roots, 170–172
- particular solution, 73, 86
- passive network, 53, 245, 247, 254, 262, 271
- peak overshoot, 92
- periodic signals, 203
  - power, 226–229
- phase current, 141
- phase voltage, 141
- phasor, 114, 115
  - addition of, 116
  - Cartesian form, 127
  - circuit analysis with, 125
  - convention, 131
  - diagram, 120, 121, 134
  - KVL, KCL, 116
  - maximum power transfer, 135
  - network theorems, 129
  - polar form, 127
  - power, 132
  - Tellegen's theorem, 117
- Π-network, 245, 246
- planar network, 27
- pole, 90, 167, 182, 183, 185, 186
- port, 44, 241
- power
  - active, *see* active power
  - apparent, *see* apparent power
  - complex, *see* complex power
  - conservation of, 26
  - instantaneous, *see* instantaneous power
  - reactive, *see* reactive power
- power factor, 132, 133
  - lagging, 133
  - leading, 133, 134, 138
- power triangle, 134
- practical sources, 50
- probe compensation, 223
- prototype filter, 284
- quality factor, 190, 192

- ramp function, 161
- reactance, 118
- reactive power, 132, 133, 137, 138
- reciprocal network, *see* passive network
- reciprocity theorem, 53, 54, 129, 131
- region of convergence, 157
- residue integration, 167
- resistor, 4
- resonance, 188
  - parallel, 188, 191, 193
  - series, 188–190
- response
  - critically damped, 88, 90
  - over-damped, 87, 90
  - under-damped, 89, 90
- rise time, 77, 92
- RoC, 157

- second order circuit, 84
- series RC, 74
- series resonance, 188–190
- series RL, 78
- series RLC, 84
- settling time, 92
- simulation, 41
- sinc(x), 219
- source transformation, 49
- SPICE, 65
- square wave, 209, 217, 218, 220
  - duty cycle, 221, 224
  - frequency, 218
  - scaling, 218
  - shifting, 218
- star-connected load, 139
- star-delta conversion, 248
- state variable, 79, 81, 84
- steady state, 109
- step function, 161

step response, 91
super-mesh, 19
super-node, 12, 14
superposition, 41, 42, 113, 129, 131, 222
susceptance, 118
system, 43
  frequency response, 186, 187
  impulse response, 112
  linear, 44, 111
  linear time invariant, *see* LTI
  LTI, 44, 113
  sinusoidal steady state, 184, 185
  stability, 183, 186
  time invariant, 44, 112
system transfer function, 258, 264

T-network, 244, 247
T-parameters, 261
tableau analysis, 39, 41
tableau equation, 39, 112
Tellegen's theorem, 26, 117, 129, 263
Thévenin's theorem, 45, 129
theorem
  compensation, 51, 129
  final value, 166
  Heaviside expansion, 170
  incidence matrix, 24
  initial value, 165
  Laplace transform linearity, 158
  maximum power transfer, 57, 59, 129, 135
  Norton's, 46, 129
  reciprocity, 53, 54, 129
  reduced incidence matrix, 24
  sinusoidal steady state analysis, 113
  superposition, 41, 42
  Tellegen's, 26, 117, 129, 263
  Thévenin's, 45, 129
three-phase circuits, 138, 139
time constant, 76, 79–82, 95
time invariance, 44, 111, 112
transfer function, 180–182
transformer, 124, 125
transmission parameters, 261
trigonometric Fourier series, 205
  coefficients, 205
  discontinuous waveform, 209, 210
  even symmetry, 211
  half-wave symmetry, 215
  odd symmetry, 212, 213
  quarter-wave symmetry, 215
  square wave, 209
trigonometric system, 203
  orthogonality, 203
  period, 204
turns ratio, 125
two-port network, 241
  cascade of, 263, 264
  series-series, 265
  series-shunt, 266
  shunt-series, 267
  shunt-shunt, 268

unbalanced load, 143
under-damped response, 89
under-damped system, 91

var, 134
VCCS, 7, 13
VCVS, 7, 13
voltage gain, 259
voltage source, 6

waveform symmetries, 211

Y-parameters, 245

Z-parameters, 243
zero, 182, 185, 186
zero input, 85
zero-input response, 72, 173
zero-state response, 73, 86, 173